Microbiological Methods for Assessing Soil Quality

Microbiological Methods for Assessing Soil Quality

Edited by

Jaap Bloem
Alterra, Wageningen, The Netherlands

David W. Hopkins
University of Stirling, UK

and

Anna Benedetti
Istituto Sperimentale per la Nutrizione delle Piante, Rome, Italy

CABI Publishing

CABI is a trading name of CAB International

CABI Head Office
Nosworthy Way
Wallingford
Oxfordshire OX10 8DE
UK

Tel: +44 (0)1491 832111
Fax: +44 (0)1491 833508
Email: cabi@cabi.org
Web site: www.cabi.org

CABI North American Office
875 Massachusetts Avenue
7th Floor
Cambridge, MA 02139
USA

Tel: +1 617 395 4056
Fax: +1 617 354 6875
Email: cabi-nao@cabi.org

A catalogue record for this book is available from the British Library, London, UK.

The Library of Congress has cataloged the hardcover edition as follows
Microbiological methods for assessing soil quality / edited by Jaap Bloem, David W. Hopkins and Anna Benedetti
p. cm.
Includes index.
ISBN 0-85199-098-3 (alk. paper)
1. Soil microbiology. 2. Soils—Quality. 3. Soils—Analysis.
I. Bloem, Jaap, 1958- II. Hopkins, David W., Dr. III. Benedetti, Anna, Dr.
IV. Title

QR111.M39 2005
579′.1757—dc22

2005001632

ISBN-13: 978-0-85199-098-9 (hardback edition)
ISBN-13: 978-1-84593-500-9 (paperback edition)

First published 2006
Paperback edition 2008

Typset by Columns Design Ltd, Reading
Printed and bound in the UK by Biddles Ltd, King's Lynn.

Contents

Editors

Editors-in-Chief

Jaap Bloem
Department of Soil Sciences, Alterra, PO Box 47, NL-6700 AA Wageningen, The Netherlands

David W. Hopkins
School of Biological and Environmental Sciences, University of Stirling, Stirling FK9 4LA, UK

Anna Benedetti
Consiglio per la ricerca e la sperimentazione in Agricoltura, Istituto Sperimentale per la Nutrizione delle Piante, Via della Navicella, 2, 00184 Rome, Italy

Editorial Board

Richard G. Burns
School of Land and Food Sciences, The University of Queensland, Brisbane, Queensland 4072, Australia

Oliver Dilly
Lehrstuhl für Bodenschutz und Rekultivierung, Brandenburgische Technische Universität, Postfach 101344, D-03013 Cottbus, Germany

Andreas Fließbach
Research Institute of Organic Agriculture (FiBL), Ackerstrasse, CH-5070 Frick, Switzerland

Philippe Lemanceau
UMR 1229 INRA/Université de Bourgogne, 'Microbiologie et Géochimie des Sols', INRA-CMSE, BP 86510 21065, Dijon cedex, France

James M. Lynch
Forest Research, Alice Holt Lodge, Farnham GU10 4LH, UK

Yvan Moënne-Loccoz
UMR CNRS 5557 Ecologie Microbienne, Université Claude Bernard (Lyon 1), 43 bd du 11 Novembre, 69622 Villeurbanne cedex, France

Paolo Nannipieri
Dipartimento de Nutrizione delle Pianta e Scienza del Suolo, Università di Firenze, Piazzale di Cascine 28, Florence, Italy

Fabio Tittarelli
Consiglio per la ricerca e la sperimentazione in Agricoltura, Istituto Sperimentale per la Nutrizione delle Piante, Via della Navicella, 2, 00184 Rome, Italy

Jan Dirk van Elsas
Department of Microbial Ecology, Groningen University, Kerklaan 30, NL-9750 RA Haren, The Netherlands

Abbreviations

AIM	acetylene inhibition method
AM	arbuscular mycorrhiza
AO	acridine orange
AODC	acridine orange direct count
APS	ammonium persulphate
ARDRA	amplified ribosomal DNA restriction analysis
ATP	adenosine 5'-triphosphate
AUDPC	area under the disease progress curve
AWCD	average well colour development
BAS	basal respiration
(B_C/TOC)	C biomass/total organic C (ratio)
BNF	biological nitrogen fixation
CAC	citric acid cycle
CEC	cation exchange capacity
CEN	Comité Européen de Normalisation
CFE	chloroform fumigation–extraction
CFU	colony-forming units
CLPP	community-level physiological profiles
C_{mic}	microbial biomass
COST	COopération dans le domaine de la recherche Scientifique et Technique
CSLM	confocal scanning laser microscopy
CV	coefficient of variation
CWDEs	cell-wall-degrading enzymes
DAPI	4',6-diamidino-2-phenylindole-dihydrochloride
DEPC	diethyl-pyrocarbonate
DFS	differential fluorescent stain
DGGE	denaturing gradient gel electrophoresis
DMSO	dimethyl sulphoxide

dNTP	deoxynucleoside 5'-triphosphate
dpm	disintegrations per minute
dps	disintegrations per second
dsDNA	double-stranded DNA
DTAF	5-(4,6-dichlorotriazin-2-yl) aminofluorescein
dTTP	deoxythymidine triphosphate
DW	dry weight
EAP	Environmental Action Programme
EL	ester-linked
EPA	Environmental Protection Agency
EU	European Union
FAME	fatty acid methyl ester (analysis)
FAO	Food and Agriculture Organization
FB	Fluorescent Brightener
FDA	fluorescein diacetate
FID	flame ionization detector
FISH	fluorescence *in situ* hybridization
GC	gas chromatograph
GC–MS	GC coupled with a mass spectrometer
GM	genetically modified
HPLC	high-pressure liquid chromatograph
IC	ion chromatograph
INT	iodonitrotetrazolium chloride
INTF	iodonitrotetrazolium formazan
IPP	intact phospholipid profiling
IR	infrared
ISO	International Organization for Standardization
KAc	potassium acetate
MIC50	mean inhibitory concentration at 50%
MIDI or MIS	Microbial Identification System
MN_{Bas}	basal nitrogen mineralization
MPN	most probable number
MST	mean survival time
MUB	modified universal buffer
MUF	methylumbelliferyl
Nbio	N in microbial biomass
Ndfa	nitrogen derived from the atmosphere
NMDS	non-metric multidimensional scaling
Ntot	total soil nitrogen
OECD	Organization for Economic Co-operation and Development
p.a.	pro analysis (reagent purity)
PBS	phosphate-buffered saline
PCA	principal components analysis
PCR	polymerase chain reaction
PGPF	plant-growth-promoting fungi
PGPR	plant-growth-promoting rhizobacteria
PLFA	phospholipid fatty acid (analysis)

p-NPP	*p*-nitrophenyl phosphate
qCO_2	metabolic quotient
Q_N	nitrogen mineralization quotient
RAPD	random amplified polymorphic DNA
RCF	relative centrifugal force
rpm	revolutions/minute
RQ	respiratory quotient
RS	ripper subsoiling
SDS	sodium dodecyl sulphate
SEM	scanning electron microscopy
SINDI	Soil Indicators (New Zealand)
SIR	substrate-induced respiration
SOM	soil organic matter
SQI	Soil Quality Index
SSC	standard saline citrate
SSCP	single-strand conformation polymorphism
SSSA	Soil Science Society of America
TEM	transmission electron microscopy
TGGE	temperature gradient gel electrophoresis
T-RFLP	terminal restriction fragment length polymorphism
TY	tryptone-yeast extract
UV	ultraviolet
v/v	volume in volume
WHC	water-holding capacity
w/v	weight in volume

I

Approaches to Defining, Monitoring, Evaluating and Managing Soil Quality

1 Introduction

Anna Benedetti[1] and Oliver Dilly[2]

[1]*Consiglio per la ricerca e la sperimentazione in Agricoltura, Istituto Sperimentale per la Nutrizione delle Piante, Via della Navicella, 2, 00184 Rome, Italy;* [2]*Lehrstuhl für Bodenschutz und Rekultivierung, Brandenburgische Technische Universität, Postfach 101344, D-03013 Cottbus, Germany*

Introduction

Having adopted the Treaty on Biological Diversity of Rio de Janeiro (UNCED, 1992), many governments are becoming increasingly concerned about sustaining biodiversity and maintaining life support functions. In several countries, national or regional programmes have been established to monitor soil quality and/or the state of biodiversity. Most monitoring programmes include microbiological indicators, because soil microorganisms have key functions in decomposition and nutrient cycling, respond promptly to changes in the environment and reflect the sum of all factors regulating nutrient cycling (see also Chapter 3). Currently the European Union (EU) and many countries all over the world are working on legislation for the protection of soil quality and biodiversity. Policy makers, as well as land users, need indicators and monitoring systems to enable them to report on trends for the future and to evaluate the effects of soil management. This book details approaches and microbiological methods for assessing soil quality.

The European Commission has been promoting cooperation and the coordination of nationally funded research through so-called COST actions ('COopération dans le domaine de la recherche Scientifique et Technique'; http://cost.cordis.lu/src/whatiscost.cfm, accessed 27 April 2004). COST Action 831 'Biotechnology of Soil: Monitoring, Conservation and Remediation' started in October 1997 and ended in December 2002. An important aim of COST Action 831 was the development of a handbook on microbiological methods for assessing soil quality. COST Action 831 has enabled working groups of European soil microbiologists to discuss and evaluate the potential use of microbiological, biochemical and molecular tools to assess soil quality. The scientific community is constantly challenged by operative institutions, such as national and local authorities, state boards, private boards, consultants and standardization agencies, to deliver feasible methods for acquiring representative biological data on soil quality. This is extremely difficult, since soil microorganisms respond and adapt rapidly to environmental conditions. In addition, the impacts caused by human activities may be barely distinguishable from natural fluctuations, especially when changes are detected late and comparison with historical data or unaffected control sites is not possible.

Various authors have made numerous suggestions. For instance, Domsch (1980) and Domsch *et al.* (1983) proposed that any alteration, caused by either natural agents or pollutants, which returns to normal microbiological values within 30 days should be considered normal fluctuation; alterations lasting for 60 days can be regarded as tolerable, whereas those persisting for over 90 days are stress agents. Brookes (1995) suggested that no parameter should be used alone, but that related parameters should be identified and utilized together as an 'internal control', e.g. biomass carbon (C) and total soil organic C. In general, there is an approximate linear relationship between these two variables, so when soils show marked variations from what is considered to be the normal ratio between biomass C and total organic C in a particular soil management system, climate and soil type, this ratio becomes an indicator of deterioration and change in soil ecosystem functions.

Criteria for Indicators of Soil Quality

Criteria for indicators of soil quality relate mainly to: (i) their utility in defining ecosystem processes; (ii) their ability to integrate physical, chemical and biological properties; and (iii) their sensitivity to management and climatic variations (Doran, 2000). These criteria apply to soil organisms, which are thus useful indicators of sustainable land management. Ideally, soil organisms and ecological indicators should be:

1. Sensitive to variations in management;
2. Well correlated with beneficial soil functions;
3. Useful for elucidating ecosystem processes;
4. Comprehensible and useful to land managers;
5. Easy and inexpensive to measure.

Brookes (1995) proposed the following criteria for selecting a microbiological parameter as an indicator of soil pollution.

1. It should be possible to determine the property of interest accurately and precisely in a wide range of soil types and conditions.
2. Determination should be easy and of low cost, as many samples must be analysed.
3. The nature of the parameter must be such that control determinations are also possible, so that the effect of the pollutant can be assessed exactly.
4. The parameter must be sensitive enough to detect pollution, but also stable enough to avoid false alarms.

5. The parameter must have general scientific validity based on reliable scientific knowledge.
6. If the reliability of a single parameter is limited, two or more independent parameters should be selected. In this case their interrelations in unpolluted areas must also be known.

These two approaches are synergetic, as the criteria proposed by Doran (2000) focus on the sphere of interest, while Brookes' (1995) criteria identify the requisites of an indicator.

Two crucial points had to be clarified by the working groups of COST Action 831 before any choice of, or suggestion about, microbial indicators of soil quality was made:

1. Who is the handbook for?
2. How do we define soil quality?

Potential Users of this Handbook

This handbook is aimed at professionals, students and organizations working in the field of agriculture and the environment, such as:

- soil scientists, colleges, universities, libraries;
- consultants in environmental risk assessment and soil management;
- analysis laboratories, e.g. those involved in ecological monitoring;
- international (e.g. EU, Organization for Economic Cooperation and Development (OECD), Food and Agriculture Organization (FAO)), national, regional and local authorities involved in soil protection and management;
- international (e.g. ISO and CEN) and national standardization agencies.

It aims to provide clear instructions to technicians operating outside of the scientific research sector, and is meant to provide a seamless link between science and application. In contrast to earlier books on microbiological methods (for instance Alef and Nannipieri, 1995), this handbook focuses on a limited number of methods which are applicable, or already applied, in regional or national soil quality monitoring programmes. It also provides an overview of monitoring programmes implemented in several countries.

The people who create, study and assess innovative solutions using scientific methods are seldom involved directly in transferring information to end-users. This can create a knowledge gap that often leads to misinformation or poor information. The purpose of this book is to provide applicable microbiological methods for assessing soil quality. Part I provides an overview of approaches to defining, monitoring, evaluating and managing soil quality. In Part II, methods are described in sufficient detail to enable this handbook to be used as a practical guide in the laboratory. Finally, Chapter 10 gives a census of the main methods used in over 30 European soil microbiological laboratories.

Defining Soil Quality

During a COST 831 Joint Working Groups meeting on 'Defining soil quality', held in Rome in December 1998 (Benedetti *et al.*, 2000), there was broad discussion about the criteria for the definition of 'soil quality and/or qualities of soils'. This can be applied to a wide range of agricultural soils, forestry soils, grazing pastures, natural environment soils, etc., and may include different climate zones. However, the focus of our activities is in the COST domain of agriculture and biotechnology. An overview on defining soil quality is given in Chapter 2.

Evaluating Soil Quality

Once the aim and the potential users of the handbook had been defined, the next step was to establish how to evaluate soil quality, and which parameters and methods to adopt. Many questions had to be answered and were debated during a Joint Working Groups meeting on 'Evaluating soil quality', in Kiel, Germany (May 2000). The issues ranged from problems related to sampling, storage and pre-incubation of soil samples for microbiological analyses, to the choice of the most efficient methods and indicators (Bloem and Breure, 2003). An overview on evaluating soil quality is given in Chapter 3.

The methods can be divided into four groups, depending on the information they can provide:

1. Soil microbial biomass and number.
2. Soil microbial activity.
3. Soil microbial diversity and community structure.
4. Plant–microbe interactions.

Soil microbial biomass and activity are relatively easy to determine using routine methods, and are used to assess soil quality. For monitoring programmes where large amounts of samples have to be processed, often the soil is sieved, mixed and pre-incubated under standardized conditions in the laboratory to reduce variation and to facilitate comparison between different locations and different sampling dates. Direct analyses of microbial biomass and activity of field samples are also possible, and are often performed in more fundamental research. However, the higher variation found in direct analysis usually requires more replicates in space and time than with pre-incubated samples. Compared to biomass and activity, soil microbial diversity and community structure is more complicated to measure, and requires more specialized techniques, which are less easy to standardize. However, molecular techniques for their assessment are rapidly improving. The study of plant–microbe interactions is also relatively specialized and time consuming, and often requires *in situ* determinations that are rarely performed in optimal conditions. Field temperature and humidity can vary greatly and also reach extreme values which are very unfavourable for microbiological activity. Moreover, substrate concentration and pH values are seldom optimal.

Methods

Once the parameters and methods for assessing soil quality had been selected, the detailed protocol for each method was proposed and discussed during a combined meeting of working groups on 'Microbiological methods for soil quality', in Wageningen, The Netherlands (November 2001). Here, the preparation of the methods section of the handbook was initiated.

1. Soil microbial biomass and number

All the methods capable of defining the weight and number of soil microorganisms in a soil sample are included. The conventional methods for determining numbers of microbes living in soil are based on viable or direct counting procedures (Zuberer, 1994; Alef, 1995; Alef and Nannipieri, 1995; Dobereiner, 1995; Lorch *et al.*, 1995). Viable counting procedures require culturable cells and comprise two approaches: the plate count technique and the most probable number (MPN) technique. Some unculturable soil microorganisms may be potentially culturable if adequate nutritional conditions for their growth could be provided. However, many remain unculturable because they are dormant and require special resuscitation before regaining the ability to grow; or they are non-viable but still intact and detectable by microscopy (Madsen, 1996). Using specific culture media, specific functional groups of microbes can be counted. However, even with general growth media, the numbers of microbes detected are usually at least an order of magnitude lower than those obtained by direct microscopy.

Direct enumeration techniques allow the counting of total numbers of both bacteria and fungi, but usually give no indication of the composition of the respective communities. Generally, with these techniques, a known amount of homogenized soil suspension is placed on a known area of a microscope slide, the microorganisms are then stained with a fluorescent dye and are counted using a microscope (Bloem *et al.*, 1995). A disadvantage of microscopic counts is that visual counting is subjective and relatively time consuming. Therefore biochemical and physiological methods, e.g. chloroform fumigation extraction of microbial carbon and nitrogen, and substrate-induced respiration, are most commonly used (Chapter 6).

2. Soil microbial activity

Biochemical techniques are described that reveal information about the metabolic processes of microbial communities, both in their entirety (e.g.

respiration and mineralization) and according to functional groups (e.g. nitrification and denitrification).

Microbial activity can be divided into potential and actual activity. *Actual activity* means the activity microorganisms develop when conditions necessary for metabolism are less than optimal, as occurs in the open field. This activity can be determined using field sensors, but to date no serial and routine methods are available. Therefore potential activity is usually determined. *Potential activity* means metabolic activity, including enzymatic activities, that soil microorganisms are capable of developing under optimal conditions of, for example, temperature, humidity, nutrients and substrates.

Biochemical methods can be divided into two subgroups. The first includes the methods that measure active populations in their entirety, usually without adding substrates. The second contains methods that are able to define the activity and potential activity of specific organisms or metabolic groups, usually after adding specific substrates; for example, respirometric tests with specific carbon sources, and potential nitrification after addition of ammonium. A selection of commonly used methods is given in Chapter 7.

3. Soil microbial diversity and community structure

This group of methods includes the most up-to-date techniques for acquiring ecological and molecular data.

Traditionally, culturing techniques have been used for the analysis of soil microbial communities. However, only a small fraction (< 0.1%) of the soil microbial community has been determined using this approach. A number of methods are currently available for studies on soil microbial communities. The use of molecular techniques for investigating microbial diversity in soil communities continues to provide new understanding of the distribution and diversity of organisms in soil habitats. The use of RNA or DNA sequences, combined with fluorescent oligonucleotide probes, provides a powerful approach for the characterization and study of soil microbes that cannot currently be cultured. Among the most useful of these methods are those in which small subunit RNA genes are amplified from soil-extracted nucleic acids. Using these techniques, microbial RNA genes can be detected directly from soil samples and sequenced. These sequences can then be compared with those from known microorganisms. Additionally, group- and taxon-specific oligonucleotide probes can be developed from these sequences, making direct determination of microorganisms in soil habitats possible.

Phospholipid fatty acid analysis and community-level physiological profiles have also been utilized successfully by soil scientists, to access a greater proportion of the soil microbial community than can be obtained using culturing techniques. In recent years, molecular methods for soil microbial community analysis have provided new understanding of the phylogenetic diversity of microbial communities in soil (Insam *et al.*, 1997; Loczko *et al.*, 1997; Hill *et al.*, 2000); Chapter 8 describes a selection of these methods.

4. Plant–microbe interactions

The rhizosphere is recognized as the zone of influence of plant roots on the associated biota and soil (Lynch, 1998). Most studies to date have involved an ecophysiological description of this region, with emphasis on the influx of nutrients to plants, including nutrient supply mediated by symbionts (e.g. mycorrhizal fungi and nitrogen-fixing *Rhizobium* bacteria) and free-living microorganisms (e.g. plant-growth-promoting bacteria), and the efflux of photosynthetic carbon compounds, which provide essential substrates for the associated biota, from plant roots (rhizodeposition products). These qualitative and quantitative studies have been very valuable for generating energy budgets of plant and crop productivity.

Some of the methods used in the rhizosphere are the same as those used in bulk soil for determination of biomass, activity and diversity (as described in Chapters 6, 7 and 8). In addition, there are more specific techniques; for example, those for evaluating soil-nodulating potential (of nitrogen fixers), bioassays using arbuscular mycorrhizal fungi, bio-indicators for assessing phytosanitary soil quality and assessment of indigenous free-living plant-beneficial bacteria. Chapter 9 provides a selection of methods that relate soil microbial activity to plants.

Relationships Between Different Parameters and Evaluation of Results

None of the four method groups stands alone (Fig. 1.1). They can often be interfaced, and the decision to include one method in a given category rather than another is a consequence of the type of interpretation one wants to give to the results obtained. For instance, the soil adenosine triphosphate (ATP) content has been used as an indicator of both biomass (group 1) and activity (group 2). The use of ATP as an index of microbial biomass is based on the assumption that ATP is present as a relatively constant component of microbial cells, and that it is not associated with dead cells nor adsorbed to soil particles. A significant correlation was found between the ATP content and the microbial biomass of different soils (Jenkinson, 1988). However, the linear relationship between ATP and microbial biomass only holds when both are determined after soil pre-incubation at constant temperature and moisture conditions. The ATP content changes rapidly, depending on the physiological state of the cell. Therefore, it was hypothesized that ATP content measured immediately after sampling reflects microbial activity rather than biomass (Jenkinson *et al.*, 1979). The accuracy of interpretation and comparison of ATP values in different soils depends on the methods used to extract ATP from soil as well as on soil handling (Nannipieri *et al.*, 1990); for this reason, ATP determination was not included in our selection of methods.

The substrate-induced respiration (SIR) method, introduced by Anderson and Domsch (1975, 1978), depends on microbial biomass as well as activity, and reflects the metabolically active component of the microbial

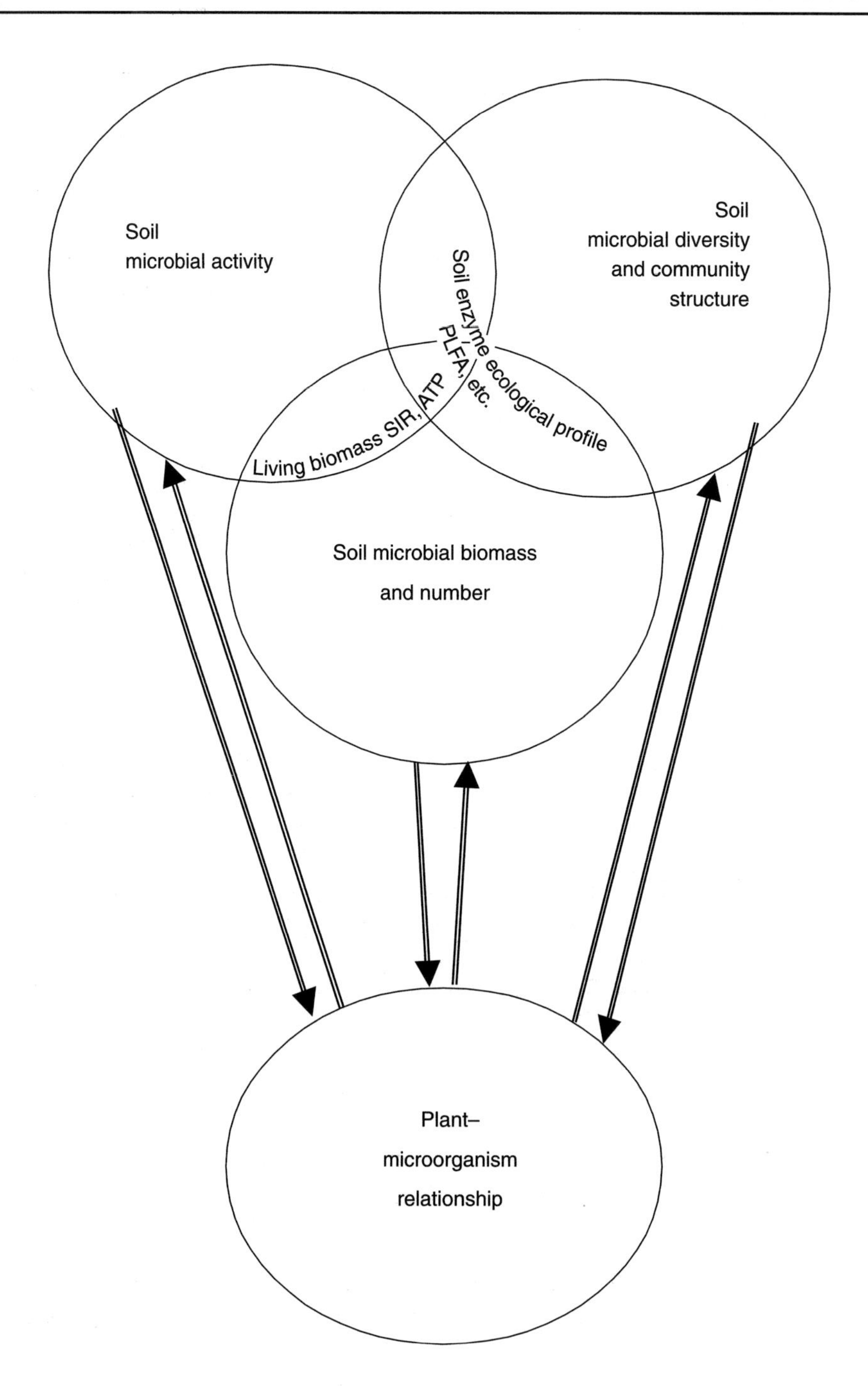

Fig. 1.1. Relationships between different soil microbiological parameters. SIR, substrate-induced respiration; ATP, adenosine triphosphate; PLFA, phospholipid fatty acid.

biomass. The microbial respiratory activity (usually determined as
lution) of a glucose-amended soil is stimulated to a maximum wi
minutes after adding saturating amounts of substrate. The enhanc
respiration is usually stable for 6–8 h and is assumed to depend on the level of microbial biomass of the soil (Sparling, 1995). Thus, the initial respiratory response to glucose is taken as an index of the soil microbial biomass before the start of microbial growth (Howarth and Paul, 1994; Sparling, 1995). After about 8 h, an increase in the respiratory activity up to a plateau phase reflects microbial growth.

Similar considerations apply to the community-level physiological profile method (CLPP or Biolog^{TM}), which provides information about: (i) the structure of the microbial community (group 3); (ii) the efficiency of specific functional groups of microorganisms in metabolizing specific substrates (groups 2 and 3); or (iii) enzymatic activities (group 2). Functional diversity, as determined by CLPP, reflects both the genetic diversity and the physiological activity of organisms inhabiting the system, and is more important for the long-term stability of an ecosystem than diversity at the taxonomical level (Garland and Mills, 1991). These so-called 'multifunctional methods' may be considered as complementary to Brookes' (1995) concept of 'internal control', which describes biochemical and chemical parameters as being interrelated.

In fact, one of the most complex parts of the soil microbiologist's work is in the assessment of relationships between different parameters, as it embraces the choice of the appropriate monitoring techniques and consideration of the interpretative criteria of results obtained by the previously mentioned methods.

How should the analytically acquired results be evaluated, which information should be deduced and what strategies should be adopted? Luckily, the literature comes to our rescue and proposes approaches for the integrated processing of results, such as amoeba, star or cobweb diagrams and the use of a Soil Quality Index (SQI) to summarize large amounts of data (Chapter 3). Recently, Herrick (2000) affirmed that soil quality appears to be an ideal indicator of sustainable land management, provided that:

1. Causal relationships between soil quality and ecosystem functions are demonstrated, including biodiversity conservation, biomass production and conservation of soil and water resources.
2. The power of soil quality indicators to predict response to disturbance is increased.
3. Accessibility of monitoring systems to land managers is increased.
4. Soil quality is integrated with other biophysical and socio-economic indicators.
5. Soil quality is placed in a landscape context.

Table 1.1. Groups of microbiological, biochemical and molecular methods.

Method groups			
1. Soil microbial biomass and number	2. Soil microbial activity	3. Soil microbial diversity and community structure	4. Plant–microbe interactions
Chloroform fumigation extraction Substrate-induced respiration Direct microscopic counts	Without substrate Soil respiration N mineralization With substrate or tracer Nitrification Thymidine and leucine incorporation N_2O emission and denitrification	Molecular methods based on microbial DNA or RNA Community-level physiological profiles (BIOLOG) Phospholipid fatty acid analysis	Nodulating symbiotic bacteria Arbuscular mycorrhiza Phytosanitary soil quality Free-living plant-beneficial microorganisms

Handbook Contents

In conclusion, the first three chapters of this handbook introduce the three main topics that are decisive factors leading to the choice and subsequent use of some selected microbiological parameters as environmental indicators:

Defining soil quality → Monitoring and evaluating soil quality → Managing soil quality

Issues related to 'Managing soil quality' (Chapter 4) were presented and discussed during the final COST 831 Joint Working Groups meeting in Budapest, Hungary (September 2002).

The general section is followed by a technical section where the methods are set out into four groups, according to the classification above (Table 1.1).

A brief description is given of the potential of each method group (Chapters 6–9), with a selection of only some of the parameters available, i.e. the ones having the requisites set down in the introduction. The selected parameters are accompanied by a detailed description of the method according to the design used for ISO standardization.

The final chapter gives the results of a census of the main methods used in over 30 European laboratories which have participated in COST Action 831.

References

Alef, K. (1995) Nutrient sterilization, aerobic and anaerobic culture technique. In: Alef, K. and Nannipieri, P. (eds) *Methods in Applied Soil Microbiology and Biochemistry*. Academic Press, New York, pp. 123–133.

Alef, K. and Nannipieri, P. (1995) *Methods in Applied Soil Microbiology and Biochemistry*. Academic Press, New York.

Anderson, J.P.E. and Domsch, K.H. (1975) Measurement of bacterial and fungal contributions to soil respiration of selected agricultural and forest soil. *Canadian Journal of Microbiology* 21, 314–322.

Anderson, J.P.E. and Domsch, K.H. (1978) A physiological method for the quantitative measurement of microbial biomass in soil. *Soil Biology and Biochemistry* 10, 215–221.

Benedetti, A., Tittarelli, F., Pinzari, F. and De Bertoldi, S. (2000) *Proceedings of the Joint WGs Meeting of the Cost Action 831 Biotechnology of Soil: Monitoring, Conservation and Remediation*, 10–11 December 1998, Rome. European Communities, Luxembourg.

Bloem, J. and Breure, A.M. (2003) Microbial indicators. In: Markert, B.A., Breure A.M. and Zechmeister, H.G. (eds) *Bioindicators/Biomonitors – Principles, Assessment, Concept*. Elsevier, Amsterdam, pp. 259–282.

Bloem, J., Bolhuis, P.R., Veninga, M.R. and Wieringa, J. (1995) Microscopic methods for counting bacteria and fungi in soil. In: Alef, K. and Nannipieri, P. (eds) *Methods in Applied Soil Microbiology and Biochemistry*. Academic Press, New York, pp. 162–173.

Brookes, P.C. (1995) The use of microbial parameters in monitoring soil pollution by heavy metals. *Biology and Fertility of Soil* 19, 269–279.

Dobereiner, J. (1995) Isolation and identification of nitrogen fixing bacteria from soil and plants. In: Alef, K. and Nannipieri, P. (eds) *Methods in Applied Soil Microbiology and Biochemistry*. Academic Press, New York, pp. 134–135.

Domsch, K.H. (1980) Interpretation and evaluation of data. *Recommended Tests for Assessing the Side-effects of Pesticides on the Soil Microflora*. Weed Research Organization Technical Report No. 59, pp. 6–8.

Domsch, K.H., Jagnow, G. and Anderson, T.H. (1983) An ecological concept for the assessment of side-effects of agrochemicals on soil micro-organisms. *Residue Reviews* 86, 65–105.

Doran, J.W. (2000) Soil health and sustainability: managing the biotic component of soil quality. *Applied Soil Ecology* 15, 3–11.

Garland, J.L. and Mills, A.L. (1991) Classification and characterization of heterotrophic microbial communities on the basis of patterns of community-level sole-carbon-source utilization patterns. *Applied and Environmental Microbiology* 57, 2351–2359.

Herrick, J.E. (2000) Soil quality: an indicator of sustainable land management? *Applied Soil Ecology* 15, 75–83.

Hill, G.T., Mitkowski, N.A., Aldrich-Wolfe, L., Emele, L.R., Jurkonie, D.D., Ficke, A., Maldonado-Ramirez, S., Lynch, S.T. and Nelson, E.B. (2000) Methods for assessing the composition and diversity of soil microbial communities. *Applied Soil Ecology* 15, 25–36.

Howarth, W.R. and Paul, E.A. (1994) Microbial biomass. In: Weaver, R.W., Angle, S., Bottomley, P., Bezdicet, D., Smith, S., Tabatabai, A. and Woollen, A. (eds) *Methods of Soil Analysis. Part 2: Microbiological and Biochemical Properties*. Soil Science Society of America, Madison, Wisconsin, pp. 753–773.

Insam, H., Amor, K., Renner, M. and Crepaz, C. (1997) Changes in functional abilities of the microbial community during composting of manure. *Microbial Ecology* 31, 77–87.

Jenkinson, D.S. (1988) Determination of microbial biomass carbon and nitrogen in soil. In: Wilson, J.K. (ed.) *Advances in Nitrogen Cycling in Agricultural Ecosystems*. CAB International, Wallingford, UK, pp. 368–386.

Jenkinson, D.S., Davidson, S.A. and

Powlson, D.S. (1979) Adenosine triphosphate and microbial biomass in soil. *Soil Biology and Biochemistry* 11, 521–527.

Loczko, E., Rudaz, A. and Aragno, M. (1997) Diversity of anthropogenically influenced or disturbed soil microbial communities. In: Insam, H. and Rangger, A. (eds) *Microbial Communities Functional Versus Structural Approaches*. Springer-Verlag, Berlin, pp. 57–67.

Lorch, H.J., Benckieser, G. and Ottow, J.C.G. (1995) Basic methods for counting microorganisms in soil and water. In: Alef, K. and Nannipieri, P. (eds) *Methods in Applied Soil Microbiology and Biochemistry*. Academic Press, New York, pp. 136–161.

Lynch, J.M. (1998) What is the rhizosphere? In: Atkinsons, D. (ed.) *Proceedings of Inter Cost Actions 821, 830, 831 Meeting*. Agricultural School, 17–19 September, Edinburgh.

Madsen, E.L. (1996) A critical analysis of methods for determining the composition and biogeochemical activities of soil microbial communities *in situ*. In: Stotzky, G. and Bollag, J.M. (eds) *Soil Biochemistry*, vol. 9, Marcel Dekker, New York, pp. 287–370.

Nannipieri, P., Ceccanti, B. and Grego, S. (1990) Ecological significance of the biological activity in soil. In: Bollag, J.M. and Stotzky, G. (eds) *Soil Biochemistry*, vol. 6, Marcel Dekker, New York, pp. 293–355.

Sparling, G.P. (1995) The substrate-induced respiration method. In: Alef, K. and Nannipieri, P. (eds) *Methods in Applied Soil Microbiology and Biochemistry*. Academic Press, New York, pp. 397–404.

UNCED (United Nations Conference on Environment and Development) (1992) Agenda 21. June, Rio de Janeiro.

Zuberer, D.A. (1994) Recovery and enumeration of viable bacteria. In: Weaver, R.W., Angle, S. and Bottomley, P. (eds) *Methods of Soil Analysis. Part 2: Microbiological and Biochemical Properties*. Soil Science Society of America Book Series, No. 5, Madison, Wisconsin, pp. 119–144.

2 Defining Soil Quality

RICHARD G. BURNS,[1] PAOLO NANNIPIERI,[2] ANNA BENEDETTI[3] AND DAVID W. HOPKINS[4]

[1]School of Land and Food Sciences, The University of Queensland, Brisbane, Queensland 4072, Australia; [2]Dipartimento de Nutrizione delle Pianta e Scienza del Suolo, Università di Firenze, Piazzale di Cascine 28, Florence, Italy; [3]Consiglio per la ricerca e la sperimentazione in Agricoltura, Istituto Sperimentale per la Nutrizione delle Piante, Via della Navicella, 2, 00184 Rome, Italy; [4]School of Biological and Environmental Sciences, University of Stirling, Stirling FK9 4LA, UK

Abstract

Environmental quality is a complex concept. Defining one component of it, soil quality, is therefore usually attempted using indicators that represent, with differing levels of approximation, particular constituents, processes or conditions. In this chapter, we review briefly the ideal characteristics of a soil quality indicator and then outline some of the national frameworks for assessing soil quality that have been proposed. A recurrent theme of the existing frameworks is the use of parameters that individually give useful information, but which can be aggregated to provide an overall indicator or index of soil fertility.

Introduction

Environmental quality is a composite of the desirable properties of soil, air and water. For water and air, where relatively precise analyses can be reported, analytical data do not necessarily provide an holistic assessment of the quality of these components of the biosphere. Soils represent an even more complex environment because they are an intimate mixture of the living and non-living components and because they vary naturally in both space and time over a range of scales. Defining soil quality is, therefore, usually attempted using somewhat arbitrarily chosen chemical, biological and physical indicators which represent particular constituents, processes or conditions. A good indicator of quality must have several characteristics. It must be representative of the sites to which it is being applied; it must be accessible both in terms of the availability of the methods required to measure it and the ease with which the measurements can be interpreted by the end-user; and it must be reliable, meaning that it must be reproducible and applicable to a range of sites. Since soil quality cannot be summarized by a single component or process, its assessment must include information about several indicators. These, depending on the stated purpose, may have different scales of measurement (e.g. aggregates, horizons, profiles, catchments) and make a different proportional contribution to the evaluation of fertility. For example, in order to describe the extent, nature and likely impact of a pollutant in a particular soil, it is necessary to employ a range of indicators. These will include the concentration of the pollutants and their vertical and horizontal distribution across the site, and intrinsic soil factors such as pH, clay and organic matter content, and ion exchange capacity. All these must be considered because they will influence the bioavailability of the pollutant and, therefore, its persistence, movement and effect on selected important processes. On the other hand, if the objective is prediction of plant nutrient availability, legume nodulation or natural biological control, other factors will assume importance. From the extensive literature, it is possible to deduce several characteristics that might contribute to an ideal indicator of soil fertility, and many of these have been summarized in the literature; for example, the Organization for Economic Co-operation and Development (OECD, 1999), recognized seven categories, as follows.

1. Political relevance and user benefits; indicators should:

- provide a representative picture of the environmental conditions and of the societal pressures or reactions to the changing state of the environment;
- be simple, easy to interpret and able to indicate temporal trends;
- be reactive to environmental changes and to related human activity;
- provide a basis for international comparisons;
- have national worth and be applicable to nationally relevant regional themes;
- have threshold or reference values, such that users can evaluate the significance of the indicator values.

2. Analytical validity; analyses should:

- be well founded, both technically and scientifically;
- be based, where possible, on international standards and have international consensus in terms of validity;
- be easily applied to economic models, forecast estimates and information systems.

3. Measurability; measurements should:

- be easily available or made available at a reasonable cost:benefit ratio;
- be adequately documented and of verified quality;
- be able to be updated at regular intervals according to well-defined procedures.

4. Representativeness; indicators should:

- correlate with a specific phenomenon or characteristic;

- correlate with previously reported effects with the minimum of statistical dispersion;
- not be easily obscured by profile factors;
- have sufficient general validity to many analogous, non-identical situations.

5. Accessibility; indicators should:

- be easily measurable;
- offer the possibility of being monitored automatically;
- be easy to sample and have a threshold of analytical detection which is accessible by standard techniques.

6. Reliability; indicators should:

- have minimum systematic errors.

7. Operativeness; indicators should:

- be easily and directly utilizable for quantifying acts of intervention, costs and benefits.

The above list serves as a guide to the selection of useful indicators, but it should be recognized that no single indicator can meet all requirements. Furthermore, a major problem in the use of any indicator is the establishment of threshold or reference values. This is only possible if many data are available and, even then, is a somewhat subjective choice, based on the current and projected use of the land (see also Chapter 3).

According to the OECD (1999) the 'definitions of indicators as a concept (let alone specific indicators), vary widely' and, furthermore, different agencies and authors use their own terms and definitions. These include: variables, parameters, measures, statistical measures, proxy measures, values, measuring instruments, fractions, indices, a piece of information, empirical models of reality and signs! The OECD (1993) bravely attempted to define a few of these terms, thus:

- *parameter* – a property that is measured or observed;
- *indicator* – a value derived from parameters, which points to or provides information about, or describes, the state with a significance extending beyond that directly associated with a parameter value; and
- *index* – a set of aggregated or weighted parameters or indicators.

Any parameter that gives useful information on soil quality can be used as an indicator, and a set of indicators can be aggregated into an index. However, in order to correctly apply the terminology it is necessary to understand the meaning of each term. For example, soil organic matter content is universally recognized as an *indicator* of soil quality and, in general, the organic carbon content of an agricultural soil is a *parameter* closely correlated with the organic matter content. At temperate latitudes, the average organic matter content falls between 1% and 5%. A reduction of this quantity over a period of time is likely to be an *index* of impoverishment and could be a strong *indicator* of the loss of soil fertility and deterioration of other soil properties – such as structural stability and water retention. Similarly, the effectiveness of an amendment can be evaluated by measuring the increase in organic carbon, which then becomes an *indicator* of effectiveness of the fertilizer used. Organic matter, together with other *parameters*, can thus become an *index* of soil quality. An example of a soil quality index is given in Chapter 3. Thus, a change in the value of a single *parameter* outside a certain range can be the important *sign* of quality improvement or reduction.

International Indicators of Soil Quality

Since the beginning of the 1980s, a decrease in soil productive capacity has been observed in more than 10% of cultivated land worldwide, as evidenced by soil erosion, atmospheric pollution, amount of land in farming, excessive grazing, salinization and desertification (Francaviglia, 2004; Van-Camp *et al.*, 2004).

A definitive set of basic indicators for the evaluation of soil quality has not yet been provided, despite various international proposals, including that published by the Soil Science Society of America (SSSA). This is due mainly to the continuing difficulty in defining soil quality and how it can be assessed. Many definitions have been suggested in recent years, but one that best represents the concept was given by Doran and Parkin (1994): 'The capacity of the soil to interact with the ecosystem in order to maintain biological productivity, environmental quality and to promote animal and plant health.' This definition is similar to the three essential criteria for soil quality that were identified by the Rodale Institute (1991), namely:

- *productivity* – the soil's capacity to increase plant biological productivity;
- *environmental quality* – the soil's capacity to attenuate environmental contamination, pathogens and external damage; and
- *health of living organisms* – the interrelation between soil quality and animal, plant and human health.

The parameters for the evaluation of soil quality can be subdivided into those that are physical, chemical and biological. However, integration among them is fundamental to our understanding. Currently, definitions of soil quality standards are being discussed within international regulatory bodies. The US Environmental Protection Agency (EPA), for example, has proposed over 1800 parameters as chemical indicators of soil quality. Within the OECD, agroenvironmental indicators (there are approximately 250), including those related to soil quality, are currently being defined. So far, 58 have been proposed as soil quality indicators, but some of them are different approaches to assessment of the same indicator, for example organic matter content estimated by modelling and by analysis. The indicators and parameters in Table 2.1 have been proposed for soil and site assessment by the ISO Technical Committee 190 on 'Soil Quality' and correspond to the physical, chemical and biological parameters essential in the consideration of soil restoration.

Although approaches to assessing soil quality have developed independently in several countries, there is considerable overlap between the parameters listed in Table 2.1, those proposed by the SSSA (Table 2.2), and those used in New Zealand (Table 2.3). In the case of both the USA and

Table 2.1. Parameters proposed in a working document of ISO Technical Commitee 190 'Soil Quality', Sub-Committee 7 'Soil and Site Assessment'.

Parameters	International standards
Physical parameters	
Petrographic features	
Mineralogy	
Nature of the mother rock	
Soil profile	
Texture	
Water content	ISO 10537
Presence of roots	
Hydraulic conductivity	DIS 11275–1/DIS 11275–2
Pore water pressure	CD 15048/ISO 11276
Plasticity index	
Consistency	
Structure stability	
Degree of infiltration	
Particle size distribution	ISO 11277
Aggregation state	DIS 11273–1
Skeleton	CD 11273–2
Apparent density	FDIS 11272
Chemical parameters	
pH	ISO 10390
Redox potential	ISO 11271
Salinity	
Sodium	
Total organic carbon	ISO 10694
Carbon dioxide losses at specific temperatures	
Cation exchange capacity	ISO 11260/ISO 13526
Dry matter content	ISO 11465
Carbonates	ISO 10693
Specific electric conductivity	ISO 11265
Exchange acidity	DIS 14254
Biological indicators	
Microbial activity	ISO 14239/ISO 11266/ISO 14238/NP 15473
Harmful plant species	
Toxicity for plants	ISO 11269
Toxicity for microorganisms	
Presence of pathogens	
Microbial biomass	ISO 14240
Toxicity for macrofauna	ISO 11268

New Zealand, there is acceptance of a range of complementary parameters. The New Zealand Soil Indicators (SINDI) approach relies on a small set of indicators matched to particular national issues, so that Olsen P is, for example, prioritized as the principal indicator of soil fertility, and direct biological assessment is limited to a nitrogen mineralization assay which simultaneously provides a soil fertility indicator and acts as a surrogate for microbial biomass. Clearly, this approach reduces the demand for time-consuming and technically complex laboratory analyses. A useful feature of SINDI is that it is supported by an on-line assessment framework (http://sindi.landcare.cri.nz, accessed 25 November 2004), in which the values for the different indicators can be compared with the expected norms for particular soil types, and in which there are links to management information and advice. Although there is no minimum dataset recognized for the assessment of soil quality in Canada, the same multifaceted approach was adopted by Agriculture and Agri-Food Canada, which included assessment of the soil organic resources, structural condition, contamination and hydrological conditions (Acton and Gregorich, 1995).

Soil quality depends on several biological, chemical and physical soil properties and, theoretically, its definition should require the determination

Table 2.2. Physical, chemical and biological features proposed as basic indicators of soil quality and based on the definition of Doran and Parkin (1994).

Soil features	Methodology
Physical indicators	
Soil texture	Water-gauge method
Depth of the soil and root systems	Soil excavation and extraction
Apparent density and infiltration	Field determination with the use of infiltration rings
Water retention features	Water content at pressures of 33 kPa and 1500 kPa
Water content	Gravimetrical analysis (weight loss over 24 h at 105°C)
Soil temperature	Thermometer
Chemical indicators	
Total organic C and N	Combustion (volumetric method)
pH	Field and laboratory determinations with pH meter
Electrical conductivity	Field and laboratory determinations with a conductometer
Inorganic N (NH_4^+ and NO_3^-), P and K concentrations	Field and laboratory determinations (volumetric method)
Biological indicators	
C and N from microbial biomass	Fumigation/incubation with chloroform (volumetric method)
Potentially mineralizable N	Anaerobic incubation (volumetric method)
Soil respiration	Field determination by means of covered infiltration rings, and in the laboratory by measuring the biomass

Table 2.3. Indicators of soil quality for New Zealand used in the SINDI (Soil Indicators) scheme (http://sindi.landcare.cri.nz).

Soil property	Comments
Soil fertility indicator	
Olsen P	Plant-available phosphorus
Soil pH	Acidity or alkalinity of soil
Organic resources	
Anaerobic nitrogen mineralization	Availability of the nitrogen reserve to plants and a surrogate measure of microbial biomass
Total (organic) C	Organic matter reserves, which is also positively related to soil structure and ability to retain water
Total N	Organic N reserves
Soil physical quality	
Bulk density	Soil compaction, physical environment for roots and soil organisms
Macroporosity	Availability of water and air, retention of water, drainage properties

of these properties. Biological parameters have assumed particular importance in the assessment of soil quality because organisms respond more rapidly than most chemical and physical parameters to changes in land use, environmental condition or contamination (Doran and Parkin, 1994; Nannipieri *et al.*, 2001; Nannipieri and Badalucco, 2002; Gil-Sotres *et al.*, 2005). It is equally well established that soil organisms play crucial roles in many processes that underpin soil quality, such as organic matter decomposition and nutrient cycling, nitrogen fixation and aggregate formation and stabilization. For this reason, the size of the soil microbial biomass, respiration, potential nitrogen (N) mineralization, enzyme activities, abundance of fungi, nematodes and earthworms have all been used as indicators of soil quality (Lee, 1985; Doran, 1987; Dick *et al.*, 1988; Kennedy and Papendick, 1995; Wall and Moore, 1999). The following chapters present some of the methods commonly used as indicators and critically evaluate their contribution to soil quality.

References

Acton, D.F. and Gregorich, L.J. (1995) *The Health of Our Soils – Towards Sustainable Agriculture in Canada*. Centre for Land and Biological Resources Research, Research Branch, Agriculture and Agri-Food Canada, Ottawa.

Dick, R.P., Rasmunssen, P.E. and Kerle, E.A. (1988) Influence of long term residue management on soil enzyme activities in relation to soil chemical properties of a wheat-fallow system. *Biology and Fertility of Soils* 6, 159–164.

Doran, J.W. (1987) Microbial biomass and mineralizable nitrogen distrubution in no-tillage and plowed soils. *Biology and Fertility of Soils* 5, 68–75.

Doran, J.W. and Parkin, T.B. (1994) Defining and assessing soil quality. In: *Defining Soil*

Quality for a Sustainable Environment, Soil Science Society of America Special Publication no. 35. SSSA, Madison, Wisconsin.

Francaviglia, R. (ed.) (2004) Agricultural impacts on soil erosion and soil biodiversity: developing indicators for policy analysis. *Proceedings of the OECD Expert Meeting on Soil Erosion and Soil Biodiversity Indicators, 25–28 March 2003, Rome, Italy.* OECD, Paris. Available at: http://webdomino1.oecd.org/comnet/agr/soil_ero_bio.nsf (accessed 17 December 2004).

Gil-Sotres, F., Trasar-Cepeda, C., Leiros, M.C. and Seoane, S. (2005) Different approaches to evaluating soil quality using biochemical properties. *Soil Biology & Biochemistry* 37, 877–887.

Kennedy, A.C. and Papendick, R.I. (1995) Microbial characteristics of soil quality. *Journal of Soil and Water Conservation* 50, 243–248.

Lee, K.E. (1985) *Earthworms: Their Ecology and Relationship with Soil and Land Use.* Academic Press, London.

Nannipieri, P. and Badalucco, L. (2002) Biological processes. In: Kenbi, D.K. and Nieder, R. (eds) *Handbook of Processes and Modelling in the Soil–Plant System.* The Haworth Press Inc., Binghampton, New York, pp. 57–82.

Nannipieri, P., Kandeler, E. and Ruggiero, P. (2001) Enzyme activities and microbiological and biochemical processes in soil. In: Burns, R.G. and Dick, R. (eds) *Enzymes in the Environment.* Marcel Dekker, New York, pp. 1–33.

OECD (1993) *OECD Core Set of Indicators for Environmental Performance Reviews – A synthesis report by the group on the state of the environment.* Environmental Monographs no. 83. OECD/GD(93)179. Organization for Economic Cooperation and Development, Paris.

OECD (1999) *Environmental Indicators for Agriculture*, vol. 2, *Issues and Design.* The York Workshop, Organization for Economic Cooperation and Development, Paris.

Rodale Institute (1991) *Conference Report and Abstract, International Conference on the Assessment and Monitoring of Soil Quality. Emmaus, Pennsylvania, 11–13 July 1991.* Rodale Press, Emmaus, Pennsylvania.

Van-Camp, L., Bujarrabal, B., Gentile, A.-R., Jones, R.J.A., Montanarella, L., Olazabal, C. and Selvaradjou, S.-K. (2004) Reports of the Technical Working Groups Established under the Thematic Strategy for Soil Protection. EUR 21319 EN/3, 872 pp. Office for Official Publications of the European Communities, Luxembourg. Available at: http://eusoils.jrc.it/ESDB_Archive/eusoils_docs/doc.html#OtherReports (accessed on 17 December 2004).

Wall, D.H. and Moore, J.C. (1999) Interactions underground-soil biodiversity, mutualism, and ecosystem processes. *BioScience* 49, 109–117.

3 Monitoring and Evaluating So Quality

JAAP BLOEM,[1] ANTON J. SCHOUTEN,[2] SØREN J. SØRENSEN,[3] MICHIEL RUTGERS,[2] ADRI VAN DER WERF [4] AND ANTON M. BREURE[2]

[1]*Department of Soil Sciences, Alterra, PO Box 47, NL-6700 AA Wageningen, The Netherlands;* [2]*Laboratory for Ecological Risk Assessment, RIVM, PO Box 1, NL-3720 BA Bilthoven, The Netherlands;* [3]*Department of General Microbiology, Institute of Molecular Biology, University of Copenhagen, Sølvgade 83H, DK 1307K Copenhagen, Denmark;* [4]*Plant Research International, PO Box 16, NL-6700 AA Wageningen, The Netherlands*

Abstract

Soil quality influences agricultural sustainability, environmental quality and, consequently, plant, animal and human health. Microorganisms are useful indicators of soil quality because they have key functions in the decomposition of organic matter, nutrient cycling and maintenance of soil structure. We summarize methods used for monitoring biomass, activity and diversity of soil organisms and show some results of the Dutch Soil Quality Network.

In contaminated soils microbial community structure was changed, but diversity was not always reduced. In contrast, microbial biomass and activity were reduced markedly. In agricultural soils there were large differences between different categories of soil type and land use. Organic management resulted in an increased role of soil organisms, as indicated by higher numbers and activity. Replacement of mineral fertilizers by farmyard manure stimulated the bacterial branch of the soil food web. Reduced availability of mineral nutrients appeared to increase fungi, presumably mycorrhizas. Bacterial DNA profiles did not indicate low genetic diversity in agricultural soils, compared with some acid and contaminated soils. Organic farms did not show higher genetic diversity than intensively farmed areas. At extensive grassland farms and organic grassland farms nitrogen mineralization was about 50% higher than on intensively farmed areas. Also, microbial biomass and activity, and different groups of soil invertebrates, tended to be higher.

Soil biodiversity cannot be monitored meaningfully with only a few simple tools. Extensive and long-term monitoring is probably the most realistic approach to obtain objective information on differences between, changes within, and human impact on, ecosystems. In most countries, microbial biomass, respiration and potential nitrogen

(N) mineralization are regarded as part of a minimum data set. Adding the main functional groups of the soil food web brings us closer to understanding biodiversity, potentially enabling us to relate the structure of the soil community to functions.

Soil Quality Monitoring and Microbiological Indicators

Following adoption of the Treaty on Biological Diversity of Rio de Janeiro (UNCED, 1992), participating governments have been concerned about the protection of endangered species, mainly plants and larger animals. Viable nature conservation areas, consisting of core areas linked by transition zones, have been developed. In addition, there is increasing concern, at both the national and the international level, about sustainable use of biodiversity and maintenance of life support functions such as decomposition and nutrient cycling (FAO, 1999; OECD, 2003; Schloter *et al.*, Chapter 4, this volume). In all soils, these vital ecosystem processes depend largely on the activities of microorganisms and small soil invertebrates that are rarely visible with the naked eye (also called 'cryptobiota').

Soil quality determines agricultural sustainability, environmental quality and, consequently, plant, animal and human health (Doran and Parkin, 1996). This chapter provides an introduction to biological approaches presently used in different countries to monitor and evaluate soil quality. Monitoring was initiated in several countries in 1992, but little information has been exchanged or published in the international literature so far. This chapter is based mainly on the experience of the Dutch Soil Quality Network (Schouten *et al.*, 2000; Bloem and Breure, 2003; Bloem *et al.*, 2004), and discussions of the working groups of EU COST Action 831 'Biotechnology of Soil, Monitoring, Conservation and Remediation'.

Soil quality has been defined as 'the capacity of a soil to function within ecosystem boundaries to sustain biological productivity, maintain environmental quality, and promote plant and animal health' (Doran and Parkin, 1994; Stenberg, 1999). The phrase 'within ecosystem boundaries' implies that each soil is different. There are no absolute quality estimates and each soil must be evaluated in relation to natural differences such as soil type, land use and climate. The term 'soil quality' is often used to describe the fitness of a soil for (agricultural) use, while the term 'soil health' is seen more as an inherent attribute regardless of land use. Often these terms are used as synonyms. There are many definitions of soil quality and soil health (Burns *et al.*, Chapter 2, this volume). Quality or health of an ecosystem is a value judgement. Although ecological health has been criticized as a nebulous concept in a scientific context, a useful consequence of that notion is that environmental monitoring programmes need to adopt a holistic ecosystem approach (Lancaster, 2000). Many different aspects need to be measured, including physical, chemical and biological characteristics. Here we focus on soil organisms and the processes they mediate.

An agricultural soil usually contains about 3000 kg (fresh weight) of soil organisms per hectare. This is equivalent to 5 cows, 60 sheep or 35 farmers living under the surface. Many thousands of species (or genotype contribute to a huge below-ground biodiversity. Soil invertebrates fragment dead organic matter and thus facilitate decomposition. Their direct contribution to the biochemical modification or flux of organic residue is usually small compared with the contribution of bacteria and fungi. Decomposition by bacteria and fungi causes release of mineral nutrients (mineralization) essential for plant growth. Mineralization is further performed by organisms that feed on bacteria and fungi (bacterivores and fungivores), such as protozoa and nematodes. Some small soil invertebrates (e.g. nematodes) feed directly on plant roots (herbivores). Predators eat other, usually smaller, soil invertebrates, and omnivores feed on different food sources. All these trophic interactions in the soil food web contribute to the flow of energy and nutrients through the ecosystem (Hunt *et al.*, 1987). Models predict that the abundances of the different functional groups of organisms, i.e. the structure of the soil food web, affect the stability of the soil ecosystem (De Ruiter *et al.*, 1995). Mycorrhizal fungi that live in symbiosis with plant roots promote the uptake of mineral nutrients by plants. Bacteria, fungi and invertebrates glue soil particles together, form stable aggregates and thus improve soil structure. Invertebrates also improve soil structure by mixing the soil (bioturbation).

The following comprise the main functional groups of the soil food web:

- Earthworms consume plant residues and soil, including (micro)organisms. Often they form the major part of the soil fauna biomass, with maximally 1000 individuals/m^2, 3000 kg fresh biomass/ha, or a few hundred kg of carbon (C) per hectare.
- Enchytraeids are relatives of earthworms with a much smaller size and a similar diet. Their population densities are between 10^2 and $10^6/m^2$, with a biomass up to 1 kg C/ha.
- Mites (fungivores, bacterivores, predators) have a size of about 1 mm, population densities of 10^4–$10^5/m^2$, and a biomass of up to 0.1 kg C/ha.
- Springtails (fungivores, omnivores) also have a size of about 1 mm. They reach population densities of 10^3–$10^5/m^2$ and a biomass of up to 1 kg C/ha.
- Nematodes (bacterivores, herbivores, fungivores, predators/omnivores) have a size of about 500 μm, population densities of 10–50/g soil, and a biomass up to 1 kg C/ha.
- Protozoa (amoebae, flagellates, ciliates) are unicellular animals with a size of 2–200 μm, population densities of about 10^6 cells/g soil, and a biomass of about 10 kg C/ha.
- Bacteria are usually smaller than 2 μm, with population densities of about 10^9 cells/g soil, and a biomass of 50–500 kg C/ha.
- Fungal hyphae usually have diameters from 2 μm to 10 μm, and reach total lengths of 10–1000 m/g soil, and a biomass of 1–500 kg C/ha.

These cryptobiota (hidden soil life) play a key role in life support functions (Bloem *et al.*, 1997; Brussaard *et al.*, 1997, 2003; Bloem and Breure, 2003), but are not part of any recognized list of endangered species. It is questionable whether a species-based approach is sufficient to attain a sustainable use of ecosystems inside, and especially outside, protected areas. Therefore, research networks have been initiated to monitor large areas, including agricultural soils.

Since about 1993, national or regional programmes have been established in several countries to monitor soil quality and/or the state of biodiversity (Stenberg, 1999; Nielsen and Winding, 2002). These include: Canada (23 sites), France, parts (Bundesländer) of Germany (about 350 sites; Höper, 1999; Oberholzer and Höper, 2000), parts (cantons) of Switzerland (Maurer-Troxler, 1999), the Czech Republic, the UK, Austria, the USA (21 sites; Robertson *et al.*, 1999) and New Zealand (500 sites; G. Sparling *et al.*, http://www.landcareresearch.co.nz, accessed 30 January 2004).

In The Netherlands, 200 sites are part of the Dutch Soil Quality Network, consisting of ten categories of a specific soil type with a specific land use, with 20 replicates per category (Schouten *et al.*, 2000). The replicates are mainly conventional farms. The 200 sites are representative of 70% of the surface area of The Netherlands. In addition, 50–100 sites from outside this network are sampled; for instance, organic farms or polluted areas which are supposed to be good and bad references, respectively. Each year two types of soil and land use are sampled (40 sites plus reference sites). Thus, it takes 5 years to complete one round of monitoring the whole network of 200 sites plus references. In 1993, the Dutch network started to obtain policy information on abiotic soil status. The aim was to measure changes over time and finally to evaluate the actual soil quality. A set of biological indicators has been included since 1997, consisting of microbiological indicators and several soil fauna groups, in order to take a cross-section through the soil ecosystem.

In most countries one or more microbiological indicators have been included. As part of a monitoring system, microorganisms are useful indicators of soil quality because they have key functions in decomposition of organic matter and nutrient cycling, they respond promptly to changes in the environment and they reflect the sum of all factors regulating the degradation and transformation of organic matter and nutrients (Stenberg, 1999; Bloem and Breure, 2003).

Sampling

For the application of microbiological indicators a lot of methodological choices have to be made.

How can variation in space and time be accounted for?

Mainly by taking many replicates and aiming at long-term monitoring. Samples can be taken from replicated field plots or can be pooled from larger areas. In The Netherlands, about 20 farms (replicates) spread over the country are sampled per category of soil type and land use. One mixed sample per farm (about 10–100 ha) is made up of 320 cores. These mixed samples are used for chemical, microbiological and nematode analyses. Separate soil cores or blocks (six replicates per site) are taken for analysis of mites, enchytraeids and earthworms. Some reference sites consist of smaller contaminated areas or experimental fields. Here replicated field plots (about 10 m × 10 m) are sampled.

Sampling depth is best decided by considering soil horizons and tillage depth (Stenberg, 1999). In a ploughed arable field, 0–25 cm would be appropriate; in grassland, and especially in forest, higher numbers of thinner layers would be better. However, this would result in a variable sampling depth or an increase in the number of samples by taking more than one layer. Given the large number of samples, analysing more than one depth would be too time consuming and expensive. Sampling 0–25 cm would dilute microbial activity considerably in some grassland and forest soils, where life is concentrated closer to the surface. Therefore, in the Dutch monitoring network, samples are taken from 0–10 cm depth and litter is removed before sampling. To reduce variation caused by variable weather conditions, samples are pre-incubated for 4 weeks at constant temperature (12°C) and moisture content (50% of water-holding capacity) before microbiological analyses are performed. Since each soil and land-use type in the monitoring network is analysed once every 5 years, effects of a dry summer, for instance, should be minimized.

Samples can be sieved through 2 mm or 5 mm mesh-size, or not at all. In The Netherlands and in Sweden, soil is sieved through 5 mm and 4 mm mesh sizes, respectively (it is practically impossible to pass a heavy clay soil through a 2 mm sieve). Sieving is useful to reduce variation in process rate measurements, such as respiration and mineralization, to facilitate mixing and to allow identical subsamples to be sent to different laboratories. However, sampling and sieving are major disturbances, which generally increase microbial activity and also reduce soil structure. Therefore, the results of the first week of incubation are not used for calculation of process rates.

When should samples be taken?

For microbiological parameters, early spring or late autumn is the best time. Then soil conditions are relatively mild and stable, and short-term effects of the crop are avoided. These periods are proposed in Sweden (Stenberg, 1999). In The Netherlands, for practical reasons, samples are taken from March to June. The land must be dry enough to access, and farmers prefer sampling of arable land before soil tillage and sowing new crops. Sampling of about 50 farms takes 2–3 months.

Storage and Pre-incubation of Soil

How should samples be stored: at –20°C, 1–5°C or field temperature? For how long? Obviously it is best to perform soil biological analyses soon after sampling. On the other hand, storage is inevitable when large amounts of samples from many sites have to be handled. The preferred method for storing soil samples in different countries appears to be related to the climate. In Sweden and Finland, freezing at –20°C for at most 1 year is practised. Stenberg *et al.* (1998) found that the effects of freezing were generally smaller than those of refrigeration. They suggested that microflora in northern soils, subjected annually to several freeze and thaw cycles, may have adapted to this stress factor. In the UK, Denmark, Germany and Switzerland soil samples are stored at 4°C, and in Italy samples are air-dried. It is generally recommended that samples for microbial analysis are stored at 2–4°C (Wollum, 1994; Nielsen and Winding, 2002). Biomass and activity usually tend to decrease during storage because available organic substrates are slowly depleted. This decrease is supposed to be slower at 4°C in a refrigerator, and may be stopped by freezing. However, in frozen samples we have found more than 50% reduction in bacterial cell numbers as counted by direct microscopy.

Using sandy soil from arable fields and grassland (Korthals *et al.*, 1996), we investigated the effects of storage for 6 months at 12°C, 2°C and –20°C. After storage, the soil was pre-incubated for 4 weeks at 12°C and 50% water-holding capacity, and subsequently analysed. The samples were taken in May, when the moisture content in the field was between 14% and 20% (w/w), corresponding to 47–67% of the water-holding capacity of the soil. The soil was a fimic anthrosol (FAO classification) with a texture of 3% clay, 10% silt, 87% sand, an organic carbon content of 2–3% (w/w) and a pH(-KCl) of 5. The results of the stored samples were compared to results of microbiological analyses started 1 day after sampling. With all storage methods, bacterial biomass, as determined by microscopy and image analysis, did not decline in grassland soil but was strongly reduced (–70%) in arable soil (Fig. 3.1).

Bacterial growth rate (thymidine incorporation) showed the opposite: it remained high in arable soil but was strongly reduced in grassland soil during storage at 12°C and 2°C. Thus, grassland bacteria (apparently k-strategists or persisters) survived better than arable soil bacteria (apparently r-strategists or colonizers). Grassland may select for persisters because it is a more stable environment with a relatively constant food supply from grass roots, whereas arable soils may favour colonizers because substrate inputs are highly seasonal. Thus, effects of storage may be different for different microbial communities (e.g. from grassland versus arable land) and parameters (e.g. biomass versus growth rate). After freezing, growth rate had doubled in arable soil. With all storage methods, respiration (CO_2 evolution) decreased by at least 40%. N mineralization was strongly reduced after storage at –20°C and in the arable soil also at 2°C. This may have been caused by N immobilization during re-growth of bacteria when the temperature was increased. Reduction in N mineralization was less at 12°C but

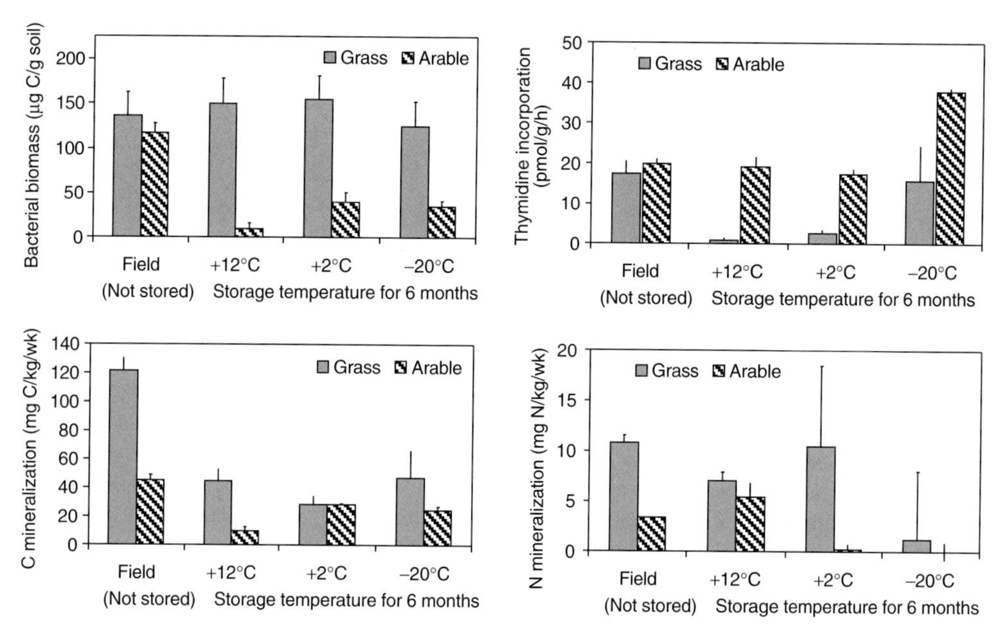

Fig. 3.1. Effect of storage of soil samples on bacterial biomass and growth rate (thymidine incorporation), soil respiration (C mineralization) and N mineralization. Error bars indicate standard error (SE), $n = 3$.

here it had decreased in grassland soil and increased in arable soil. Stenberg *et al.* (1998) reported that N mineralization capacity was greatly influenced by freezing, but that other parameters, such as basal respiration and microbial biomass, were only a little affected. After storage for 6 months, followed by 1 month pre-incubation, the number of bacterial DNA bands, as obtained by denaturing gradient gel electrophoresis, was reduced by about 20%, and some qualitative changes had occurred in the DNA banding pattern, regardless of the storage temperature. Our results support the view that soil samples for microbiological analyses should be stored for as short a time as possible (Anderson, 1987; Zelles *et al.*, 1991).

In the Dutch monitoring programme, storage of soil samples for 1–2 months is inevitable. This will cause extra variation in the results. In Germany and Switzerland, soil samples are stored for 6 months at most. In The Netherlands, a storage temperature of 12°C was chosen, which is close to the average annual soil temperature. The optimum storage method may differ for different microbiological parameters. However, using more than one storage method increases handling time and cost of monitoring. Moreover, microbiological parameters can best be related to each other when they are measured in the same portion of soil. In the Dutch soil monitoring network, after storage at 12°C, the samples are pre-incubated for 4 weeks at 12°C and 50% of the water-holding capacity. In Sweden, after freezing, samples are pre-incubated for a few days. In the UK, Germany and Italy, after storage soils are pre-incubated for 1–2 weeks. Also, the optimum time of pre-incubation may depend on the parameters to be measured. Measurement of potential microbial activities in slurries at 37°C, as in Sweden (Torstensson *et al.*, 1998), may require a shorter pre-incubation than measurements in soil under conditions similar to those in the field, as in The Netherlands.

Methods and Choice of Indicators

A range of methods is used to assess the amount (biomass), activity and diversity of soil organisms (Akkermans *et al.*, 1995; Alef and Nannipieri, 1995). No one method is best suited for all purposes and they need calibration and standardization before use (Paul *et al.*, 1999). A selection of methods is described in this book. Here, we summarize methods that are applicable to relatively large amounts of samples and which are already used in monitoring programmes.

Biomass

For methods see Chapter 6.

- Chloroform fumigation extraction (CFE): the soil is fumigated with chloroform, which permeabilizes cell membranes. The increase in extractable organic carbon (and nitrogen), compared to an unfumigated control, is a measure of the total microbial biomass (C and N).
- Substrate-induced respiration (SIR): a substrate (glucose) is added to soil at a saturating concentration and is utilized by microorganisms. The increased CO_2 evolution in the first few hours before a growth response occurs, compared to that of an unamended control, is a measure of the (responsive) microbial biomass.
- Direct microscopy can be combined with automatic image analysis (Bloem *et al.* 1995, Paul *et al.*, 1999): number and body size are determined and biomass is calculated. This can be done for different groups, e.g. fungi and bacteria. Fungi and bacteria are counted directly in soil smears after fluorescent staining. Since their numbers are much lower, soil invertebrates (nematodes, springtails, mites, etc.) are extracted from the soil before counting and identification.

Activity

For methods see Chapter 7.

- Respiration: CO_2 evolution under standardized conditions in the laboratory, without addition of substrates (basal respiration).

- Bacterial growth rate: incorporation rate of [^{3}H]thymidine and [^{14}C]leucine into bacterial DNA and proteins during a short incubation (1 h).
- Potential N mineralization: increase in mineral N under standardized moisture content and temperature in the laboratory, without addition of substrate.
- Potential nitrification: conversion of added NH_4^+ via NO_2^- to NO_3^- under optimal conditions.
- Enzyme activities, e.g. dehydrogenase, phosphatase, cellulase.

Diversity/community structure

For methods see Chapter 8.

- DNA profiles obtained by DGGE or TGGE (denaturing or temperature gradient gel electrophoresis): DNA is extracted from soil, amplified by PCR (polymerase chain reaction) and separated by gel electrophoresis. This results in a banding pattern where the number of DNA bands reflects the dominant genotypes and genetic diversity (Fig. 3.2).
- Community-level physiological profiles (CLPP): the ability to utilize a range of (31 or 95) sole-carbon-source substrates is tested in Biolog™ multiwell plates. Colour development in a well indicates utilization of a specific substrate. The pattern of colour development characterizes the functional diversity, if equal amounts of bacterial cells are added. If a fixed amount of soil is added, it reflects the number of active bacteria (Fig. 3.3).
- Phospholipid fatty acid (PLFA) analysis: PLFAs are essential membrane components of living cells. Specific PLFAs predominate in certain taxonomic groups and are relatively conservative in their concentrations within them. Measuring the concentrations of different PLFAs extracted from soils can, therefore, provide a biochemical fingerprint of the soil microbial community. The PLFA profiles reflect the community structure and show which groups are dominant. PLFAs do not, however, give any quantitative information about the number of species.
- Soil fauna: (usually microscopic) enumeration and identification of functional groups or species.

All these methods measure different aspects of microbial communities, and a combination of methods is needed for monitoring the diversity and functioning of soil microorganisms.

The choice of methods depends on the questions asked and both the expertise and budget available. Microbial biomass, respiration per unit of biomass (qCO_2) and also biodiversity are regarded as the most sensitive parameters, especially to assess the effects of soil contamination (Brookes, 1995; Giller *et al.*, 1998). The following is a list of the methods used for monitoring in several countries.

- Germany: microbial biomass (SIR), respiration, soil enzymes (Höper, 1999).

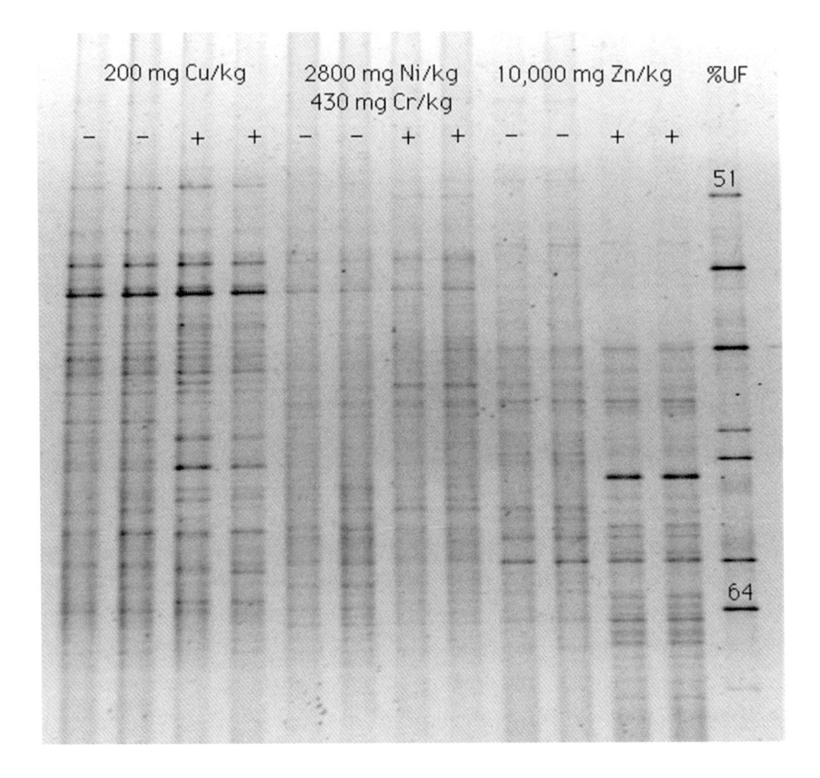

Fig. 3.2. Denaturing gradient gel electrophoresis (DGGE) DNA profiles of soils contaminated with heavy metals (+), compared to uncontaminated controls (–).

- Switzerland: earthworms, microbial biomass (CFE), respiration, N mineralization (Maurer-Troxler, 1999).
- Czech Republic: microbial biomass (SIR), respiration, N mineralization, nitrification, soil enzymes.
- United Kingdom: microbial biomass (CFE), respiration, microbial diversity (CLPP, PLFA).
- New Zealand: microbial biomass (CFE), respiration, N mineralization (G. Sparling *et al.*, available at: http://www.landcareresearch.co.nz/research/rurallanduse/soilquality/Soil_Quality_Indicators_Home.asp, accessed 30 January 2004).

We do not pretend that this list is complete. In New Zealand, the interpretation of microbial biomass and respiration measures was found to be too problematic for practical application (Carter *et al.*, 1999; Schipper and Sparling, 2000). Therefore, seven mainly abiotic soil properties were selected as core indicators of soil quality: total C, total N, mineralizable N, pH, Olsen P, bulk density and macroporosity. The only microbiological indicator used in the 500 soils project is mineralizable N as determined by anaerobic incubation under waterlogged conditions for 7 days at 40°C. The method is relatively simple (see also Canali and Benedetti, Chapter 7.3, this volume).

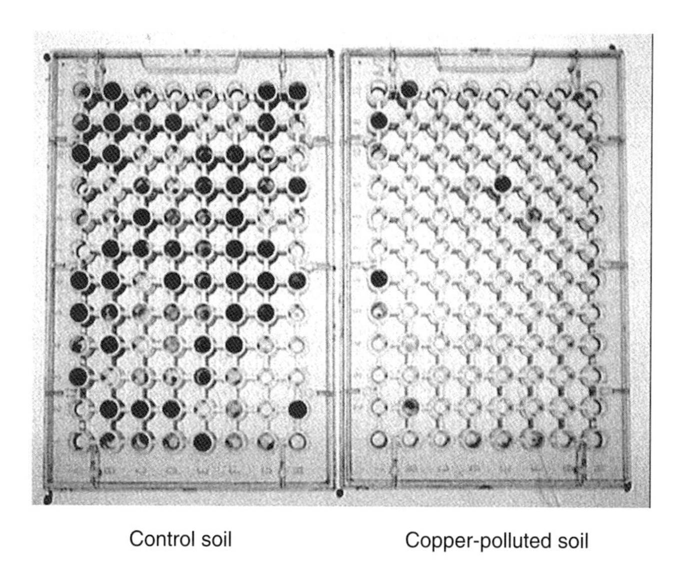

Control soil Copper-polluted soil

Fig. 3.3. Community-level physiological profiles: utilization of different substrates in Biolog^{TM} plates.

Mineralizable N is regarded as a measure of readily decomposed organic N, and as a measure of biological activity (Hill *et al.*, 2004). Thus, in most countries a relatively small number of variables is monitored, usually at a relative large number of sites (up to 500 in New Zealand). In the USA, in contrast, a large number of methods and variables are used at a limited number of 21 long-term ecological research sites (Robertson *et al.*, 1999).

In the Dutch Soil Quality Network, besides chemical and physical variables, nematodes have been measured since 1993. Since 1997 a wider range of biological variables has been included (Schouten *et al.*, 2000):

- bacterial biomass (microscopy and image analysis);
- bacterial growth rate (thymidine and leucine incorporation into DNA and proteins);
- bacterial functional diversity (substrate utilization profiles using Biolog^{TM});
- bacterial genetic diversity (DNA profiles using DGGE);
- potential carbon and nitrogen mineralization (6-week laboratory incubations);
- nematodes, abundance and diversity;
- mites, abundance and diversity;
- enchytraeids, abundance and diversity;
- earthworms, abundance and diversity.

Thus, important processes and functional groups of the soil food web are included. A relatively large number of biological variables was chosen because monitoring of soil biodiversity had a high priority. Each variable is measured in a single specialized laboratory, using a single method.

Earthworms, enchytraeids, mites and nematodes are included as important functional groups of soil fauna. Microbial indicators currently used are bacterial biomass, growth rate, functional diversity and genetic diversity (Bloem and Breure, 2003). These different microbial indicators are measured in the same subsamples of soil after pre-incubation for 4 weeks at 12°C and 50% of the water-holding capacity (WHC). Potential carbon and nitrogen mineralization are determined in soil incubated for 6 weeks at 20°C and 50% WHC. C and N mineralization rates are calculated from differences in CO_2 and mineral N concentrations between week six and week one. Results of the first week are not used, to avoid disturbance effects of sample handling.

Results from Contaminated and Experimental Reference Sites

In the Dutch Soil Quality Network, large areas of about 50 ha (mainly farms) are sampled and the replicates within categories of soil type and land use are spread all over the country. Some of the reference sites were smaller long-term experimental fields and contaminated sites. Such sites are expected to show the most clear and contrasting results.

In a heavily contaminated field soil (10,000 mg zinc/kg), bacterial DGGE DNA profiles showed a significantly reduced diversity compared to remediated plots (31 versus 50 DNA bands). Also, in a slightly contaminated soil (160 mg copper/kg), the number of DNA bands was reduced from 50 to 42. However, in a soil contaminated with nickel and chromium (2800 mg/kg and 430 mg/kg, respectively) there was no reduction in the number of DNA bands. In all cases contamination caused significant qualitative changes in the DNA profiles (Fig. 3.2). Also, community-level physiological profiles (Biolog™) indicated changes in the community structure (results not shown). Thus, community structure changed, but diversity was not always reduced in seriously contaminated soils. Biomass contents for different groups of organisms, respiration and mineralization were much more reduced than diversity. Bacterial growth rate (thymidine incorporation) was the most sensitive, with decreases of more than 70% (Bloem and Breure, 2003). Thus, biological indicators showed significant reductions in contaminated soils.

Less extreme effects may be expected in agricultural soils. Sustainable agriculture aims at maintaining good crop yields with minimal impact on the environment, while at least avoiding deterioration in soil fertility and providing essential nutrients for plant growth. Further, sustainable agriculture supports a diverse and active community of soil organisms, exhibits a good soil structure and allows for undisturbed decomposition (Mäder *et al.*, 2002). Therefore, agricultural practices are adjusted to integrate organic, or more extensive, management. The main principles are to restrict stocking densities, avoid synthetic pesticides, avoid mineral fertilizers and use organic manure (Hansen *et al.*, 2001). Finally, this will result in an increased role of soil organisms, e.g. decomposers, nitrogen fixers and mycorrhizas, in plant nutrition and disease suppression.

Indeed, larger microbial biomass, diversity (Biolog™), enzyme ties, more mycorrhizas and higher earthworm abundance have observed in organic farming during a 20-year experiment in Switzerland (Mäder *et al.*, 2002). Application of farmyard manure, instead of mineral fertilizer, stimulates the bacterial branch of the soil food web. In fields that had received farmyard manure for 45 years (Marstorp *et al.*, 2000), we found a fourfold greater bacterial biomass than in fields that received only mineral nitrogen (Fig. 3.4). Bacterial biomass was lower (twofold) with sewage sludge. However, a larger input of organic matter does not necessarily lead to more bacterial biomass. Arable fields under integrated farming showed greater bacterial activity but not a significantly larger bacterial biomass (Bloem *et al.*, 1994); this was attributed to bacterivorous protozoa and nematodes, which reached 20–60% greater densities. N mineralization by the soil food web was 30% higher than in conventional fields. This compensated for a 35% smaller input of mineral fertilizer in the integrated fields and supported a crop yield of 90% of that in the conventional fields.

In grassland plots that received no (O) or incomplete (N only, or PK) mineral fertilization for 50 years, there was up to 2.5 times more fungal biomass than in plots with full mineral fertilization (NPK) (Fig. 3.5). Liming (Ca) had the same effect as fertilization because this stimulated mineralization of the high amount of organic matter (20%) in this clay soil. Part of the fungal biomass may have been mycorrhizas, which support plant nutrition at low availability of mineral nutrients. The grassland plots that received no mineral nitrogen had high plant diversity (40 species) whereas N-fertilized plots contained low plant diversity (20 species). There was no simple quantitative relationship between above-ground and below-ground diversity. The numbers of bacterial DNA bands (61–72) were not significantly different. Nevertheless, there are qualitative relationships between above- and below-ground diversity. Principal component analysis clearly separated soil bacterial DNA profiles of grassland plots from those of arable plots on the

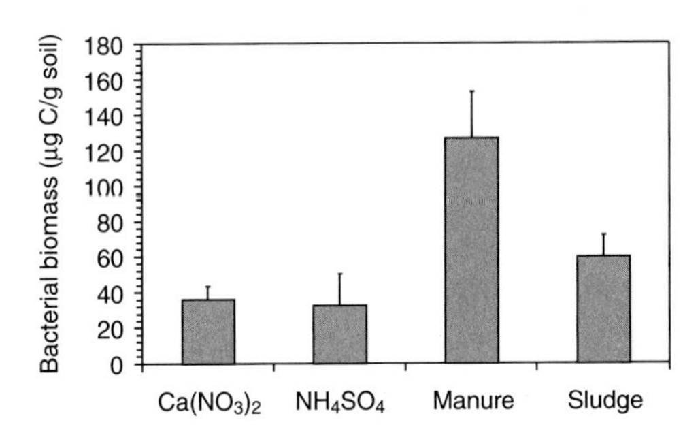

Fig. 3.4. Effect of fertilization on bacteria in arable fields, Ultuna, Sweden; experiment began 1956. Error bars indicate SE, $n = 3$.

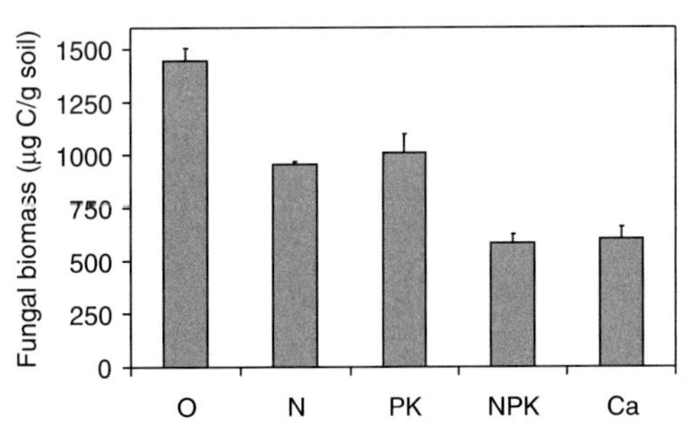

Fig. 3.5. Effect of fertilization on fungal hyphae in grassland, Ossekampen, The Netherlands; experiment began 1950. Error bars indicate SE, $n = 2$. O, no NPK.

same soil sampled on the same date. However, with samples of four different seasons, temporal variation in the DNA profiles impeded a clear separation. Spatial variation in DNA profiles was small between replicate plots on the same field, but large between different sites. The variation between 10–20 farms of the same category was as large as the differences between categories of intensive, organic and extensive grassland farms on sand. Therefore, qualitative information (community structure) is more difficult to handle than quantitative information, such as the number of species (richness) and relative abundances (evenness). Richness and evenness are combined in the Shannon diversity index (Atlas and Bartha, 1993).

The Dutch Soil Quality Network: Results of the First Year

In the first year (1997), a pilot study was performed to test whether the selected biological variables were sufficiently reproducible when applied in large-scale monitoring using mixed samples of whole farms. In that year, mites were included at only two reference farms, where the total food web structure was determined, including important functional groups such as fungi and protozoa. Thus, for these two farms, C and N mineralization were calculated using a food web model, where mineralization rates are calculated from the observed biomasses of different functional groups of soil organisms. Food web modelling was also used to predict food web stability (De Ruiter *et al.*, 1995). The budget was not sufficient to adopt this fundamental and comprehensive approach in the routine programme. Originally, potential nitrification was included in the programme, but in 1999 this indicator was replaced by potential C and N mineralization. Also in 1999, genetic diversity of bacteria was included because of increased political and scientific interest in biodiversity and because molecular techniques became available for routine use.

Results of the first year (1997) of the Dutch Soil Quality Network are summarized in Table 3.1. In this year agricultural grasslands on sea-clay and horticultural farms (bulbs and vegetables) on various soils were sampled. Sea-clay (Dutch classification) encompasses soils in loam and clay deposits of marine origin. Both types of soil are managed intensively, but frequencies of soil tillage and pesticide application are higher in horticulture. Soil pH-KCl was 6.5 in both categories. The grassland soils had a significantly higher content of clay (24% versus 7%) and organic matter (6.2% versus 2.8%) than the horticultural soils. Correspondingly, the cation exchange capacity (CEC) was significantly lower in the horticultural soils (6.5 cmol+/kg versus 22 cmol+/kg). In spite of the lower P-binding capacity, the horticultural soils had a higher phosphate content (0.77 mg P_2O_5/g versus 0.41 mg P_2O_5/g soil), which is attributed to a higher degree of fertilization. Contents of heavy metals (cadmium, copper, lead and zinc), some formerly used persistent pesticides (DDT, HCB, dieldrin and lindane) and polyaromatic hydrocarbons (phenanthrene and benzo(a)pyrene) were at acceptable concentrations and not significantly different between the categories.

Table 3.1. Indicators measured in 20 agricultural grasslands on sea-clay and 17 horticultural farms on various soils.

Soil biota	Indicators	Grassland on sea-clay ($n = 20$)	Horticulture ($n = 17$)	Statistical significance of difference Grl. – Hort.
Bacteria	Thymidine incorp. (pmol/g/h)	179	108	***
	Leucine incorp. (pmol/g/h)	847	392	***
	Bacterial biomass (mg C/g)	232.4	56	***
	Colony-forming units (10^7/g)	17.1	2.6	***
	Potential nitrification (mg NO_3-N/kg/week)	93	74	***
Biolog™	LogCFU-50 (number of CFU corresponding with 50% activity)	3.73	2.87	***
	H-coefficient (evenness of decomposition of 31 substrates)	0.39	0.60	***
	Gg50 (μg soil with 50% activity)	95	44	*
Nematodes	Abundance (number/100 g)	4,629	2,069	***
	Number of taxa	26.1	21.8	*
	Maturity index	1.77	1.47	***
	Trophic diversity index	2.12	1.51	***
	Number spec. bacterial feeding	11.4	13.3	*
	Number spec. carnivores	0.4	0.6	ns
	Number spec. hyphal feeding	2.1	2.1	ns
	Number spec. omnivores	1.0	1.2	ns
	Number spec. plant feeding	11.4	4.5	***
	Number of functional groups	3.9	4.3	ns
Enchytraeids	Abundance (number/m^2)	24,908	16,096	**
	Number of taxa	8.2	5.5	***
	Biomass (g/m^2)	5.6	1.1	***
	Number of Friderica (number/m^2)	8,654	1,300	***
Earthworms	Abundance (number/m^2)	317	40	***
	Biomass (g/m^2)	70.1	3.8	***
	Endogé-species (deeper in soil)	2.10	0.82	***
	Epigé-species (closer to surface)	1.2	0.06	***
		(n = 1)	(n = 1)	
Mites	Abundance (number/m^2)	37,900	18,100	
	Number of species	23	20	
	Number of functional groups	8	10	
Food web (model-calculations)	N-mineralization (kg N/ha/yr)	335	115	
	C-mineralization (kg C/ha/yr)	6,150	1,750	
	Stability	0.47	0.61	

CFU, colony-forming units; Grl., grassland; Hort., horticulture; ns, not significant.

In a single soil, thymidine incorporation, leucine incorporation and bacterial biomass usually show coefficients of variation (CV) of 10%, 5% and 30%, respectively. In the national monitoring programme, variation is much larger because of differences between farms. The variation between 20 replicate farms of the same category was about 30% for the growth rate measurements and about 60% for the biomass measurements. With 20 replicates, this results in standard errors of about 7% and 13% of the mean. This is sufficient to establish statistical differences between categories. Analyses of variance of (log transformed) data yielded a significance (P) < 0.001 for most indicators (Table 3.1), thus demonstrating reproducibility and discriminative power.

Thymidine and leucine incorporation indicated a 50% lower bacterial growth rate in horticulture compared to grassland. Bacterial biomass in horticulture was only 25% of the biomass in the grasslands. Thus, the specific growth rate per unit of bacterial biomass was twofold higher in the horticultural soils. Also, the results of the Biolog™ assay indicated a greater specific activity of bacteria in the horticultural soils. Using Biolog™ ECO-plates, the decomposition of 31 different carbon sources can be tested (in triplicate) in 96 wells of a microtitre plate. Of each soil suspension, four different dilutions were incubated in Biolog™ ECO-plates, and colony-forming units (CFU) were counted in parallel (for methods see Rutgers and Breure, 1999; Breure and Rutgers, 2000; Chapter 8, this volume). LogCFU-50, the number of CFU giving 50% activity, was smaller in the horticultural soils, and thus the activity per CFU greater. The activity in Biolog™ plates is measured as average colour development in the wells, reflecting respiration and growth. Potential nitrification was high in all grasslands and in most horticultural soils. On average, the rate of potential nitrification was 20% smaller in horticulture. In three (out of 17) horticultural farms, potential nitrification rate was at least 50% slower than in the other soils. Most indicators were lower in the horticultural soils. Food web modelling, applied to only two farms, indicated lower carbon and nitrogen mineralization in the soil of a horticultural farm compared to a grassland farm. Predicted soil food web stability was higher on the horticultural farm. It may be speculated that under intensive management more stress-resistant organisms are selected.

Presentation of results

The large amount of data in Table 3.1 is not suitable for presentation to either policy makers or the public. The results can be presented in a more clear and illustrative way for diagnostic purposes using the AMOEBA method (Ten Brink *et al.*, 1991), or similar cobweb or star diagrams (Stenberg, 1999; Mäder *et al.*, 2002). The former method results in an amoeba-like graphical representation of all indicator values, scaled against a historical, undisturbed or desired situation. As an example, the indicator data from one organic grassland farm were used as a reference for the intensively managed farms (Fig. 3.6).

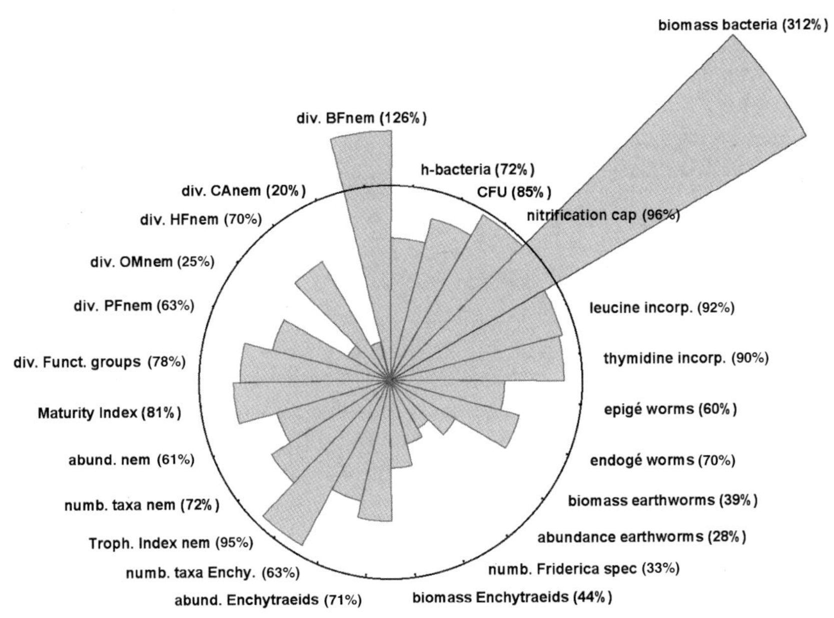

Fig. 3.6. AMOEBA presentation of indicator values from the grassland farms on sea-clay, relative to those in the reference (circle), which is set to 100%. In this case the reference is only one organic farm (from Schouten *et al.*, 2000). The graph facilitates a clear overview of the differences between categories of soil type and land use. The indicators are described in more detail in Table 3.1.

The value of each variable in the reference was scaled as 100%. This yields a circle of 100% values for the reference. In the example, mite fauna and total food web structure were omitted due to lack of data from the reference. Almost all the indicative parameters of the 20 grassland farms appeared within the 100% circle. Apparently, biodiversity within functional groups, and process rates, were lower in intensive than in organic grassland. Only bacterial biomass was much greater (312%), which was reflected in the diversity of bacterial-feeding nematodes (126%). Despite the large bacterial biomass, its activity (indicated by the number of colony-forming units, potential nitrification and [^{14}C]leucine and [^{3}H]thymidine incorporation) was approximately 90% of the reference, so the specific activity was smaller.

The indexed indicator values used to construct the AMOEBA can be further condensed into a Soil Quality Index (SQI), using the average factorial deviation from the reference value (Ten Brink *et al.*, 1991). The index is calculated as follows:

$$\mathrm{SQI} = 10^{\log m - \frac{\sum_{i=1}^{n} |\log m - \log n_i|}{n}}$$

where m is the reference (set to 100%) and n are the measured values as percentages of the reference.

For example:

	Reference	Sample
Indicator 1	100	50
Indicator 2	100	200

log reference = 2, log sample is 1.7 and 2.3, respectively.
The (absolute) difference between reference and sample is:

For indicator 1: $|2 - 1.7| = 0.3$
For indicator 2: $|2 - 2.3| = 0.3$

The sum of the differences is 0.6 and the average difference is 0.3.
The average value of the sample is 2 – 0.3 = 1.7 on the log scale. Back transformed this gives an SQI of $10^{1.7} = 50\%$.

Thus, in this SQI a value of 50% has the same weight as one of 200% of the reference value (both a factor two). For the AMOEBA in Fig. 3.6, this exercise resulted in an SQI value of 65% for the intensively managed farms versus the single organic farm. AMOEBA-like figures and derived indices can be used as tools for comparison (in space and time) and for relatively simple presentation of complicated results. It must be realized that they are only simplified reflections of complex ecosystems and should not be taken as absolute values. The SQI approach has been applied mainly on sites with local contamination, where an uncontaminated reference is available. It is more difficult to define a proper reference for larger areas with diffuse contamination, and for the different categories of soil type and land use of the Dutch Soil Quality Network.

The Dutch Soil Quality Network: Results of a 5-year Cycle

After the pilot study in 1997, the programme was evaluated in 1998 and continued in 1999 by sampling of grassland farms on sand with different management intensities: extensive, organic and intensive. The categories are based on the number of livestock units per hectare. One livestock unit is defined as the amount of cattle, pigs and/or poultry excreting an average of 41 kg N/ha/yr. The organic farms had been certified for at least 5 years before sampling, used compost and/or farmyard manure and no biocides, and had on average 1.6 livestock units/ha. Extensive farms used mineral fertilizer and less farmyard manure and had 2.3 livestock units/ha. Intensive farms used mineral fertilizer and farmyard manure and had 3.0 livestock units/ha. In 2000, highly intensive (Int+) farms with pigs (5.1 livestock units/ha) on sand, and forest on sand were sampled. Intensive arable farms on sand followed in 2001, when further sampling on farms was prevented by an outbreak of foot-and-mouth disease. Therefore, organic arable farms on sand and grassland farms on peat could not be sampled as planned in 2001. In 2002, sampling organic arable farms on sand was completed. Also in 2002, arable farms on sea-clay (both intensive and organic) and grassland farms on river clay (intensive) were sampled. River-clay (Dutch classification) encompasses loam and clay soils in river deposits.

Bacterial DNA profiles reflecting the dominant species or genotypes did not indicate low genetic diversity in agricultural soils. Organic farms did not show higher genetic diversity of bacteria than intensive farms. Between 48 and 69 DNA bands were found, with Shannon diversity indices from 3.57 ± 0.03 to 3.85 ± 0.02 (± SE). There may be up to 10,000 bacterial genotypes/g soil, but most species occur in very low numbers. If a species has a density below 10^6 cells/g soil (about 1/1000 of the total number), it is below the detection limit of our DGGE method (Dilly *et al.*, 2004). Thus, this method reflects the composition of most of the bacterial biomass, but not the total number of species. Low bacterial diversity (25 DNA bands, Shannon index 2.37 ± 0.22) was only found in the usually acid forest soils, which also contained a very small bacterial biomass (Fig. 3.7).

The bacterial biomass was also very low in horticultural soils, higher in arable soils and high in grassland. Grassland contained a higher bacterial biomass on clay than on sand. The biomass appears to reflect management intensity, which is higher in arable land (tillage, pesticides, etc.) than in grassland. Similar to the biomass, potential nitrogen mineralization was very low in forest, higher in arable land and highest in grassland (Fig. 3.8).

Potential N mineralization was about 50% lower in clay than in sandy soils. At extensive and organic grassland farms on sand, N mineralization (i.e. soil fertility) was about 50% higher than at intensive farms. At the highly intensive farms, mineralization was almost as high as on the organic farms, probably because a lot of (pig) manure had been applied. Not only microbial biomass and activity, but also different groups of soil fauna, tended to be higher at organic and extensive farms (Fig. 3.9).

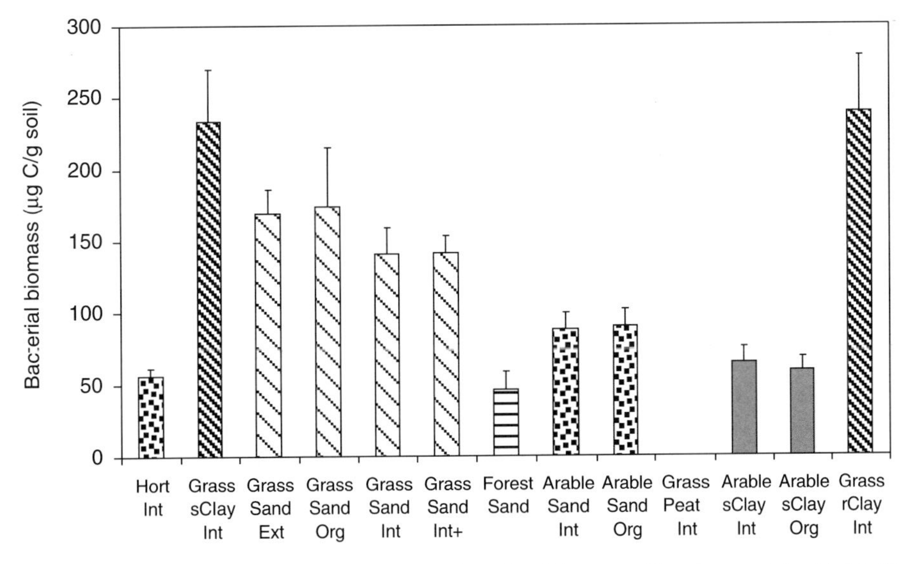

Fig. 3.7. Bacterial biomass in different categories of soil type and land use of the Dutch Soil Quality Network. Hort, horticultural; Grass, grassland; sClay, sea-clay; rClay, river-clay; Int, intensive farms; Ext, extensive farms; Org, organic farms; Int+, highly intensive farms. Error bars indicate SE, $n = 20$ (10 for organic farms).

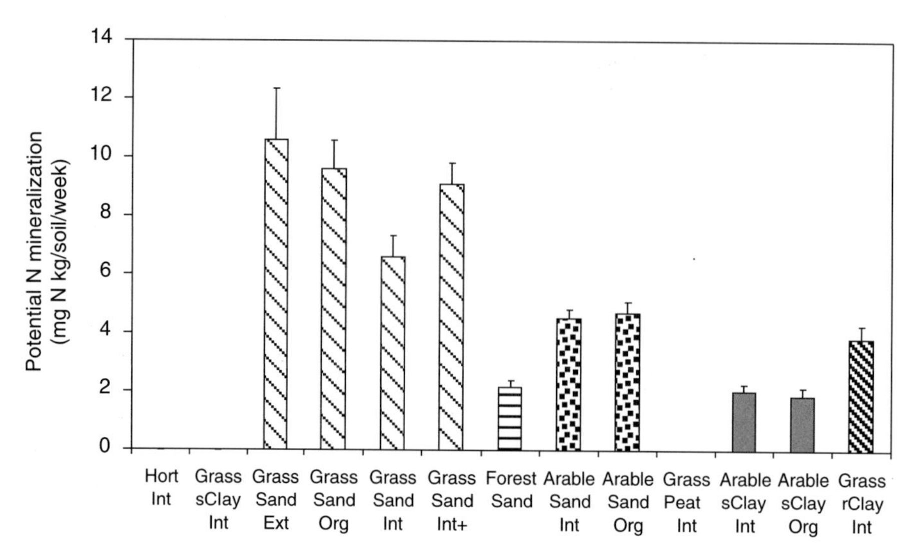

Fig. 3.8. Potential N mineralization in different categories of soil type and land use of the Dutch Soil Quality Network. Hort, horticultural; Grass, grassland; sClay, sea-clay; rClay, river-clay; Int, intensive farms; Ext, extensive farms; Org, organic farms; Int+, highly intensive farms. Error bars indicate SE, $n = 20$ (10 for organic farms).

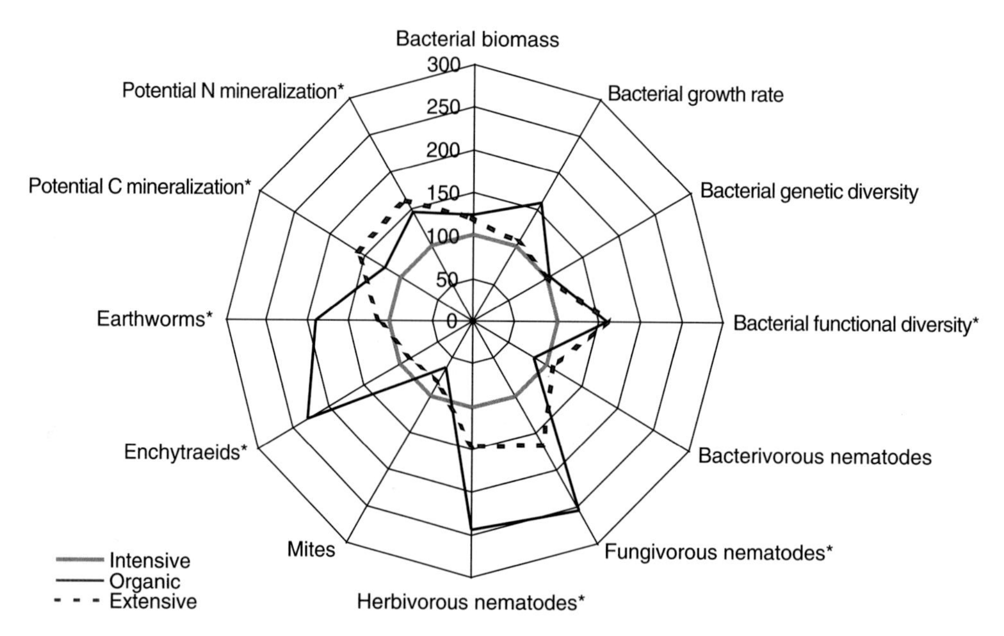

Fig. 3.9. Soil biological variables at organic, extensive and intensive grassland farms on sand. Intensive is set to 100%; * indicates a statistically significant difference ($P \leq 0.05$, analysis of variance) between categories.

Analysis of variance indicated statistically significant differe rial functional diversity (colour development in Biolog^{TM} plates, activity), fungivorous nematodes, herbivorous nematodes, earthworms, potential C mineralization (respiration) and potent ization. Bacterial growth rate (thymidine incorporation) was 60% higher in organic than in intensive grassland on sand (23.6 ± 4.8 versus 14.8 ± 2.9 pmol thymidine/g soil/h, ± SE, n = 10 for organic and 20 for intensive farms). Similarly, bacterial growth rate was 200% higher in organic than in intensive arable land on sand (114 ± 18 pmol/g/h versus 46.6 ± 3.9 pmol/g/h), and 33% higher in organic than in intensive arable land on clay (153 ± 26 pmol/g/h versus 115 ± 11 pmol/g/h). Both on sand and on clay there was no difference in N mineralization between intensive and organic arable farms (Fig. 3.8). The differences between categories of farms are less significant than differences between treatments in long-term experimental fields. This may be caused by several reasons. Sampling at the farm level causes greater variation than sampling in replicated experimental fields. Furthermore, it may take a long time, perhaps decades rather than 5 years, before soil organisms increase significantly after a change in management. The categories used here are rather broad, and there are relatively large differences between farms within one category. With more detailed analysis, using narrow clusters instead of broad categories of farms, trends can be detected. Bacterial biomass and bacterivorous nematodes tend to decrease with increasing stocking density on grassland farms on sand (Fig. 3.10).

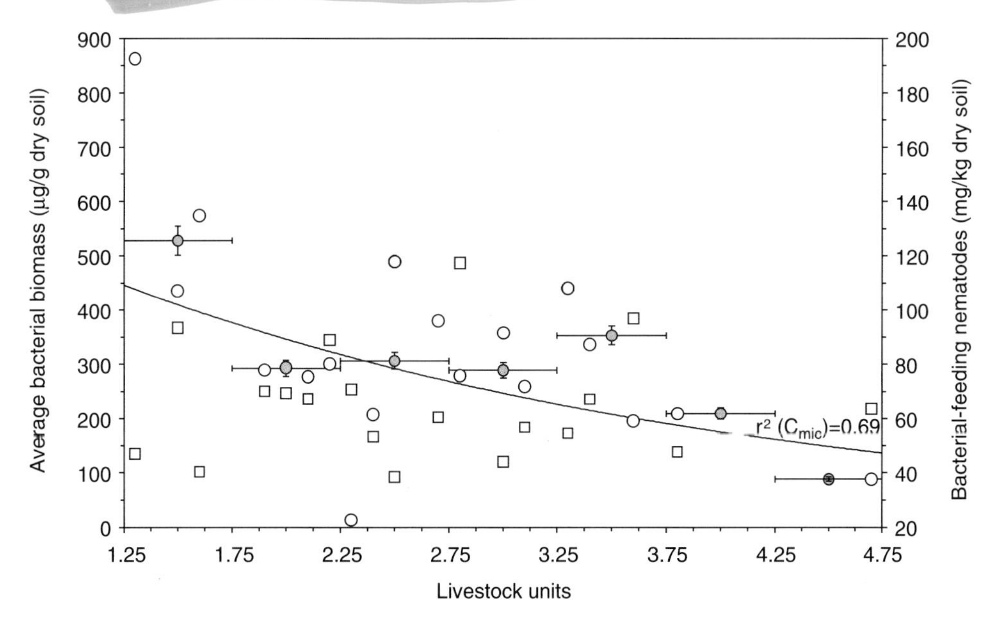

Fig. 3.10. Scatterplot of the average bacterial biomass (circles) and the bacterial-feeding nematode biomass (squares) along an increasing farming intensity gradient (livestock units/ha).The open symbols represent narrow clusters. The exponential trendline for bacterial biomass is fitted on wider clusters (closed circles) of 0.5 livestock unit width (n = 9, α = 0.05; vertical error bars for 5% bacterial biomass). (From Mulder *et al.*, 2003.)

When sufficient data (> 200 sites) become available after completing one 5-year cycle of monitoring, habitat response models can be developed to predict the value of biological variables from environmental variables such as stocking density, pH, organic matter content, clay, etc. (Oberholzer and Höper, 2000; Mulder *et al.*, 2003).

Thus, experience with monitoring soil life shows large differences between categories of soil type and land use, and significant effects of management on most groups of organisms and potential C and N mineralization.

Measuring Soil Biodiversity: Impediments and Research Needs

Soil ecosystems are complex. Therefore, many different aspects need to be measured (Lancaster, 2000). It is important to use a set of various indicators, and not a few indicators selected *a priori*, which are supposed to be the most sensitive. Some indicators are more sensitive to contamination (e.g. bacterial growth rate); others are more sensitive to differences in soil fertility and agricultural management (e.g. N mineralization).

The complexity of biodiversity implies that time and money are major impediments for thorough monitoring. Techniques are available, but extensive monitoring is expensive. The costs of sampling and analysing one site (mixed sample of one farm) of the Dutch Soil Quality Network in 2002 were €5500. This amounts to €330,000 per year for 60 sites. Still, the important functional groups of fungi and protozoa are not (yet) included in the routine programme. Protozoa are the major bacterivores, but showed strong temporal variation, and fungi usually had a very low biomass in Dutch agricultural soils (Bloem *et al.*, 1994; Velvis, 1997). However, recently fungal dominance was found in a few agricultural soils, especially at low mineral N fertilization (Fig. 3.5). This indicates that the role of fungi may increase when fertilization is reduced, e.g. in sustainable agriculture and set-aside land. Therefore, the fungal to bacterial ratio deserves more research as a potentially simple indicator of efficient nutrient use and sustainability (Bardgett and McAlister, 1999; Mulder *et al.*, 2003). Based on the results obtained, in 2004 fungi were included in the indicator set. On the other hand, bacterial genetic diversity based on DGGE DNA profiles was omitted, because quantitative differences appeared to be small and qualitative results appeared difficult to interpret, especially in large-scale monitoring.

Relationships between below-ground biodiversity, in terms of species richness and function, are still not clear. Experimental reduction of biodiversity in soil did not significantly affect functioning in terms of decomposition and mineralization (Griffiths *et al.*, 2001). Nevertheless, biodiversity may play a role in stability against perturbation and stress, in nutrient retention and in suppression of plant diseases (Griffiths *et al.*, 2000). Fundamental research is still needed to clarify the importance of soil biodiversity. Species identity (community structure) is probably more important than species richness, but is more difficult to analyse. New techniques, such as microarrays (DNA chips), are needed to monitor large numbers of species, genes and functions, and may offer new opportunities to link diversity to function.

Using approaches as described here, biodiversity and functioning of soil ecosystems can be monitored. In pollution gradients, it is possible to use a local unpolluted control (Bloem and Breure, 2003). However, in many cases such a reference is not available. Generally, the value of an indicator is affected not only by stress factors, but also by soil type, land use and vegetation. Therefore, reference values for specific soil types have to be deduced from many observations, e.g. 20 replicates per type. The choice of a desired reference is a political rather than a scientific issue, and depends on the aims of land use. A biologically active and fertile soil is needed in (organic) farming, but a high mineralization of nutrients from organic matter may hamper conversion of agricultural land to a species-rich natural vegetation. For a specific soil and land-use type, the reference could be the current average of 20 conventional farms, or the average of 20 organic farms. Soils showing very low or very high indicator values may be suspect and need further examination. Sufficient data and experience are needed to make judgements of desirable reference values. Monitoring changes of indicators over time can reduce the importance of (subjective) reference values. Such changes may be easier to interpret than momentary values (Lancaster, 2000). Spatially extensive and long-term monitoring may be not ideal, but it is probably the most realistic approach to obtain objective information on differences between, temporal changes within, and human impact on ecosystems.

To make progress with monitoring and understanding the complexity of soil life it is necessary to start. Methods are already available. Differences between methods and laboratories occur with any variable, and can be minimized by using a limited number of specialized laboratories for each variable. For any purpose, standardization and intercalibration are necessary. Learning by doing is inevitable, thus long-term and extensive monitoring programmes need to be flexible. To begin with, monitoring programmes may be relatively simple with a limited number of variables, which should be increased when possible. Better one indicator than nothing, but the more the better. Choosing a minimum data set and ranking of indicators remains a subjective exercise. Looking at monitoring programmes established in different countries, it appears that microbial biomass, respiration and potential N mineralization are commonly regarded as part of a minimum data set. These are very useful, but more data are needed if we aim at monitoring biodiversity. Adding the main functional groups of the soil food web brings us closer to understanding biodiversity and provides the potential to relate the structure of the soil community to ecosystem functioning and environmental stress (Mulder *et al.*, 2005).

Acknowledgements

We thank many colleagues for their invaluable contribution to the results: An Vos, Popko Bolhuis, Meint Veninga, Gerard Jagers op Akkerhuis, Henk Siepel and Wim Dimmers, Alterra, Wageningen, The Netherlands; Christian Mulder, Marja Wouterse, Rob Baerselman, Niels Masselink, Margot Groot, Ruud Jeths and Hans Bronswijk, RIVM, Bilthoven, The Netherlands; Wim Didden, Sub-department of Soil Quality, Wageningen University, The Netherlands; Harm Keidel, Christel Siepman and others, Blgg bv, Oosterbeek, The Netherlands; and Ernst Witter, Department of Soil Sciences, Swedish University of Agricultural Sciences, Uppsala, Sweden, for providing samples from Ultuna (Fig. 3.4).

References

Akkermans, A.D.L., van Elsas, J.D. and De Bruijn, F.J. (eds) (1995) *Molecular Microbial Ecology Manual.* Kluwer Academic Publishers, Dordrecht, The Netherlands.

Alef, K. and Nannipieri, P. (eds) (1995) *Methods in Applied Soil Microbiology and Biochemistry*. Academic Press, London.

Anderson, J.P.E. (1987) Handling and storage of soils for pesticide experiments. In: Somerville, L. and Greaves, M.P. (eds) *Pesticide Effects on Soil Microflora*. Taylor & Francis, London, pp. 45–60.

Atlas, R.M. and Bartha, R. (1993) *Microbial Ecology. Fundamentals and Applications.* The Benjamin/Cummings Publishing Company, Redwood City, California, pp. 140–145.

Bardgett, R.D. and McAlister, E. (1999) The measurement of soil fungal:bacterial biomass ratios as an indicator of ecosystem self-regulation in temperate meadow grasslands. *Biology and Fertility of Soils* 29, 282–290.

Bloem, J. and Breure, A.M. (2003) Microbial indicators. In: Markert, B., Breure, A.M. and Zechmeister, H. (eds) *Bioindicators/Biomonitors – Principles, Assessment, Concepts*. Elsevier, Amsterdam, pp. 259–282.

Bloem, J., Lebbink, G., Zwart, K.B., Bouwman, L.A., Burgers, S.L.G.E., De Vos, J.A. and De Ruiter, P.C. (1994) Dynamics of microorganisms, microbivores and nitrogen mineralisation in winter wheat fields under conventional and integrated management. *Agriculture, Ecosystems and Environment* 51, 129–143.

Bloem, J., Veninga, M. and Shepherd, J. (1995) Fully automatic determination of soil bacterium numbers, cell volumes and frequencies of dividing cells by confocal laser scanning microscopy and image analysis. *Applied and Environmental Microbiology* 61, 926–936.

Bloem, J., De Ruiter, P.C. and Bouwman, L.A. (1997) Food webs and nutrient cycling in agro-ecosystems. In: van Elsas, J.D., Trevors, J.T. and Wellington, E. (eds) *Modern Soil Microbiology*. Marcel Dekker, New York, pp. 245–278.

Bloem, J., Schouten, A.J., Didden, W., Jagers op Akkerhuis, G., Keidel, H., Rutgers, M. and Breure, A.M. (2004) Measuring soil biodiversity: experiences, impediments and research needs. In: Francaviglia, R. (ed.) *Agricultural impacts on soil erosion and soil biodiversity: developing indicators for policy analysis*, Proceedings of an OECD expert meeting, 25–28 March 2003, Rome, Italy. OECD, Paris, p. 109–129. Available at: http://webdomino1.oecd.org/comnet/agr/soil_ero_bio.nsf (accessed 26 October 2004).

Breure, A.M. and Rutgers, M. (2000) The application of Biolog plates to characterize microbial communities. In: Benedetti, A., Tittarelli, F., De Bertoldi, S. and Pinzari, F. (eds) *Biotechnology of Soil: Monitoring, Conservation and Bioremediation*, Proceedings of the COST Action 831 joint working group meeting 10–11 December 1998, Rome, Italy. (EUR 19548), EU, Brussels, pp. 179–185.

Brookes, P.C. (1995) The use of microbial parameters in monitoring soil pollution by heavy metals. *Biology and Fertility of Soils* 19, 269–279.

Brussaard, L., Behan-Pelletier, V.M., Bignell, D.E., Brown, V.K., Didden, W., Folgarait, P., Fragoso, C., Freckman, D.W., Gupta, V.V.S.R., Hattori, T., Hawksworth, D.L., Klopatek, C., Lavelle, P., Malloch, D.W., Rusek, J., Söderström, B., Tiedje, J.M. and Virginia, R.A. (1997) Biodiversity and ecosystem functioning in soil. *Ambio* 26, 563–570.

Brussaard, L., Kuyper, T.W., Didden, W.A.M., De Goede, R.G.M. and Bloem, J. (2003) Biological soil quality from biomass to biodiversity – importance and resilience to management stress and disturbance. In: Schjønning, P., Christensen, B.T. and Elmholt, S. (eds) *Managing Soil Quality – Challenges in Modern Agriculture.* CAB International, Wallingford, UK, pp. 139–161.

Carter, M.R., Gregorich, E.G., Angers, D.A., Beare, M.H., Sparling, G.P., Wardle, D.A. and Voroney, R.P. (1999) Interpretation of microbial biomass measurements for soil quality assessment in humid temperate regions. *Canadian Journal of Soil Science* 79, 507–520.

De Ruiter, P.C., Neutel, A.-M. and Moore, J.C. (1995) Energetics, patterns of interaction strengths, and stability in real ecosystems. *Science* 269, 1257–1260.

Dilly, O., Bloem, J., Vos, A. and Munch, J.C. (2004) Bacterial diversity during litter decomposition in agricultural soils. *Applied and Environmental Microbiology* 70, 468–474.

Doran, J.W. and Parkin, T.B. (1994) Defining and assessing soil quality. In: Doran, J.W., Coleman, D.C., Bezdicek, D.F. and Stewart, B.A. (eds) *Defining Soil Quality for a Sustainable Environment*, Special Publication 35. American Society of Agronomy, Madison, Wisconsin, pp. 3–21.

Doran, J.W. and Parkin, Quantitative indicators of s minimum data set. In: Do Jones, A.J. (eds) *Methods for Assessing* *Quality*, Special Publication 49. Soil Science Society of America, Madison, Wisconsin, pp. 25–37.

FAO (1999) *Sustaining Agricultural Biodiversity and Agro-ecosystem Functions.* Report available at: http://www.fao.org/sd/epdirect/EPre0065.htm (accessed 9 April 2004).

Giller, K.E., Witter, E. and McGrath, S.P. (1998) Toxicity of heavy metals to microorganisms and microbial processes in agricultural soils: a review. *Soil Biology and Biochemistry* 20, 1389–1414.

Griffiths, B.S., Ritz, K., Bardgett, R.D., Cook, R., Christensen, S., Ekelund, F., Sørensen, S., Bååth, E., Bloem, J., De Ruiter, P., Dolfing, J. and Nicolardot, B. (2000) Ecosystem response of pasture soil communities to fumigation-induced microbial diversity reductions: an examination of the biodiversity–ecosystem function relationship. *Oikos* 90, 279–294.

Griffiths, B.S., Ritz, K., Wheatly, R., Kuan, H.L., Boag, B., Christensen, S., Ekelund, F., Sørensen, S.J., Muller, S. and Bloem, J. (2001) An examination of the biodiversity–ecosystem function relationship in arable soil microbial communities. *Soil Biology and Biochemistry* 33, 1713–1722.

Hansen, B., Alroe, H.F. and Kristensen, E.S. (2001) Approaches to assess the environmental impact of organic farming with particular regard to Denmark. *Agriculture Ecosystems and Environment* 83, 11–26.

Hill, R., Frampton, C., Cuff, J. and Sparling, G. (2004) National soil quality review and programme design, unpublished report.

Höper, H. (1999) Bodenmikrobiologische Untersuchungen in der Bodendauerbeobachtung in Deutschland. *VBB-Bulletin* 3, 13–14. Arbeitsgruppe Vollzug Bodenbiologie, FiBL, CH-5070 Frick, Switzerland [in German].

Hunt, H.W., Coleman, D.C., Ingham, E.R., Ingham, R.E., Elliott, E.T., Moore, J.C., Rose, S.L., Reid, C.F.F. and Morley, C.R. (1987) The detrital food web in a short-grass prairie. *Biology and Fertility of Soils* 3, 57–68.

Korthals, G.W., Alexiev, A.D., Lexmond, T.M., Kammenga, J.E. and Bongers, T. (1996) Long-term effects of copper and pH on the nematode community in an agroecosystem. *Environmental Toxicology and Chemistry* 15, 979–985.

Lancaster, J. (2000) The ridiculous notion of assessing ecological health and identifying the useful concepts underneath. *Human and Ecological Risk Assessment* 6, 213–222.

Mäder, P., Fliessbach, A., Dubois, D., Gunst, L., Fried, P. and Niggli, U. (2002) Soil fertility and biodiversity in organic farming. *Science* 296, 1694–1697.

Marstorp, H., Guan, X. and Gong, P. (2000) Relationship between dsDNA, chloroform labile C and ergosterol in soils of different organic matter contents and pH. *Soil Biology and Biochemistry* 32, 879–882.

Maurer-Troxler, C. (1999) Einsatz bodenbiologischer Parameter in der langfristigen Bodenbeobachtung des Kantons Bern. *VBB-Bulletin* 3, 11–13. Arbeitsgruppe Vollzug Bodenbiologie. FiBL, CH-5070 Frick, Switzerland [in German].

Mulder, Ch., De Zwart, D., Van Wijnen, H.J., Schouten, A.J. and Breure, A.M. (2003) Observational and simulated evidence of ecological shifts within the soil nematode community of agroecosystems under conventional and organic farming. *Functional Ecology* 17, 516–525.

Mulder, Ch., Cohen, J.E., Setälä, H., Bloem, J. and Breure, A.M. (2005) Bacterial traits, organism mass, and numerical abundance in the detrital soil food web of Dutch agricultural grasslands. *Ecology Letters* 8, 80–90.

Nielsen, M.N. and Winding, A. (2002) Microorganisms as indicators of soil health. National Environmental Research Institute (NERI), Denmark, Technical Report no. 388. Available at: http://www.dmu.dk (accessed 30 January 2004).

Oberholzer, H.R. and Höper, H. (2000) Reference systems for the microbiological evaluation of soils. *Verband Deutscher Landwirtschaftlicher Untersuchungs- und Forschungsanstalten* 55, 19–34.

OECD (2003) *Agriculture and Biodiversity: Developing Indicators for Policy Analysis.* Proceedings from an OECD expert meeting Zurich, Switzerland, November 2001. OECD, Paris.

Paul, E.A., Harris, D., Klug, M. and Ruess, R. (1999) The determination of microbial biomass. In: Robertson, G.P., Coleman, D.C., Bledsoe, C.S. and Sollins, P. (eds) *Standard Soil Methods for Long-term Ecological Research.* Oxford University Press, New York, pp. 291–317.

Robertson, G.P., Coleman, D.C., Bledsoe, C.S. and Sollins, P. (eds) (1999) *Standard Soil Methods for Long-term Ecological Research.* Oxford University Press, New York.

Rutgers, M. and Breure, A.M. (1999) Risk assessment, microbial communities, and pollution induced community tolerance. *Human and Ecological Risk Assessment* 5, 661–670.

Schipper, L.A. and Sparling, G.P. (2000) Performance of soil condition indicators across taxonomic groups and land uses. *Soil Science Society of America Journal* 64, 300–311.

Schouten, A.J., Bloem, J., Didden, W.A.M., Rutgers, M., Siepel, H., Posthuma, L. and Breure, A.M. (2000) Development of a biological indicator for soil quality. *SETAC Globe* 1, 30–32.

Stenberg, B. (1999) Monitoring soil quality of arable land: microbiological indicators. *Acta Agriculturae Scandinavica, Section B, Soil and Plant Science* 49, 1–24.

Stenberg, B., Johansson, M., Pell, M., Sjödahl-Svensson, K., Stenström, J. and Torstensson, L. (1998) Microbial biomass and activities in soil as affected by frozen and cold storage. *Soil Biology and Biochemistry* 30, 393–402.

Ten Brink, B.J.E., Hosper, S.H. and Colijn, F. (1991) A quantitative method for description and assessment of ecosystems: the AMOEBA-approach. *Marine Pollution Bulletin* 23, 265–270.

Torstensson, L., Pell, M. and Stenberg, B. (1998) Need of a strategy for evaluation of arable soil quality. *Ambio* 27, 1–77.

UNCED (United Nations Conference on Environment and Development) (1992) Agenda 21. June 1992, Rio de Janeiro.

Velvis, H. (1997) Evaluation of the selective respiratory inhibition method for measuring the ratio of fungal:bacterial activity in acid agricultural soils. *Biology and Fertility of Soils* 25, 354–360.

Wollum, A.G. (1994) Soil sampling for microbiological analysis. In: Weaver, R.W., Angle, S., Bottomley, P., Bezdicek, D.F., Smith, S., Tabatabai, A. and Wollum, A.G. (eds) *Methods of Soil Analysis. Part 2: Microbiological and Biochemical Properties.* Soil Science Society of America, Madison, Wisconsin, pp. 1–14.

Zelles, L., Adrian, P., Bai, Q.Y., Stepper, K., Adrian, M.V., Fischer, K., Maier, A. and Ziegler, A. (1991) Microbial activity measured in soils stored under different temperature and humidity conditions. *Soil Biology and Biochemistry* 23, 955–962.

4 Managing Soil Quality

Michael Schloter,[1] Jean Charles Munch[1] and Fabio Tittarelli[2]

[1]*GSF Research Centre for Environment and Health, Institute of Soil Ecology, PO Box 1129, D-85764 Neuherberg, Germany;* [2]*Consiglio per la ricerca e la sperimentazione in Agricottura, Istituto Sperimentale per la Nutrizione delle Piante, Via della Navicella, 2, 00184 Rome, Italy*

Abstract

Soil quality is defined as the 'continued capacity of soil to function as a vital living system, within ecosystem and land use boundaries, to sustain biological productivity, promote the quality of air and water environments, and maintain plant, animal and human health'. Therefore it is not surprising that the sustainable conservation of soil quality is a key issue not only in scientific discussions, but also in political controversies. However, conservation of agricultural soils is often a topic of conflict as, on the one hand, it is a major goal to produce crops with high yields and, on the other hand, soil quality should be maintained. In this chapter this clash will be described in detail and some examples of how different farming management systems affect soil quality will be given. Furthermore, how to measure soil quality easily and whether sensitive indicators are available are still open questions. Therefore, another part of this chapter will focus on questions about indicator development and use in practice.

Introduction

During recent years the European Commission has given increasing importance to the role played by soil as a strategic compartment for many of the environmental issues affecting our countries and the planet as a whole. In 2001, publishing its Sustainable Development Strategy (COM 2001, 264), the European Commission noted that soil loss and declining fertility were eroding agricultural land viability. In 2002 (Decision 1600/2002), the European Parliament and the Council laid down the Sixth Environmental Action Programme (Sixth EAP), which covers a period of 10 years, starting from 22 July 2002. The Programme addresses the key environmental objectives and priorities of the Community, which will be met through a range of measures, including legislation and strategic approaches. In Article 6,

'Objectives and priority areas for action on nature and biodiversity', the Sixth EAP foresees the development of a thematic strategy on soil protection 'addressing the prevention of, *inter alia*, pollution, erosion, desertification, land degradation, and hydrogeological risks taking into account regional diversity, including specificities of mountain and arid areas'. On the basis of this input, the European Commission published a Communication entitled 'Towards a thematic strategy for soil protection' (COM 2002, 179), which is intended as a first contribution for building on the political commitment to soil protection. The Communication listed the following as main threats to soil in the EU and candidate countries: erosion, decline in organic matter, soil contamination, soil sealing, soil compaction, decline in soil biodiversity, salinization, floods and landslides. All these threats degrade soil processes, and are considered to be driven and exacerbated by human activity. In this context, soil microbiologists have the expertise to contribute to the ongoing discussion on soil protection through the elaboration of the concept of soil quality and, in particular, of soil quality management. This is the main objective of the following chapter, which, after a short theoretical dissertation on soil quality and soil quality indicators, provides a few case studies to demonstrate the effects of different agricultural production methods on soil microbial community structure and functions.

Soil Quality and Agricultural Sustainability

Sustainable agriculture is based on the conservation of natural resources and on the concept of productivity linked closely to the maintenance of a system aimed at saving energy and resources in the mid to long term, through optimizing recycling and enhancing biodiversity, and through biological synergy. The emphasis on 'sustainable agriculture' and, more generally, on 'sustainable land use', initiated the development of the soil quality concept during the 1990s. Despite the fact that during the past 10 years many different definitions of soil quality have been proposed, the following seems to be one of the most widely accepted, and is suitable for the purpose of sustainable soil management: 'the capacity of a soil to function within ecosystem boundaries to sustain biological productivity, maintain environmental quality, and promote plant and animal health'. In a more simplistic way, this definition of soil quality, like those proposed by other authors and not reported in this chapter, refers to the capacity of a soil to function at present and in the future, for an indefinite period of time. The concept of soil quality has become a tool for assessing the sustainability of soil management systems and has been adopted by users at different educational levels, such as policy makers, land managers and farmers. According to Karlen *et al.* (2001), the use of soil quality as a tool for the assessment of human impact on natural resources is based on the distinction between inherent and dynamic characteristics of soil. Inherent characteristics are those determined by parent material, climate, vegetation and so on, which are not influenced by human activity. They contribute to an inherent soil quality, which is meaningful in determining the capacity of a soil for a specific land use. On the other hand, dynamic characteristics are those subjected to change as a consequence of a specific soil management system. So, dynamic soil quality can be measured and used to compare different agricultural practices and/or farming systems on similar soils, or in the same soil over a period of time. To succeed in soil quality assessment, the issue of which types of measurements should be made in order to evaluate the effect of management on soil function must be solved. Soils have physical, chemical and biological properties that interact in a complex way to give them their quality or capacity to function (Seybold *et al.*, 1997). So, soil quality cannot be measured directly, but can be assessed through the measurement of changes of its attributes, or attributes of the ecosystem which are considered as indicators. Of course, we need standards for the evaluation of management systems that allow an assessment of their sustainability. Two different approaches are employed in the evaluation of sustainable management systems: the comparative assessment and the dynamic assessment.

Comparative and Dynamic Assessment of Sustainable Management Systems

Comparative and dynamic assessment are the two main approaches employed in the evaluation of sustainable management systems. In the comparative approach, the performance of a system is determined in relation to alternatives. On the other hand, in the dynamic approach a management system is assessed in terms of its performance determined over a period of time. According to the comparative approach, the characteristics, biotic and abiotic soil attributes of alternative systems are compared at time *t* and a decision about the relative sustainability of each system is based on the magnitude of the measured parameters. The main limit of this approach is that, if only outputs are measured, it provides little information about the process that created the measured condition. On the other hand, the main disadvantage of the dynamic approach is that it needs measurements of indicators for at least two points in time and consequently does not provide an immediate assessment of soil quality (Seybold *et al.*, 1997). Moreover, it can be misleading in the case of a soil that functions at its highest attainable level, which it cannot improve, or when it is functioning at its lowest attainable level and cannot go lower; both these cases would show a static trend, indicating sustaining systems, but they would have completely different quality.

In our opinion, the two approaches to assessment are complementary, since they allow different scales of evaluation. While monitoring trends is more useful for evaluation at the level of the farm, the comparative assessment seems to be more suitable to a wider scale of evaluation (on a regional scale).

Indicators of Soil Quality

According to the above discussion, in order to be useful as an indicator of the sustainability of agricultural practices and of land management, indicators of soil quality should give some measure of the capacity of a soil to function in terms of plant and biological productivity and environmental quality (Seybold *et al.*, 1997). As reported in the introduction, indicators of soil quality should be:

- sensitive to long-term change in soil management and climate, but sufficiently robust not to change as a consequence of short-term changes in weather conditions;
- well correlated with beneficial soil functions;
- useful for understanding why a soil will or will not function as desired;
- comprehensible and useful to land managers;
- easy, and not expensive, to measure;
- where possible, components of existing soil databases.

Due to the complexity of the system, there cannot be a single indicator of soil quality to assess a specific soil management system, but a minimum data set of attributes regarding soil physical, chemical and biological properties must be selected. Furthermore, the suitability of soil quality indicators depends on the kind of land, land use and scale of assessment. Different land uses may require different properties of soil and, consequently, some soil quality indicators in a given situation can be more helpful than others for the purpose of the assessment (Karlen *et al.*, 2001). Another aspect, closely related to the previous one, which should always be stressed, is that the final aim of soil quality assessment is different in an agricultural soil and in a natural ecosystem. As stated by Singer and Ewing (2000), 'in an agricultural context soil quality may be managed in order to maximize production without adverse environmental effects, while in a natural ecosystem soil quality may be observed as a baseline value or set of values against which future changes in the system may be compared'. As a general consideration, each combination of soil type, land use and climate calls for a different set of practices to enhance soil quality. Whereas, within the same country, different pedoclimatic conditions require different management practices in order to reach the same sustainability goals, even more complex issues are related to soil quality management in tropical areas or, in general, in less-developed countries.

Soil Organic Carbon Pools and Processes

Since organic matter, or more specifically organic carbon and the carbon (C) cycle as a whole, can have an important effect on soil functioning, all the attributes linked to the soil C cycle are usually recommended as components in any minimum data set for soil quality evaluation.

According to a simplified scheme, soil organic matter (SOM) can be divided into two pools: non-living and living. Non-living SOM includes materials of different age and origin, which can be further divided into pools or fractions as a function of their turnover characteristics. For example, the humified fraction is more resistant to decay. The stability and longevity of this pool is a consequence of chemical structure and organo-mineral association. This pool of soil organic matter influences different aspects of soil quality, such as the fate of ionic and non-ionic compounds, the increase of soil cation exchange capacity and the long-term stability of microaggregates (Herrick and Wander, 1997). The interpretation guideline for the sake of soil quality assessment is that the higher the humified fraction of SOM, the higher is its contribution to soil quality.

Living soil organic matter represents only a small percentage of total soil organic carbon, and includes soil micro-, meso- and macroorganisms. In particular, soil microbial biomass is regarded as the most active and dynamic pool of SOM, and plays an important role in driving soil mineralization processes. Three main aspects of soil microbial biomass are usually considered for their effect on soil functions: pool size, activity and diversity. The determination of the total amount of carbon immobilized within microbial cells permits the determination of soil microbial biomass as a pool of soil organic matter. Since this pool is responsible for the decomposition of plant and animal residues and for the immobilization and mineralization of plant nutrients, it is finally responsible for the maintenance of soil fertility (Brookes, 2000). For this reason, the concept of microbial biomass has developed as an 'early warning' of changing soil conditions and as an indicator of the direction of change. Carbon mineralization activity, as a key process of the soil C cycle, determines the rapidity of the organic matter degradation process in soil. Many studies have been carried out recently in order to verify, directly or indirectly, the potential for increasing carbon storage in soil by manipulating C inputs to minimize the rate of carbon mineralization (Jans-Hammermeister *et al.*, 1997; Fließbach and Mäder, 2000). The sensitivity of the carbon mineralization process to changes in soil management is low, because small microbial populations, in degraded soil, can mineralize organic matter to the same extent and at the same rate as large microbial populations in undegraded soils (Brookes, 1994). More sensitive to soil management changes, and more helpful as a soil quality indicator, is the combination of the two measurements, relating to both size and activity of microbial biomass. Carbon mineralization activity/unit of biomass (biomass-specific respiration) and the mineralization coefficient (respired carbon/total organic carbon) indicate efficiency in carbon utilization and energy demand. Finally, increasing importance has been given recently to soil microbial diversity measurements as indicators of community stability and of impact of stress on that community.

Interactions between the diversity of primary producers (plants) and that of decomposers (microbial communities), the two key functional groups that form the basis of all ecosystems (Loreau, 2001), have major consequences for agricultural management. Soil microorganisms control the mineralization of natural compounds and xenobiotics. Furthermore, bacteria and fungi exist at extremely high density and diversity in soils, and can react to changing environmental conditions rapidly, by adjusting: (i) activity rates; (ii) gene expression; (iii) biomass; and (iv) community structure. Some of these parameters might be perfect indicators for evaluating soil quality (Schloter *et al.*, 2003a).

Research interest in microbial biodiversity over the past 25 years has increased markedly, as microbiologists have become interested in the significance of biodiversity for ecological processes. Most of the work illustrates a dominant interest in questions concerning the effect of specific environmental factors on microbial biodiversity, the spatial and temporal heterogeneity of this biodiversity, and quantitative measures of population structure (for a review see Morris *et al.*, 2002). However, since the rise of availability of molecular genetic tools in microbial ecology in the early 1990s, it has become apparent that we know only a very small part of the diversity of the microbial world. Most of this unexplored microbial diversity seems to be hidden in yet uncultured microbes. Use of new direct methods, independent of cultivation, based on the genotype (Amann *et al.*, 1995) and phenotype (Zelles *et al.*, 1994) of the microbes, enables a deeper understanding of the composition of microbial communities. For example, using the rDNA-directed approach of dissecting bacterial communities by amplifying the 16S rDNA (*rrs*) gene from soil samples by polymerase chain reaction (PCR), and studying the diversity of the acquired *rrs* sequences, almost exclusively new sequences became apparent, which are only related to a certain degree to the well-studied bacteria in culture collections (Amann *et al.*, 1995). Based on molecular studies, it can be estimated that 1 g of soil is the habitat of more than 10^9 bacteria, belonging to about 10,000 different microbial species (Ovreas and Torsvik, 1998).

Using these new methods for studies in agricultural ecosystems, it could be clearly proven that the highest influence on microbial community structure and function, mainly in the rhizosphere, derives from used crops and applied agrochemicals. Miethling *et al.* (2000) conducted a greenhouse study with soil–plant microcosms, in order to compare the effect of the crop species lucerne (*Medicago sativa*) and rye (*Secale cereale*), soil origin (different cropping history) and a bacterial inoculant (*Sinorhizobium meliloti*) on the establishment of plant-root-colonizing microbial communities. Three community-level targeting approaches were used to characterize the variation of the extracted microbial rhizosphere consortia: (i) community-level physiological profiles (CLPP); (ii) fatty acid methyl ester (FAME) analysis; and (iii) diversity of PCR-amplified 16S rRNA target sequences from directly extracted ribosomes, determined by temperature gradient gel electrophoresis (TGGE). All approaches identified the crop species as the major determinant of microbial community characteristics. The influence of soil was consistently of minor importance, while a modification of the lucerne-associated microbial community structure after inoculation with *S. meliloti* was only consistently observed by using TGGE. In a study by Yang *et al.* (2000), the DNA sequence diversities for microbial communities in four soils affected by agricultural chemicals (mainly triadimefon and ammonium bicarbonate and their intermediates) were evaluated by random amplified polymorphic DNA (RAPD) analysis. The richness, modified richness, Shannon–Weaver index and a similarity coefficient of DNA were calculated to quantify the diversity of accessed DNA sequences. The results clearly showed that agricultural chemicals affected soil microbial community diversity at the DNA level. The four soil microbial communities were distinguishable in terms of DNA sequence richness, modified richness, Shannon–Weaver index and coefficient of DNA similarity. Analysis also showed that the amounts of organic C and microbial biomass C were low in the soil treated by pesticide (mainly triadimefon and its intermediates), but high in the soil where the chemical fertilizer (mainly ammonium bicarbonate and its intermediates) was applied. Combined, the above results may indicate that pesticide pollution caused a decrease in the soil microbial biomass but maintained a high diversity at the DNA level, compared to the control without chemical pollution. In contrast, chemical fertilizer pollution caused an increase in the soil biomass but a decrease in DNA diversity.

Managing Soil Quality: Case Studies

Managing soil quality relates to agricultural soils which are subjected to different agricultural practices as a consequence of the management system adopted. As a function of the specific objective that farmers and/or land managers and/or policy makers want to reach, relevant soil functions and/or parameters should be individualized and measured, in order to understand whether the specific soil management system adopted reaches the prefixed aims. Below are some examples of how agricultural management can influence microbial community structure and function.

Influence of precision farming and conventional agricultural management on microbial community structure and function

Precision farming summarizes cultivation practices that allow for spatial and temporal variability of soil attributes and crop parameters within an agricultural field. Distinct areas in a field are managed by applying different levels of input, depending on the yield potential of the crop in that particular area. Benefits of these actions are: reduction of the cost for crop production, thus conserving resources while maintaining high yield; and minimizing the risk of environmental pollution (Dawson, 1997).

In two studies, from Hagn *et al.* (2003a) and Schloter *et al.* (2003b), microbial community structure and function and their dynamics were investigated in relation to season, soil type and farming management practice. The research was done using soils from high- (H) and low-yield areas (L) of a field site, cultivated with winter wheat under two different farming management systems (precision farming, P; conventional farming, C) over the growing period.

It was demonstrated that the microbial biomass and the microbial community structure, measured using the phospholipid fatty acid technique and DNA-based methods in the top soils of the investigated plots, were not influenced by precision farming. Both parameters showed a typical seasonal run, which was independent of the farming management type. Microbial biomass was reduced during the summer months, due to the dry weather conditions and the hot temperatures in the top soil. The microbial community structure changed mainly after the application of fertilizers and was associated with high amounts of root exudates in late spring.

As fungal communities are essential for the reduction of soil erosion, degradation of complex organic compounds and biocontrol (Hagn *et al.*, 2003b), fungal diversity was studied in detail using cultivation-independent methods (direct extraction of DNA from soil followed by PCR amplification of a subunit of the 18S rDNA and fingerprinting (DGGE)) as well as cultivation-dependent techniques (isolation of pure cultures). Comparison of the PCR amplicons by DGGE patterns, reflecting the *total fungal community*, showed no differences between the sampling sites and no influence of the farming management systems. Only small differences were observed over the growing period. For the identification of *active hyphae*, cultivation-dependent techniques were used. The resulting isolates were subcultured and grouped by their morphology and genotype. In contrast to the cultivation-independent approaches, clear site-specific and seasonal effects on the fungal community structure could be observed. However, no effects of the different farming management techniques were seen. These results clearly indicate that the potential fungal community (including spores) is not influenced by the investigated factors, whereas active populations show a clear response to environmental changes (soil type and season). The most abundant group, consisting of *Trichoderma* species, was investigated in more detail using strain-specific genotype-based fingerprinting techniques as well as a screening for potential biocontrol activity against the wheat pathogen *Fusarium graminearum*. The genotypic distribution, as well as the potential biocontrol activity, revealed clear site-specific patterns, reflecting the soil type and the season. A clear response of the *Trichoderma* ecotypes to different farming management techniques was not seen.

Enzyme activities in the nitrogen cycle were more affected by precision farming. Proteolytic activity was significantly increased by precision farming, especially on low-yield plots; also, nitrification and denitrification activities showed a clear response to application of fertilizers. In summer, due to the low microbial biomass, all measured activities were very low and showed no sustained reaction to farming management.

These results indicated that the structure of the entire microbial community is not influenced by precision farming compared to conventional agriculture. However, the applied farming management type had a visible influence on the induction or repression of gene transcription and expression.

Influence of changes from conventional to organic farming on microbial community structure and function

The influence of two farming systems in southern Germany on aggregate greenhouse gas emission (CO_2, CH_4 and N_2O) was investigated by Flessa *et al.* (2002). One system (farm A) conformed to the principles of integrated farming (recommended by the official agricultural advisory service) and the other system (farm B) followed the principles of organic farming (neither synthetic fertilizers nor pesticides were used). Farm A consisted of 30.4 ha fields (mean fertilization rate 188 kg N/ha), 1.8 ha meadows, 12.4 ha set-aside land and 28.6 adult beef steers (year-round indoor stock keeping). Farm B consisted of 31.3 ha fields, 7 ha meadows, 18.2 ha pasture, 5.5 ha set-aside land and a herd of 35.6 adult cattle (grazing period 6 months).

The integrated assessment of greenhouse gas emissions included those from fields, pasture, cattle, cattle waste management, fertilizer production and consumption of fossil fuels. Soil N_2O emissions were estimated from 25 year-round measurements on differently managed fields. Expressed per hectare farm area, the aggregate emission of greenhouse gases was 4.2 Mg CO_2 equivalents (conventional farming) and 3.0 Mg CO_2 equivalents (organic farming). Nitrous oxide emissions (mainly from soils) contributed the major part (about 60%) of total greenhouse gas emissions in both farming systems. Methane emissions (mainly from cattle and cattle waste management) were approximately 25%, and CO_2 emissions were lowest (*c.* 15%). Mean emissions related to crop production (emissions from fields, fertilizer production, and the consumption of fossil fuels for field management and drying of crops) were 4.4 Mg CO_2 equivalents/ha and 3.2 Mg CO_2 equivalents/ha field area for farms A and B, respectively. On average, 2.53% of total N input by synthetic N fertilizers, organic fertilizers and crop residues were emitted as N_2O-nitrogen. Total annual emissions per cattle unit (live weight of 500 kg) from enteric fermentation and storage of cattle waste were about 25% higher for farm A (1.6 Mg CO_2 equivalents) than farm B (1.3 Mg CO_2 equivalents). Taken together, these results indicated that conversion from conventional to organic farming led to reduced emissions per hectare, but yield-related emissions are not reduced.

In another study by Schloter *et al.* (2004), the effects on microbial community structure of conventional farming and ecological farming were compared. Plots under ecological farming for 4–40 years were compared with plots under conventional farming practice. To characterize the soil microbial community structure under both systems, a hierarchical approach was applied (phospholipid fatty acids to describe differences in superfamilies and domains, 16S rDNA pattern to describe species variability, and enrichment of *Ochrobactrum* spp. to describe microdiversity of ecotypes). To characterize functions both at the community level and at the level of the enriched *Ochrobactrum antrophi* populations, the EcoBiolog® system was used. Although no differences in the microbial biomass were observed (the two highest biomass values were found in the soil of a conventional farming plot and a plot that had been under ecological farming practice for 40 years), the results clearly indicated that there was a significant shift in microbial community structure between the conventionally farmed plots and the plots under ecological farming practice. The use of the phospholipid fatty acid (PLFA) method and the statistic evaluation of the results by principal component analysis (PCA) revealed significant changes of the microbial community on a high taxonomic level. Just 2 years after changing farming practice, shifts in the structure of the microflora were visible. The longer the ecological farming practice was applied to the plots, the greater the differences were. The main shifts were found in the Gram-positive bacteria. After extraction of total DNA from soil, amplification of the variable region V6/V8 of the 16S rDNA and fingerprinting using DGGE, three main clusters (conventional farming, 2 years' ecological farming and 40 years' ecological farming) were visible. The main differences were not found in the total diversity (Shannon index), but in the abundance of characteristic species in each soil (Simpson index). For the soils that were under ecological farming practice for 40 years, *Actinomycetes* appeared to be highly abundant. After enrichment by specific antibodies and classifying the ecotypes by the EcoBiolog® system, 24 ecotypes could be defined. Significant differences in ecotype variability were only observed in the plots that had been under ecological farming practice for 40 years. Using the EcoBiolog®system, it could be shown that differences in the microbial community structure led to a change of function: in the conventional soil specialized microbes (which can utilize only a small part of the used C sources) were dominant, whereas in the plots under ecological farming practice generalists dominated. These results could be shown at the level of the whole microbial community and were confirmed on the microdiversity level.

Sustainable Management of Soils

Sustainable management of soils is an environmental issue worldwide, but is most urgent in less-developed countries. Many of the problems faced by tropical countries concern soil degradation. As a consequence of population pressure and the lack of new land, fallow rotation is shortened, resulting in soil erosion. Since, in many tropical soils, a large proportion of available plant nutrients are in the topsoil, soil erosion causes a substantial loss in fertility and a deterioration of soil physical structure. As the cost of reclamation of eroded soil is prohibitive for many less-developed countries, soil erosion is considered to be one of the greatest threats to the sustainability of agricultural production in tropical countries (Scholes *et al.*, 1994). Other constraints to crop productivity in tropical soils are of chemical or nutrient origin: aluminium toxicity, acidity, high phosphorus fixation with iron oxides, low nutrient reserves and high secondary salinization of irrigated land (Lal, 2002). Large areas of sub-Saharan Africa and South Asia are characterized by the constraints reported above. According to Lal (2002), the goals of sustainable agriculture in these areas can be summarized as:

- food security;
- reversal of soil degradative trends;
- improvement of surface and ground water quality;
- sequestration of C in soil in order to reduce the negative effect of emission of greenhouse gases.

The identification of the goals of sustainable agriculture in tropical countries does not, in itself, solve the issue of managing soil quality, since the same objectives can be reached following different soil management systems, as a consequence of socio-economic considerations and political decisions. A typical example regarding less-developed countries is the debate about one of the most serious problems among those listed above: food security. In order to enhance food production to meet the needs of the population, during the past 30 years some areas of the tropics have experienced a dramatic increase in per capita food production. This improvement is mainly based on the introduction of new crop varieties on fertile soils heavily supplied with water, fertilizers and pesticides. While, on a regional scale and from a macroeconomic point of view, many Asian countries can now theoretically satisfy the need of their ever-growing populations for rice and wheat, in poorer rural areas the numbers of undernourished people are still extremely high. In these areas, smallholder farming systems seem to better guarantee food security. On a farm scale, traditional cropping systems based on different food crops, rather than a single high-yield cash crop, the use of indigenous crop varieties, organic fertilizers produced on-farm and minimum tillage can be more efficient for reaching self-sufficiency. In terms of sustainability of soil management for food security on a farm level, smallholder farming systems are characterized by agricultural practices that improve soil resistance to degradation and its capacity to recover rapidly after a perturbation.

Conclusion

In contrast to the high complexity of microbial communities in soil, the ideal soil microbiological and biochemical indicator to determine soil quality should be simple to measure, should work equally well in all environments and should reliably reveal which problems exist where. It is unlikely that a sole ideal indicator can be defined with a single measure, because of the multitude of microbiological components and biochemical pathways. Therefore, a minimum data set is frequently applied (Carter *et al.*, 1997). The basic indicators and the number of measures needed are still under discussion, and depend on the aims of the investigations. National and international programmes for monitoring soil quality presently include biomass and respiration measurements, but also extend to determination of nitrogen mineralization, microbial diversity and functional groups of soil fauna.

It is still unclear whether observed changes in microbial community structure are lasting, and whether naturally occurring environmental factors can influence the genotypic ability of the soil microbiota to recover after harsh conditions and become healthy again (Sparling, 1997). Research on the resilience of soil microbiota is therefore a significant task of microbial ecology.

References

Amann, R., Ludwig, W. and Schleifer, K.H. (1995) Phylogenetic identification and in situ detection of individual microbial cells without cultivation. *Microbiological Reviews* 59, 143–149.

Brookes, P.C. (1994) The use of microbial parameters in monitoring soil pollution by heavy metals. *Biology and Fertility of Soils* 19, 269–279.

Brookes, P.C. (2000) Changes in soil microbial properties as indicators of adverse effects of heavy metals. In: Benedetti, A. (ed.) *Indicatori per la Qualità del Suolo.* Fischer, Heildelberg, pp. 205–227.

Carter, M.R., Gregorich, E.G., Anderson, D.W., Doran, J.W., Janzen, H.H. and Pierce, F.J. (1997) Concepts of soil, quality and their significance. In: Gregorich, E.G. and Carter, M.R. (eds) *Soil Quality for Crop Production and Ecosystem Health.* Elsevier, Amsterdam, pp. 1–19.

Dawson, C.J. (1997) Management for spatial variability. In: Stafford, V. (ed.) *Precision Agriculture '97.* BIOS Scientific Publishers Ltd, Oxford, pp. 45–58.

Flessa, H., Ruser, R., Dörsch, P., Kamp, T., Jimenez, M.A., Munch, J.C. and Beese, F. (2002) Integrated evaluation of greenhouse gas emissions (CO_2, CH_4, N_2O) from two farming systems in southern Germany. *Agriculture, Ecosystems and Environment* 91, 175–189.

Fließbach, A. and Mäder, P. (2000) Microbial biomass and size-density fractions differ between soils of organic and conventional agricultural systems. *Soil Biology and Biochemistry* 32, 757–768.

Hagn, A., Pritsch, K., Schloter, M. and Munch, J.C. (2003a) Fungal diversity in agricultural soil under different farming management systems with special reference to biocontrol strains of *Trichoderma* spp. *Biology and Fertility of Soils* 38, 236–244.

Hagn, A., Geue, H., Pritsch, K. and Schloter, M. (2003b) Assessment of fungal diversity and community structure in agricultural used soils. *Microbiological Post* 37, 665–680.

Herrick, J.E. and Wander, M.M. (1997) Relationship between soil organic carbon and soil quality in cropped and rangeland soils: the importance of distribution, composition and biological activity. In: Lal, R., Kimble, J.M. and Stewart, B.A. (eds) *Soil Processes and the Carbon Cycle.* CRC Press, Boca Raton, Florida, pp. 405–425.

Jans-Hammermeister, D.C., McGill, W.B. and Izaurralde, R.C. (1997) Management of soil C by manipulation of microbial metabolism: daily vs. pulsed C additions. In: Lal, R., Kimble, J.M. and Stewart, B.A. (eds) *Soil Processes and the Carbon Cycle.* CRC Press, Boca Raton, Florida, pp. 321–334.

Karlen, D.L., Andrews, S.S. and Doran, J.W. (2001) Soil quality: current concepts and applications. In: Sparks, D.L. (ed.) *Advances in Agronomy,* Vol. 7. Academic Press, London, pp. 1–39.

Lal, R. (2002) The potential of soils of the tropics to sequester carbon and mitigate the greenhouse effect. In: Sparks, D.L. (ed.) *Advances in Agronomy,* Vol. 76. Academic Press, London, pp. 1–30.

Loreau, M. (2001) Microbial diversity, producer–decomposer interactions and ecosystem processes: a theoretical model. *Proceedings of the Royal Society for Biological Science* 268, 303–309.

Miethling, R., Wieland, G., Backhaus, H. and Tebbe, C.C. (2000) Variation of microbial rhizosphere communities in response to crop species, soil origin, and inoculation with *Sinorhizobium meliloti* L33. *Microbial Ecology* 40, 43–51.

Morris, C.E., Bardin, M., Berge, O., Frey-Klett, P., Fromin, N., Girardin, H., Guinebretiere, M.H., Lebaron, P., Thiery, J.M. and Troussellier, M. (2002) Microbial biodiversity: approaches to experimental design and hypothesis testing in primary scientific literature from 1975 to 1999. *Microbiological and Molecular Biology Reviews* 66, 592–616.

Ovreas, L. and Torsvik, V.V. (1998) Microbial diversity and community structure in two different agricultural soil communities. *Microbial Ecology* 36, 303–315.

Schloter, M., Dilly, O. and Munch, J.C. (2003a) Indicators for evaluating soil quality. *Agriculture, Ecosystems and Environment* 98, 255–262.

Schloter, M., Bach, H.J., Sehy, U., Metz, S. and Munch, J.C. (2003b) Influence of precision farming on the microbial community structure and selected functions in nitrogen turnover. *Agriculture, Ecosystems and Environment* 98, 295–304.

Schloter, M., Bergmüller, C., Friedel, J., Hartmann, A. and Munch, J.C. (2005) Effects of different farming practice on the microbial community structure. *Applied and Environmental Microbiology* (submitted).

Scholes, M.C., Swift, M.J., Heal, O.W., Sanchez, P.A., Ingram, J.S.I. and Dalal, R. (1994) Soil fertility in response to the demand for sustainability. In: Woomer, P.L. and Swift, M.J. *The Biological Management of Tropical Soil Fertility*. John Wiley & Sons, New York, pp. 1–14.

Seybold, C.A., Mausbach, M.J., Karlen, D.L. and Rogers, H.H. (1997) Quantification of soil quality. In: Lal, R., Kimble, J.M. and Stewart, B.A (eds) *Soil Processes and the Carbon Cycle*. CRC Press, Boca Raton, Florida, pp. 387–404.

Singer, M.J. and Ewing, S. (2000) Soil quality. In: Sumner, M.E. (ed.) *Handbook of Soil Science*. CRC Press, Boca Raton, Florida, pp. 271–298.

Sparling, G.P. (1997) Soil microbial biomass, activity and nutrient cycling as indicators of soil health. In: Pankhurst, C., Doube, B.M. and Gupta, V. (eds) *Biological Indicators of Soil Health*. CAB International, Wallingford, UK, pp. 97–119.

Yang, Y., Yao, J., Hu, S. and Qi, Y. (2000) Effects of agricultural chemicals on DNA sequence diversity of soil microbial community: a study with RAPD marker. *Microbial Ecology* 39, 72–79.

Zelles, L., Bai, Q., Ma, X., Rackwitz, R., Winter, K. and Beese, F. (1994) Microbial biomass, metabolic activity and nutritional status determined from fatty acid patterns and polyhydroxybutyrate in agriculturally managed soils. *Soil Biology and Biochemistry* 26, 439–446.

5 Concluding Remarks

Anna Benedetti,[1] Philip C. Brookes[2] and James M. Lynch[3]

[1]*Consiglio per la ricerca e la sperimentazione in Agricoltura, Istituto Sperimentale per la Nutrizione delle Piante, Via della Navicella, 2, 00184 Rome, Italy;* [2]*Agriculture and Environment Division, Rothamsted Research, Harpenden AL5 2JQ, UK;* [3]*Forest Research, Alice Holt Lodge, Farnham GU10 4LH, UK*

Soil quality has been defined as the capacity of the soil to produce healthy and nutritious crops (Paperdick and Parr, 1992), but it is also important to consider the studies where indicators of soil quality have been sought for forestry (Moffat, 2003). The related concepts of soil resilience (the ability of the soil to recover after disturbance) and soil degradation (the loss of the soil's capacity to produce crops) also need to be considered alongside quality assessments (Elliott and Lynch, 1994). In this context, toxicity measurements are necessary, and recently a range of biosensors has become available for this purpose.

The aim of this handbook is to provide a practical guide on how to use biochemical, microbiological and molecular indicators to define soil quality. It must be stressed that non-experts cannot expect to resolve such a complex issue without the assistance of experts in the field; nevertheless, the topic will be addressed humbly. We are all aware that there are no absolute benchmarks for assessing biological soil fertility, while there are chemical or physical indicators that are easy to define, interpret and understand. On the contrary, biological indicators, e.g. soil microbial biomass concentration, require especially careful interpretation. There is no such thing as absolute high or low values, values that remain constant over time and in space; there are sets or families of similar values, and also other, and therefore interpretable, values. It must be kept in mind that a variability of 20% is common for biological populations, and this is one of the reasons why it is recommended that several indicators are utilized to produce an index to describe the behaviour of soil microorganisms.

An index can be a set of several indicators that are developed and deduced from parameters, as implicit in the definition provided by the OECD, 'Set of parameters or aggregate of weighed indicators' (1993). Moreover, biological indicators have to be related to physical, chemical and agronomic indices, etc. and be relevant to the nature of the type of study to be performed.

Selected biochemical, microbiological and molecular parameters available to the operator are given in this handbook (for instance in Chapter 3). Once the exact aim of the investigation has been defined, the list of parameters can usually be streamlined and reduced to answer the questions asked.

Activity carried out within the framework of EU COST Action 831 revealed that the selected factors need to be arranged in a hierarchical scale of indicators in function of the study goal. The topic of hierarchical scales of different indicators is widely debated, and it is evident that each single researcher is more skilled at using the parameters used in his or her laboratory. A guide to the hierarchical use of indicators is provided by the OECD requirements (1999), whereby indicators must:

- be clearly correlated with a certain phenomenon or a certain feature that is being investigated or monitored;
- be highly correlated with the above-mentioned effect with minimal statistical variability;
- be unobscured by much less significant responses;
- have a sufficiently generalized, albeit not identical, validity in many analogous situations.

It is clear that hierarchical levels can change depending upon whether the indicator is required for monitoring, for accurate characterization of a particular environment, for assessing or restoring previous changes, or for starting up research. If the aim is to study soil quality in terms of fertility, the hierarchical level represented in Fig. 5.1 could be applicable:

1. Biomass-carbon (C) and respiration rate.
2. Functional diversity.
3. Genetic diversity.
4. Case-by-case in-depth probes (heavy metals, genetically modified organisms, air pollution, erosion, etc.).

We should take the same approach we would all adopt if we were to have a medical check-up. The first thing a physician does is to carry out a series of routine basic examinations. The physician will probably only call for further tests if irregularities occur in these basic ones, which may indicate an underlying pathology. Then the physician may move to more sophisticated tests, indicated by the findings in the first set. These can then be followed by other, much more sophisticated and specific diagnostic investigations.

Likewise, the first step in assessing biological soil fertility, i.e. the expression of microbial turnover, is to perform simple biochemical tests. The same tests can be used effectively for environmental monitoring. The next step could be to study the functional diversity of the ecosystem, followed by genetic diversity, and then case-by-case in-depth probes. To date, some methodologies, for example the ecophysiological profile and bacterial and fungal DNA studies, are rarely utilized in nationwide, large-scale monitoring programmes.

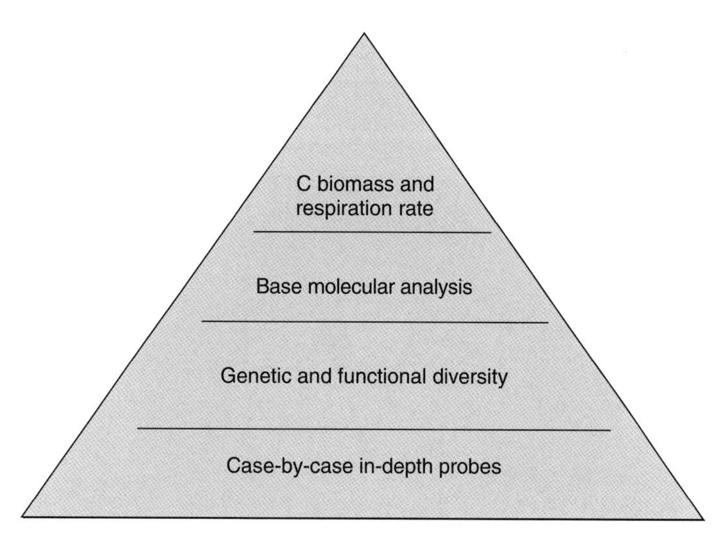

Fig. 5.1. Hierarchical scales of microbiological, biochemical and molecular parameters to define the biological fertility of a soil.

Moreover, a minimum hierarchical level must also be identified for other correlated indicators, in order to prevent false-negative and -positive results. In the case where soil fertility is related to crop yield, physical fertility is just as important as chemical and biological fertility. Obviously, however, the correct functioning of aerobic microorganisms will not occur under, for example, conditions of oxygen limitation, extremes of pH or elevated salinity. Thus, it is crucial to build other hierarchical scales, which, for chemical soil parameters, could be represented by the following:

1. Organic matter.
2. pH.
3. Available nutrients.
4. Various types of pollutants, etc.

The following parameters could be adopted for physical soil fertility indicators (Pagliai and Vignozzi, 2000; Pagliai *et al.*, 2000):

1. Porosity.
2. Aggregate stability.
3. Compactness.
4. Sealing along the profile.
5. Structure loss.
6. Superficial crusts and potential risk of their formation.
7. Fissuring.
8. Erodability.

The hierarchical scales will then be put together in an attempt to identify a minimum data set taken from the point where the different hierarchical scales overlap (Fig. 5.2).

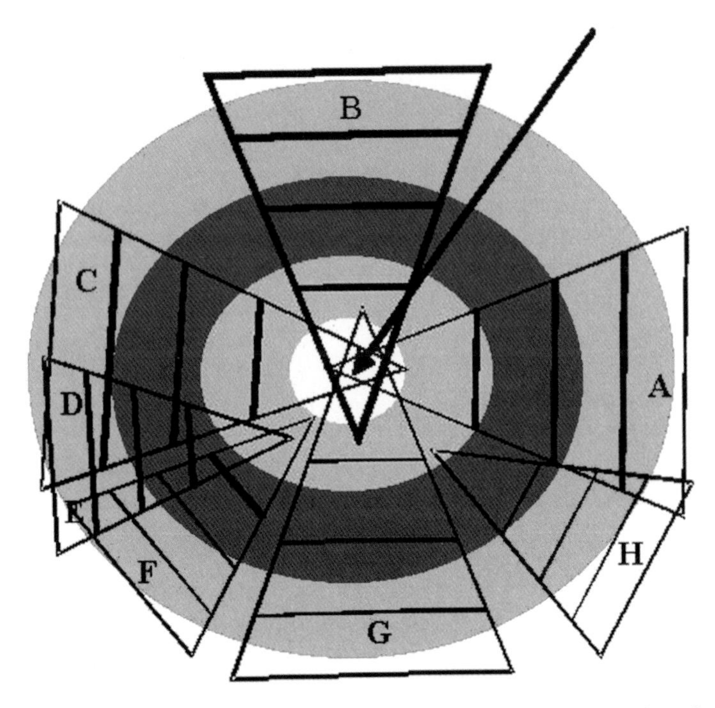

Fig. 5.2. Model showing a minimum data set for assessing soil quality and environmental sustainability. The minimum data set increases with the extent of the study (circles). A, physical parameters; B, climatic parameters; C, chemical parameters; D, microbiological parameters; E, land use; F, biological parameters; G, management; H, etc.

The extent of the study is represented by the circles in Fig. 5.2. The hierarchical scales to be used in different studies obviously differ, and range from environmental, pedological and agronomic parameters to social and economic ones. The approaches adopted in two case studies are described below.

Soil Sealing, Compaction and Erosion

Large areas of the central part of Italy are representative of the hillside environments of the Tuscan–Romagna Mountains (site 1) and of the clay hillside environments of central southern Italy (site 2). The soils, containing small amounts of organic carbon, are characterized by low structural stability and poor regeneration capacity: they must be managed correctly to minimize the potential risks of formation of surface crusts, sealed surfaces and compaction by farming machinery. The effects of such hazardous degradation in a hilly environment – the reduced rainwater infiltration rate and the creation of preferential surface runoff courses – play a role in triggering widespread and channelled erosion processes.

Soil management is crucial for the prevention and control of degradation. Different tillage systems, i.e. deep ploughing, shallow ploughing, minimum tillage and ripper subsoiling, have different effects on soil conditions. Adoption of ripper subsoiling tillage is capable of reducing structural damage caused by deep ploughing, lessening the risk of formation of surface crusts and the presence of compacted layers in the profile. This was revealed by the findings of the micromorphological analysis and quantification of the pore system in site 1. Moreover, ripper subsoiling conserves more organic carbon than does deep ploughing, especially in the topsoil layer. Also, the amounts of humified organic matter in soil managed by ripper subsoiling is greater than in soils following deep ploughing (Dell'Abate *et al.*, 2004).

Soil microbial activity has been verified by the determination of the soil microbial biomass carbon and its respiration (Table 5.1), and subsequently using the metabolic quotient (qCO_2) and the C biomass/total organic C (B_C/TOC) ratio.

Data concerning the quantity and the activity of the microbial biomass for the two different situations demonstrate that ripper subsoiling (RS) is a better management practice than deep ploughing for the maintenance of the total organic resource, and also of the living fraction. In fact, the content of microbial biomass is greater in RS for both the depths.

Results obtained for the two different management practices showed that qCO_2 was comparable in the two layers in the soil submitted to ripper subsoiling (RS), while in the soil in which the deep ploughing (DP) was practised, qCO_2 was higher at a depth of 20–40 cm.

A high value of qCO_2 in the ploughed zone (DP) highlights a situation of non-equilibrium due to the adopted practice. This conclusion can be also reached by analysing the B_C/TOC ratio: the value of microbial biomass in the deepest layer of the DP case was halved in comparison to the value found in RS to the same depth, while total organic carbon was practically constant.

In site 2, the effects of two other types of soil management (a comparison of continuous wheat and continuous lucerne) were assessed, in a field experiment established in 1994. In this case, the porosity values obtained in the 0–5 cm soil depth for the two areas on which the crops were grown show that after heavy rain, soils supporting lucerne gave a higher porosity percentage than those from the wheat-growing area. The protective action of the vegetation cover of lucerne decreased soil surface vulnerability to the

Table 5.1. Soil microbial biomass carbon (B_C), C biomass/total organic carbon ratio (B_C/TOC) and metabolic quotient (qCO_2).

Sample soil	B_C (μg/g)	B_C/TOC (%)	qCO_2 (mg CO_2-C/mg C_{mic}.h)
RS (0–20 cm)	203.8 ± 26.9	2.290 ± 0.004	0.0013 ± 0.0003
(20–40 cm)	137.7 ± 23.6	1.996 ± 0.004	0.0017 ± 0.0003
DP (0–20 cm)	121.3 ± 69.8	1.989 ± 0.012	0.0019 ± 0.0011
(20–40 cm)	73.8 ± 9.8	1.118 ± 0.01	0.0034 ± 0.0008

RS, ripper subsoiling; DP, deep ploughing.

impact of rainfall and thus lessened the risk of formation of crusts. Moreover, wheat did not seem to be the most suitable crop as it depleted the organic matter in the top horizon of the soil, while lucerne conserved the organic matter better. Removal of the finest soil particles by water erosion led to preferential loss of the most stable and most strongly sorbed organic fraction (humin), which, in contrast, accumulated in the deeper layer.

The quantity of microbial biomass in the soil cultivated with lucerne (L) was higher than that in the soil cultivated with wheat (W) (Table 5.2), and this was also confirmed by the respiration data (data not shown): compared to wheat, the lucerne crop is more effective in preserving the total soil organic carbon and also improves soil microbial life. The C biomass/total organic carbon rate had comparable values for the two crops for each depth. Also, the specific respiration (qCO_2) showed comparable values in the two cases for the two depths. It was higher under both the crops in the deepest layer, and this underlines a stress condition.

In this study, the lucerne covering showed a positive effect on the quantity of organic carbon and microbial biomass. On the other hand, microbial metabolism was not affected by the different crop (similar B_C/TOC values and metabolic quotients at the same depths). Wheat removed carbon from the soil but it did not modify microbial metabolism.

The rational answer to this environmental risk is to avoid sowing wheat in these soils, but indubitably wheat is a more profitable crop than, for example, lucerne. It is clear that the radical changes in agriculture, where the aims are to produce high yield and, at the same time, conserve 'landscaping farming' environments, must strike a balance or set a limit of degradation (Baroccio *et al.*, 2002).

Impact of Cultivation of Genetically Modified Plants

The EU Directive 18/2001 sets the general criteria on emission of genetically modified (GM) plants in Europe. For the first time, this law fixed a limit on soil, nitrogen and carbon recycling. What does this mean? Carbon and nitrogen mineralization or humification parameters should be able to detect carbon and nitrogen recycling in soil, as they are indicators of biomass activity and ecosystem functions.

Table 5.2. Soil microbial biomass carbon (B_C), C biomass/total organic carbon ratio (B_C/TOC) and metabolic quotient (qCO_2).

Sample soil	B_C (μg/g)	B_C/TOC (%)	qCO_2 (mg CO_2-C/mg C_{mic}.h)
W (0–20 cm)	153.0 ± 40.2	3.188 ± 0.004	0.0017 ± 0.0005
(20–40 cm)	89.9 ± 43.6	1.427 ± 0.008	0.0032 ± 0.0016
L (0–20 cm)	263.9 ± 144.0	3.341 ± 0.018	0.0013 ± 0.0007
(20–40 cm)	151.3 ± 63.1	1.780 ± 0.007	0.0029 ± 0.0014

W, wheat; L, lucerne.

During the past decade some authors have tested different parameters in soil growing GM and non-GM plants (Gebhard and Smalla, 1999; Lottmann *et al.*, 1999; Hopkins *et al.*, 2002; Bruinsma *et al.*, 2003; Ceccherini *et al.*, 2003). The investigation focused on soil respiration rate, biomass C, nitrification test, humification parameters, total organic carbon and the metabolic quotient to evaluate the impact of GM plants. Molecular (DGGE and PCR analysis) and ecophysiological tests (Biolog^{TM} analysis) were also carried out. The results demonstrated that biological, chemical and ecophysiological parameters alone were not able to detect the delayed impact of GM plants on diversity and microbial biomass activity, but that molecular analysis was more efficient. Baseline biological soil fertility and also microbial population management and natural fluctuations play a crucial role in this case. Molecular analysis detects possible changes in microbial diversity, such as richness and evenness, but also reveals the persistence of transgenes in soil.

In conclusion, some useful criteria that can be adopted to select soil quality indicators are as follows:

1. Correlate soil quality with land use. Define the use of the agricultural, forestry, pasture, public nature parks, unused industrial areas, industrial areas, marginal areas, etc. The choice of indicators depends on the type of soil investigated and the indicators suitable for the type of soil. For example, it may be more useful to determine substrate-induced respiration (SIR) instead of simple CO_2 evolution in forestry or pasture soils in homeostasis, and also in soils stressed by pollution and other factors.
2. Perform a case-by-case probe on the nature of the problems to be studied. Negative/positive/unknown impact, potential pollutants, recovery actions, new growing systems, GM plants, etc. When assessing the effects of heavy metals, for example, total genetic diversity and possibly even single species diversity have to be assessed along with functional diversity and the system's resistance to perturbation. The EU Directive 18/2001 on GM plants recommends assessing nitrogen and carbon recycling when evaluating impact, while it would be constructive to specify other parameters, such as mycorrhizal infection capacity, in mycorrhizic plants.
3. Identify which hierarchical scales can be related to observations.
4. Identify a minimum data set of indicators that can be correlated with each other.
5. If available, compare the observations with historical data from the monitoring framework, or repeat the same observations during the time.

Those involved in drafting this handbook aimed to ensure that potential users address the issue positively. Serial determinations are strongly recommended in order to detect the vital cycles of soil organisms, and, in turn, to understand soil functions and better manage soil conservation. The development of a new common language can be achieved thanks to the implementation of projects for monitoring biological fertility on various scales and the creation of institutional databanks and networks. Hopefully, this handbook will achieve this goal.

References

Baroccio, F., Dell'Abate, M.T. and Benedetti, A. (2002) Effect of different soil management on organic matter conservation: chemical and microbiological indicators. In: Nardi, S., Albuzio, A., Bottacin, A., Carden, D.E., Concheri, G., Ferretti, M., Chisi, R., Malagoli, M. and Masi, A. (eds) *Proceeding of XX Convegno nazionale della Società di Chimica Agraria*, Padova, Italy, pp. 29–36.

Bruinsma, M., Kowalchuk, G.A. and van Veen, J.A. (2003) Effects of genetically modified plants on microbial communities and processes in soil. *Biology and Fertility of Soils* 37, 329–337.

Ceccherini, M.T., Potè, J., Kay, E., Van, V.T., Maréchal, J., Pietramellara, G., Nannipieri, P., Vogel, T.M. and Simonet, P. (2003) Degradation and transformability of DNA from transgenic leaves. *Applied and Environmental Microbiology* 69, 673–678.

Dell'Abate, M.T., Benedetti, A. and Baroccio, F. (2005) Example n. 1: Fagna experimental site. In: Dell'Abate, M.T., Benedetti, A., Pagliai, M. and Sequi, P. (eds) *Atlas of Soil Quality Indicators*. Agriculture and Forestry Ministry, Rome (in press).

Elliott, L.F. and Lynch, J.M. (1994) Biodiversity and soil resilience. In: Greenland, D.J. and Szabolcs, I. (eds) *Soil Resilience and Sustainable Land Use*. CAB International, Wallingford, UK, pp. 353–364.

EU Council Directive 2001/18/CE (2001) Concerning the voluntary emission on environment of genetically modified organisms.

Gebhard, F. and Smalla, K. (1999) Monitoring field releases of genetically modified sugar beet for persistence of transgenic plant DNA and horizontal gene transfer. *FEMS Microbiology Ecology* 28, 261–272.

Hopkins, D.W., Marinari, S., Webster, E.A., Tilston, E.L. and Halpin, C. (2002) Decomposition in soil of residues from plants with genetic modifications to lignin biosynthesis. *Symposium on the Impact of GMOs: Soil Microbiology and Nutrient Dynamics*. IGMO, Vienna.

Lottmann, J., Heuer, H., Smalla, K. and Berg, G. (1999) Influence of transgenic T4-lysozyme-producing potato plants on potentially beneficial plant-associated bacteria. *FEMS Microbiology Ecology* 29, 365–377.

Moffat, A.J. (2003) Indicators of soil quality for UK forestry. *Forestry* 76, 547–568.

OECD (1993) Environmental monographs n. 83. *OECD Core Set of Indicators for Environmental Performance Reviews*. OECD, Paris.

OECD (1999) Environmental Indicators for Agriculture. *Issues and Design, The York Workshop*, Vol. 2. OECD, Paris.

Pagliai, M. and Vignozzi, D. (2000) Il sistema dei pori quale indicatore delle qualità strutturali dei suoli. Accademia Nazionale delle Scienze detta dei XL. *Memorie di Scienze Fisiche e naturali* 118° XXIV, 229–238.

Pagliai, M., Pellegrini, S., Vignozzi, N., Rousseva, S. and Grasselli, O. (2000) The quantification of the effect of subsoil compaction on soil porosity and related physical properties under conventional to reduced management practices. In: Horn, R., van den Akker, J.J.H. and Arvidsson, J. (eds) *Subsoil Compaction. Distribution, Processes and Consequences. Advances in Geo Ecology* 32, 305–313.

Paperdick, R.I. and Parr, J.F. (1992) Soil quality – the key to a sustainable agriculture. *American Journal of Alternative Agriculture* 7, 2–3.

II Selected Methods

6 Microbial Biomass and Numbers

6.1 Estimating Soil Microbial Biomass

ANDREAS FLIEßBACH[1] AND FRANCO WIDMER[2]

[1]*Research Institute of Organic Agriculture (FiBL), Ackerstrasse, CH-5070 Frick, Switzerland;* [2]*Agroscope FAL Reckenholz, Reckenholzstrasse 191, CH-8046 Zürich, Switzerland*

Introduction

Soils provide the base for a rich and diverse community of microorganisms. The quality of a soil can be defined by the ability to perform functions in the ecosystem of which it is a part. Soil microbiology has been a discipline covering topics such as nitrogen and organic matter transformations and nitrogen fixation. Transformation and the effects of pollutants were topics that became important with increasing awareness of environmental damage. The potential value of the soil resource for sustainable land use has been recognized and endorses the need to understand the processes involved in organic matter cycling as well as the amount of organisms in soil and their functions. The effects of land use and management on environmental quality, or even on the global carbon balance, have been highlighted in the recent past.

Most soil functions are directly or indirectly related to the soil microbiota, which may explain the effort expended in developing techniques to determine with confidence the number, volume and diversity of organisms in soil. In ecosystem research, the cycling of most elements is driven partly or completely by the amount and activity of organisms that assimilate mineral compounds or decompose organic matter.

As early as 1909, Engberding was doubtful whether microbial numbers on agar plates would give representative cell counts for agricultural soils (Engberding, 1909). But it wasn't until 1996 that researchers were able to demonstrate experimentally that plate counts show only a very small proportion of the total soil microbial community, even though such microbes exist and are viable (Torsvik *et al.*, 1996). Therefore, techniques that are based on isolating and cultivating microorganisms are considered not to reflect the whole microbial community.

Soil fertility is a central aim of sustainable land use. Organic farming systems, especially, emphasize the role of soil microorganisms in element cycling and their value for a healthy soil (Mäder *et al.*, 2002). Reliable measurement and interpretation of biological soil quality, in a world that realizes once more the potential of the soil natural resource, is an important task.

Approaches for Estimating Soil Microbial Biomass

As for most quantitative analyses, it is necessary to account for methodological, spatial and temporal variability, which can be done by analysing an appropriate number of replicates in the laboratory and in the field. For agricultural fields, four replicate bulk samples, each consisting of at least 15 single cores, may reduce spatial variability to less than 10%. Analytical variability can be checked by a representative reference soil – stored at optimum conditions for each purpose – and being analysed with each batch of analysis.

Direct cell counting

Observing and counting microorganisms in their soil habitat is considered to reliably reflect the number and volume of microbial cells in soil. Using staining techniques, it is possible to distinguish microbes from soil particles and to determine activity and biomass. *In situ* hybridization techniques using specific RNA probes may finally serve as powerful tools for counting specific organisms or functional groups in soil. Reliable techniques in microbial biovolume estimation in soil are reviewed and described by Bölter *et al.* in Section 6.4.

The microbial biomass approach

Jenkinson and Powlson (1976a) invented an approach that overcame the isolation–cultivation bias. This chloroform fumigation incubation method was originally checked by (more time-consuming) direct microscopic measurements. They pointed out that 'soil microbes are the eye of the needle, through which all organic material that enters the soil must pass', and they defined the whole microbial community as a black box – the 'holistic approach'. This is a quantitative approach of pools and fluxes and hardly considers specific organisms or functional groups of the microbial community. The microbial biomass, as such, has been inserted as an active pool of soil organic matter in recently developed models on organic matter cycling. In Section 6.2, Brookes and Joergensen explain this approach based on chloroform fumigation.

Jenkinson and Powlson's approach served as the reference for Anderson and Domsch (1978), who invented a physiological method for estimating microbial biomass in soils that has since become very popular – the substrate-induced respiration (SIR) method. Section 6.3 by Höper deals with the advantages and shortcomings of this method as it is now used, as well as the chloroform fumigation and microscopic methods, for monitoring and survey of soils.

Molecular markers

A major step towards a more detailed analysis of the microbial community was the development of molecular markers and advances in extracting them directly from soil organisms. The advantage of these markers is their specificity to living cells or even taxa. However, differing amounts in cells of different species at different physiological states may limit their use as parameters for microbial biomass estimation.

ATP

Adenosine 5′-triphosphate (ATP) occurs in living cells and does not persist in soils. The procedure for extracting ATP from soils has been validated for a wide range of soils (Contin *et al.*, 2001) and a critical review of the method is given by Martens (2001). Provided a standardized procedure is used, Jenkinson and Ladd (1981) found, and Jenkinson (1988) and Contin *et al.* (2001) confirmed, a remarkably constant ATP content of the microbial biomass of 10–12 μmol ATP/g microbial biomass C. These authors recommend the use of a strongly acidic extractant as crucial to denature soil phosphatases, whereas Martens (2001) recommends extraction with dimethyl sulphoxide (DMSO) and Na_3PO_4, as proposed by Bai *et al.* (1988).

Membrane compounds of microbial cells

Microbial membranes contain a great number of lipid compounds that can be extracted directly from soil. The variable content of signature fatty acids in microbial cells of different taxa makes it difficult to relate them to specific biomass, but as they can be extracted quantitatively from the soil, the sum of fatty acids may serve as a biomass indicator. Zelles (1999) reviews the potential and limitations of the use of fatty acids for biomass estimation.

Nucleic acid analysis of the microbial cells

Direct nucleic acid extraction from soil has gained much attention in the past few years, since it holds great potential for describing soil microorganisms that were not accessible by use of cultivation-dependent techniques (see also Chapter 8, this volume). DNA is a biomolecule, tightly associated with living organisms, and thus might serve as a biomass indicator. However, in the past it has proven very difficult to reliably extract total DNA from soils of different textures, and it became commonly accepted that DNA quantities detected in soils are very variable and barely meaningful or useful as a biomass indicator. However, recent advances in soil DNA extraction protocols have indicated that DNA, and even RNA, can be recovered efficiently from soil (Borneman and Triplett, 1997; Bundt *et al.*, 2001; Bürgmann *et al.*, 2001), and correlation of DNA quantities and biomass has been demonstrated in a few cases (Curci *et al.*, 1997; Bundt *et al.*, 2001; Macrae *et al.*, 2001; Taylor *et al.*, 2002). A broader validation of DNA extraction protocols for quantitative extraction of DNA from soils may be beneficial in at least two ways. First, it may serve as a simple and rapidly determined biomass indicator, and, secondly, it would allow further dissection of the structure of the microbial biomass by use of molecular tools (Chapter 8, this volume).

Methods Standardized for Soil Quality Determination

ISO-certified methods are described for soil microbial biomass determination: the chloroform fumigation extraction and the substrate-induced respiration technique. For comparing quantitative results originating from different laboratories, samples have to be exchanged in order to assure reliability of the absolute values. Thresholds or minimum/maximum values for microbial biomass are often discussed. For soil quality evaluation it may be interesting to see if predicted and measured values show much of a difference. If there is, this could be used as an indication of disturbance of the system. Finally, and as pointed out by most authors in soil quality research, it is recommended to use a set of methods, in particular with respect to microbial functions, that are of ecological importance. Soil microbial biomass itself may be interpreted as the mediator of some soil functions, and apparently plays key roles in physical stability and nutrient cycling.

6.2 Microbial Biomass Measurements by Fumigation–Extraction

Philip C. Brookes[1] and Rainer Georg Joergensen[2]

[1]*Soil Science Department, IACR-Rothamsted, Harpenden AL5 2JQ, UK;* [2]*Department of Soil Biology and Plant Nutrition, University of Kassel, Nordbahnhofstraße, D-37213 Witzenhausen, Germany*

Introduction

The soil microbial biomass responds much more quickly than most other soil fractions to changing environmental conditions, such as changes in substrate inputs (e.g. Powlson *et al.*, 1987) or increases in heavy metal content (Brookes and McGrath, 1984). This, and much other similar, research supports the original idea of Powlson and Jenkinson (1976) that biomass is a much more sensitive indicator of changing soil conditions than, for example, the total soil organic matter content. Thus, the biomass can serve as an 'early warning' of such changes, long before they are detectable in other ways.

Linked parameters (e.g. biomass-specific respiration or biomass as a percentage of soil organic C) are also useful as they have their own intrinsic 'internal controls' (see Barajas *et al.*, 1999 for a discussion of this). This may permit interpretation of measurements in the natural environment, where, unlike in controlled experiments, there may not be suitable non-contaminated soil (for example) to provide good 'control' or 'background' measurements.

Here we provide experimental details of two measurements of biomass which have proved useful in environmental studies, particularly at low levels (i.e. around European Union limits) of pollution by heavy metals, namely soil microbial biomass C and biomass ninhydrin N.

Principle of the Method

Following chloroform fumigation of soil, there is an increase in the amount of various components coming from the cells of soil microorganisms which are lysed by the fumigant and made partially extractable (Jenkinson and Powlson, 1976b). Organic C (Vance *et al.*, 1987), total N and NH_4-N (Brookes *et al.*, 1985), and ninhydrin-reactive N (Amato and Ladd, 1988; Joergensen and Brookes, 1990) can be measured in the same 0.5 M K_2SO_4 extract. Further information on fumigation–extraction and other microbiological methods is given by Alef and Nannipieri (1995).

Materials and Apparatus

- A room or incubator adjustable to 25°C
- An implosion-protected desiccator
- A vacuum line (water pump or electric pump)
- A horizontal or overhead shaker
- A deep-freezer at –15°C
- Folded filter papers (e.g. Whatman 42 or Schleicher & Schuell 595 1/2)
- Glass conical flasks (250 ml)

Chemicals and Reagents

- Ethanol-free chloroform ($CHCl_3$)
- Soda lime
- 0.5 M Potassium sulphate (K_2SO_4) (87.1 g/l)

Procedure

Fumigation–extraction

A moist soil sample of 50 g is divided into two subsamples of 25 g. The non-fumigated control samples are placed in 250 ml conical flasks and then immediately extracted with 100 ml 0.5 M K_2SO_4 (ratio extractant:soil is 4:1) for 30 min by oscillating shaking at 200 rpm (or 45 min overhead shaking at 40 rpm) and then filtered through a folded filter paper. For the fumigated treatment, 50 ml glass vials containing the moist soils are placed in a desiccator containing wet tissue paper and a vial of soda lime. A beaker containing 25 ml ethanol-free $CHCl_3$ and a few boiling chips is added and the desiccator evacuated until the $CHCl_3$ has boiled vigorously for 2 min. The desiccator is then incubated in the dark at 25°C for 24 h. After fumigation, $CHCl_3$ is removed by repeated (sixfold) evacuation and the soils are transferred to 250 ml bottles for extraction with 0.5 M K_2SO_4. All treatments are replicated three times. All K_2SO_4 extracts are stored at –15°C prior to analysis.

Biomass C estimated by dichromate oxidation

Principle of the method

In the presence of strong acid and dichromate, organic matter is oxidized and Cr(+VI) reduced to Cr(+III). The amount of dichromate left is back-titrated with iron II ammonium sulphate (Kalembasa and Jenkinson, 1973) and the amount of carbon oxidized is calculated.

Additional materials and apparatus

- Liebig condenser
- 250 ml round-bottom flask
- Burette

Additional chemicals and reagents

- 66.7 mM potassium chromate ($K_2Cr_2O_7$) (19.6125 g/l)
- Concentrated phosphoric acid (H_3PO_4)
- Concentrated sulphuric acid (H_2SO_4)
- 40.0 mM iron II ammonium sulphate [$(NH_4)_2[Fe(SO_4)_2] \times 6H_2O$]
- 25 mM 1.10-phenanthroline-ferrous sulphate complex solution

All chemicals are analytical reagent grade and distilled or de-ionized water is used throughout.

- *Digestion mixture*: two parts conc. H_2SO_4 are mixed with one part conc. H_3PO_4 (v/v).
- *Titration solution*: iron II ammonium sulphate (15.69 g/l) is dissolved in distilled water, acidified with 20 ml conc. H_2SO_4 and made up to 1000 ml with distilled water.

Procedure

To 8 ml soil extract in a 250 ml round-bottom flask, 2 ml of 66.7 mM (0.4 N) $K_2Cr_2O_7$ and 15 ml of the H_2SO_4/H_3PO_4 mixture are added. The mixture is gently refluxed for 30 min, allowed to cool and diluted with 20–25 ml water, added through the condenser as a rinse. The excess dichromate is measured by back-titration with 40.0 mM iron II ammonium sulphate, using 25 mM 1.10-phenanthroline-iron II sulphate complex solution as an indicator.

Calculation of results

CALCULATION OF EXTRACTABLE ORGANIC C FOLLOWING DICHROMATE DIGESTION

$$C\ (\mu g/ml) = [(HB - S)\ /\ CB] \times N \times [VD/VS] \times E \times 1000$$

where: S = consumption of titration solution by the sample (ml); HB = consumption of titration solution by the hot (refluxed) blank (ml); CB = consumption of titration solution by the cold (unrefluxed) blank (ml); N = normality of the $K_2Cr_2O_7$ solution; VD = added volume of the $K_2Cr_2O_7$ solution (ml); VS = added volume of the sample (ml); and E = 3, conversion of Cr(+VI) to Cr(+III), assuming that, on average, all organic C is as [C(0)].

$$C\ (\mu g/g\ soil) = C\ (\mu g/ml) \times (VK + SW)/DW$$

where: VK = volume of K_2SO_4 extractant (ml); SW = volume of soil water (ml); and DW = dry weight of sample (g).

CALCULATION OF BIOMASS C

Biomass C (B_C) = E_C/k_{EC}

where: E_C = (organic C extracted from fumigated soils) – (organic C extracted from non-fumigated soils) and $k_{EC} = 0.38$ (Vance *et al.*, 1987).

Biomass C by UV-persulphate oxidation

Principle of the method

In the presence of potassium persulphate ($K_2S_2O_8$), extractable soil organic carbon is oxidized by ultraviolet (UV) light to CO_2, which is measured using infrared (IR) or photo-spectrometric detection.

Additional materials and apparatus

Automatic carbon analyser with IR-detection (e.g. Dohrman DC 80) or continuous-flow systems with colourimetric detection (Skalar, Perstorp).

Additional chemicals and reagents

- $K_2S_2O_8$
- Concentrated H_3PO_4
- Sodium hexametaphosphate [$(Na(PO_4)_6)n$]
- *Oxidation reagent*: 20 g $K_2S_2O_8$ are dissolved in 900 ml distilled water, acidified to pH 2 with conc. H_3PO_4 and made up to 1000 ml
- *Acidification buffer*: 50 g sodium hexametaphosphate are dissolved in 900 ml distilled water, acidified to pH 2 with conc. H_3PO_4 and made up to 1000 ml

Procedure

For the automated UV-persulphate oxidation method, 5 ml K_2SO_4 soil extract are mixed with 5 ml acidification buffer. Any precipitate of $CaSO_4$ in the soil extracts is dissolved by this procedure. The $K_2S_2O_8$ is automatically fed into the UV oxidation chamber, where the oxidation to CO_2 is activated by UV light. The resulting CO_2 is measured by IR absorption.

Calculation of results

CALCULATION OF EXTRACTABLE ORGANIC C

C (µg/g soil) = [(S × DS) – (B × DB)] × (VK + SW)/DW

where: S = C in sample extract (μg/ml); B = C in blank extract (μg/ml); DS = dilution of sample with the acidification buffer; DB = dilution of blank with the acidification buffer; VK = volume of K_2SO_4 extractant (ml); SW = volume of soil water (ml); and DW = dry weight of sample (g).

CALCULATION OF BIOMASS C

Biomass C (B_C) = E_C/k_{EC}

where: E_C = (organic C extracted from fumigated soils) – (organic C extracted from non-fumigated soils) and k_{EC} = 0.45 (Wu *et al.*, 1990; Joergensen, 1996a).

Biomass C by oven oxidation

Extractable soil organic C is oxidized to CO_2 at 850°C in the presence of a platinum catalyser. The CO_2 is measured by infrared absorption using an automatic analyser (Shimadzu 5050, Dimatoc 100, Analytic Jena). The new oven systems use small sample volumes – so they are able to measure C in extracts containing large amounts of salts. The procedure is similar to the automated UV-persulphate oxidation method, except that the samples are diluted with water and acidified using a few drops of HCl instead of the hexametaphosphate acidification buffer. The calculations of extractable C and biomass C are identical to those used in the automated UV-persulphate oxidation method.

Determination of ninhydrin-reactive nitrogen

Principle of the method

Ninhydrin forms a purple complex with molecules containing α-amino nitrogen and with ammonium and other compounds with free α-amino groups, such as amino acids, peptides and proteins (Moore and Stein, 1948). The presence of reduced ninhydrin (hydrindantin) is essential to obtain quantitative colour development with ammonium. According to Amato and Ladd (1988), the amount of ninhydrin-reactive compounds, released from the microbial biomass during the $CHCl_3$ fumigation and extraction by 2 M KCl, is closely correlated with the initial soil microbial biomass carbon content.

Additional apparatus

- Boiling water bath
- Photo-spectrophotometer

Additional chemicals and solutions

- Ninhydrin
- Hydrindantin
- Dimethyl sulphoxide (DMSO)
- Lithium acetate dihydrate
- Acetic acid (96%)
- Citric acid
- Sodium hydroxide (NaOH)
- Ethanol (95%)
- L-Leucine
- Ammonium sulphate ($(NH_4)_2SO_4$)
- *Lithium acetate buffer*: lithium acetate (408 g) is dissolved in water (400 ml), adjusted to pH 5.2 with acetic acid and finally made up to 1 l with water
- *Ninhydrin reagent*: ninhydrin (2 g) and hydrindantin (0.3 g) are dissolved in dimethyl sulphoxide (75 ml), 25 ml of 4 M lithium acetate buffer at pH 5.2 are then added (Moore, 1968)
- *Citric acid buffer*: citric acid (42 g) and NaOH (16 g) are dissolved in water (900 ml), adjusted to pH 5 with 10 M NaOH if required, then finally made up to 1 l with water

Procedure

The procedure is described according to Joergensen and Brookes (1990) for measuring biomass C and microbial ninhydrin-reactive N in K_2SO_4 soil extracts. A 10 mM L-leucine (1.312 g/l) and a 10 mM ammonium-N [$(NH_4)_2SO_4$ 0.661 g/l] solution are prepared separately in 0.5 M K_2SO_4 and diluted within the range 0–1000 μM N. The standard solutions, K_2SO_4 soil extracts or blank (0.6 ml) and the citric acid buffer (1.4 ml) are added to 20 ml test tubes. The ninhydrin reagent (1 ml) is then added slowly, mixed thoroughly and closed with loose aluminium lids. The test tubes are then heated for 25 min in a vigorously boiling water bath. Any precipitate formed during the addition of the reagents then dissolves. After heating, an ethanol:water mixture (4 ml 1:1) is added, the solutions are thoroughly mixed again and the absorbance read at 570 nm (1 cm path length).

Calculation of results

CALCULATION OF EXTRACTED NINHYDRIN-REACTIVE N (N_{NIN})

$$N_{nin}(\mu g/g \text{ soil}) = (S - B)/L \times N \times (VK + SW)/DW$$

where: S = absorbance of the sample; B = absorbance of the blank; L = millimolar absorbance coefficient of leucine; N = 14 (atomic weight of nitrogen); VK = volume of K_2SO_4 extractant (ml); SW = volume of soil water (ml); and DW = dry weight of the sample (g).

CALCULATION OF MICROBIAL NINHYDRIN-REACTIVE N

B_{nin} = (N_{nin} extracted from the fumigated soil) – (N_{nin} extracted from the non-fumigated soil)

CALCULATION OF MICROBIAL BIOMASS CARBON

Biomass C = $B_{nin} \times 22$ (soils pH-$H_2O > 5.0$)
Biomass C = $B_{nin} \times 35$ (soils pH-$H_2O \leq 5.0$)

The conversion factors were obtained by correlating the microbial biomass C and B_{nin} in the same extracts of 110 arable, grassland and forest soils by the fumigation–extraction method (Joergensen, 1996b).

Discussion

Biomass measurements are certainly useful in studies of soil protection. They have the advantage that they are relatively cheap and simple, as well as being rapid. There is now a considerable amount of literature to show that these measurements are useful in determining effects of stresses on the soil ecosystem. Biomass ninhydrin measurements have two advantages over biomass C. First, a reflux digestion is not required for ninhydrin N. This makes it very suitable for situations with minimal laboratory facilities. Secondly, in both biomass C and N measurements the fraction coming from the biomass is determined following subtraction of an appropriate 'control'. With biomass C this value is often half of the total, while with biomass ninhydrin N it is commonly about 10 or less. This causes considerably less error in its determination. Both parameters are very closely correlated, however, so biomass C may be readily estimated from biomass ninhydrin N, as described above. One feature of the fumigation–extraction method frequently caused concern. Upon thawing of frozen K_2SO_4 soil extracts, a white precipitate of $CaSO_4$ occurs in near-neutral or alkaline soils. However, this causes no analytical problems in either method and may be safely ignored.

Acknowledgements

IACR receives grant-aided support from the Biotechnology and Biological Sciences Research Council of the United Kingdom.

6.3 Substrate-induced Respiration

Heinrich Höper

Geological Survey of Lower Saxony, Friedrich-Missler-Straße 46/48, D-28211 Bremen, Germany

Introduction

The method of substrate-induced respiration was developed by Anderson and Domsch (1978). It is based on the principle that, under standardized conditions, the metabolism of glucose added in excess is limited by the amount of active aerobic microorganisms in the soil. During the first hours after substrate addition there is no significant growth of the microbial populations, and the respiratory response is proportional to the amount of microbial biomass in the soil. Anderson and Domsch (1978) established a conversion factor by correlating the substrate-induced respiration with the microbial biomass determined by the fumigation–incubation method: 1 ml CO_2/h corresponded to 40 mg microbial biomass carbon (C_{mic}).

For the measurements of substrate-induced respiration, Anderson and Domsch (1978) used a continuously operating CO_2-analyser (Ultragas 3, Wösthoff, Bochum, Germany); however, soil samples were not permanently aerated and were flushed with CO_2-free air only 20 min before measurement, thus leading to a temporary increase in CO_2 and depletion in O_2 in the soil sample. Later, respiration was also measured by oxygen consumption (Sapromat, Voith, Germany) in closed vessels (Beck, 1984).

Heinemeyer *et al.* (1989) developed a system where soil samples are continuously aerated with ambient air and the CO_2 production is detected by an infrared gas analyser. This very sensitive system allows the detection of respiration rates at high resolution after the addition of glucose. Concentration changes of less than 1 μl/l CO_2 are detectable. The problem of CO_2 absorption in carbonate-rich soils is overcome by the continuous flow. Kaiser *et al.* (1992) compared the soil microbial biomass as analysed by the so-called Heinemeyer device with the fumigation–extraction method according to Vance *et al.* (1987) and proposed changing the conversion factor to 30 mg C_{mic}/1 ml CO_2/h. This factor can also be used for mineral horizons of forest soils, as can be derived from the data of Anderson and Joergensen (1997) (Fig. 6.1).

The substrate-induced respiration method is based on the detection of a respiratory response of soil microorganisms on supply of glucose. Thus, only glucose-responsive and active organisms are measured. Based on this principle, the substrate-induced respiration method detects predominantly bacterial biomass. In some cases, as in peat and marsh soils, differences

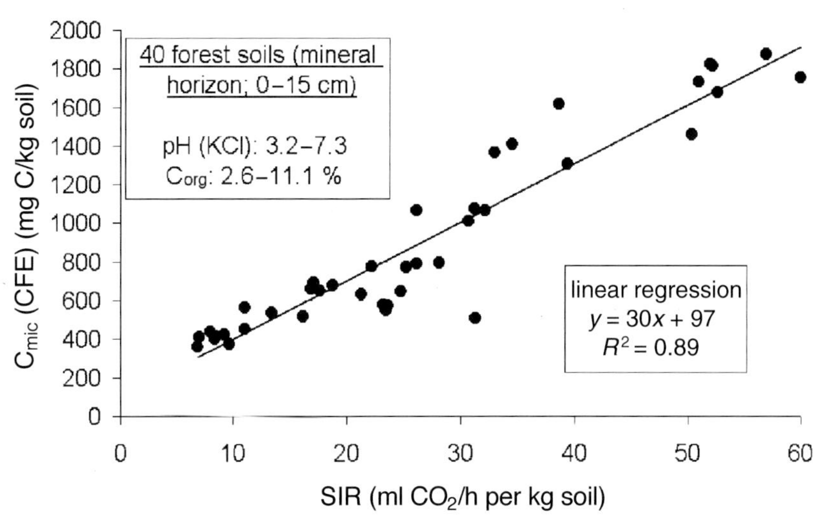

Fig. 6.1. Relation between substrate-induced respiration (SIR) and microbial biomass (C_{mic}) measured by the chloroform fumigation–extraction (CFE) method in mineral horizons of forest soils. Data were recalculated from Anderson and Joergensen (1997).

between the fumigation–extraction method and the substrate-induced respiration method were attributed to population structure, e.g. bacteria-to-fungi ratio, differing from those of common aerobic soils (Anderson and Joergensen, 1997; Brake *et al.*, 1999; Joergensen and Scheu, 1999; Chander *et al.*, 2001). Additionally, in several non-fertilized soils, nitrogen and phosphorus can become the limiting factors for microbial growth and the maximum initial respiratory response (Dilly, 1999). In this case, the amended glucose substrate should be enriched with a nitrogen, phosphorus, potassium and sulphur source (Palmborg and Nordgren, 1993).

Nevertheless, the method has been calibrated to determine the total soil microbial biomass (Anderson and Domsch, 1978) in a wide range of agricultural and forest soils. The results of the fumigation–extraction method and the substrate-induced respiration method were well correlated (Kaiser *et al.*, 1992; Anderson and Joergensen, 1997). Also, Lin and Brookes (1996) confirmed a good relationship between the fumigation–extraction method for a group of amended and unamended soils, and concluded that these soils had the same SIR response, although the activity status of the soil microorganisms and the community structure should be different.

Principle of the Method

The method of substrate-induced respiration is based on the principle that microorganisms react to the addition of glucose with an immediate respiration response that is proportional to their biomass as long as the organisms do not utilize the glucose for growth. Under standardized conditions, especially with respect to temperature and water content of the sample, and if glucose is given in excess without being inhibitory, the only factor limiting the respiratory response within the first hours is the amount of microorganisms. The optimal glucose concentration should be tested in preliminary assays. In arable soils with an expected microbial biomass below 800 mg C_{mic}/kg soil, a glucose amendment of 3000 mg/kg is often used.

A typical respiration curve is shown in Fig. 6.2, where the respiration rate is almost constant during the first 3 h after glucose amendment and increases due to microbial growth between 4 and 14 h.

The respiration rate after glucose supply can be measured by any method for measuring respiration rates in general. Nevertheless, methods that permit hourly monitoring of the respiration rates should be preferred, as the initial respiratory response of the soil sample on glucose supply can be derived from the actual curve shape (see Section 7.2). As an example of continuous measurement, the procedure using the Heinemeyer soil biomass analyser (Heinemeyer *et al.*, 1989) will be described in more detail below. Under low-budget conditions, an approach for soil respiration as developed by Jäggi (1976), based on static incubation with an alkali trap, can be used (Beck *et al.*, 1993; Anonymous, 1998; Pell *et al.*, Section 7.2).

Materials and Apparatus

General equipment

- Room or chamber at constant temperature of 22 ± 1°C
- Balance (1000 g, resolution 0.1 g)

Heinemeyer soil biomass analyser

- SIR soil biomass analyser (Heinemeyer *et al.*, 1989) with an infrared gas analyser and a flow meter for flow rates < 500 ml/min
- 24 fibreglass tubes, inner diameter 4 cm, length 25 cm. 48 sample holders of rubber foam
- Bubble meter, 100 ml, to control the flow meter
- Plastic beakers to permit homogeneous incorporation of glucose into the soil
- Hand mixer

Static incubation in airtight jars (Jäggi, 1976)

- Incubation vessels: SCHOTT-bottles with ISO thread, 250 ml
- Vessels for the soil material: flanged test tubes of polypropylene, inner diameter 29 mm × 105 mm, with lateral holes (3 cm below the fringe, 12 holes with 2 mm diameter) for gas exchange

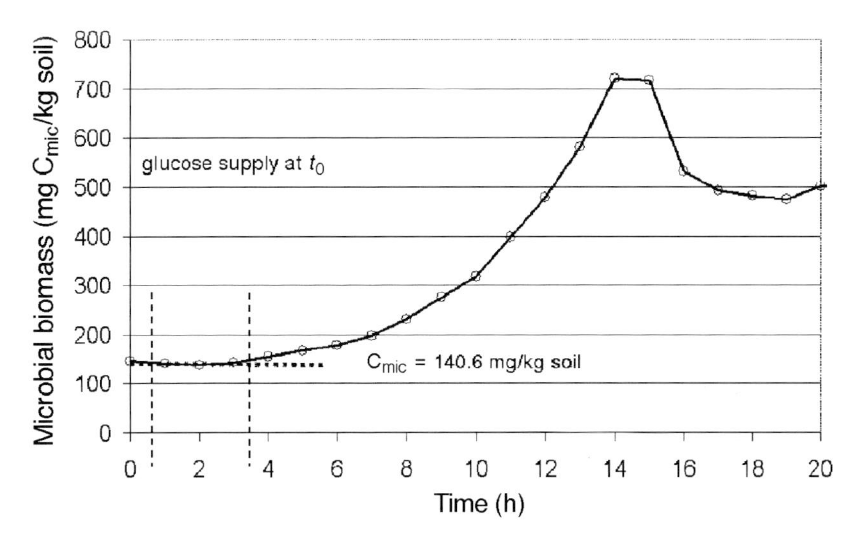

Fig. 6.2. Respiration rates converted to microbial biomass of a soil after the supply of 3000 mg glucose/kg soil.

- Rubber seal: O-ring 35 mm × 5 mm
- Burette, volume 20 ml or 10 ml, with CO_2 trap

Chemicals and solutions

General reagents

- Glucose–talcum mixture, ratio 3:7; add 150 g glucose p.a. to 350 g talcum

Heinemeyer soil biomass analyser

- Reference gases 330 μl/l CO_2 and 400 μl/l CO_2 in synthetic air
- Soda lime with indicator to produce CO_2-free air
- Demineralized water with a small amount of phosphoric acid to reduce the pH to about 5.0

Static incubation in airtight jars (Jäggi, 1976)

- Demineralized water (H_2O, electric conductivity < 5 μS/cm)
- *Sodium hydroxide solution* 0.025 M: sodium hydroxide solution 0.1 mol/l Titrisol diluted in 4 l H_2O
- *Hydrochloric acid* 0.025 M: hydrochloric acid 0.1 mol/l Titrisol diluted in 4 l H_2O

- *Phenolphthalein solution*: 0.2 g phenolphthalein ($C_{20}H_{14}O_4$, M = 318.33 g/mol) dissolved in 200 ml ethanol (60%)
- *Barium chloride solution* 0.5 M: 12.22 g barium chloride ($BaCl_2$ $2H_2O$, M = 244.28 g/mol, p.a.) dissolved in 100 ml H_2O

Procedure

General procedure

Sample preparation

Soils are partially dried or rewetted to 40–60% of the maximum water-holding capacity and passed through a 2 mm sieve. The water content of the soil should be as wet as possible but allowing the soil to be sieved and maintain the aggregate structure. Until the analysis, soil samples should be stored at 4°C. In order to reduce the effects of fresh organic matter amendments to the soil just before sampling, a pre-incubation of 1 week at about 22°C before the measurement is advised.

Determination of the optimal glucose concentration

Glucose is added in excess to the soil in order to get the maximum initial respiratory response. It has to be ascertained that the glucose does not inhibit microbial activity. At least five different concentrations should be tested. In agricultural soils, concentrations between 500 mg/kg and 6000 mg/kg soil are appropriate. Replicates are not necessary at this stage of assessment. The glucose concentration leading to the maximum initial respiratory response is optimal for measuring microbial biomass.

Measurement of microbial biomass

The measurement should last 24 h to control the exponential growth and to be sure that glucose was the only limiting factor for microbial respiration.

Heinemeyer soil biomass analyser

Glucose incorporation and measurement of respiration rate

At least three replicates should be analysed. To establish a concentration of 3000 mg glucose/kg soil, 0.5 g of the glucose–talcum mixture is added to humid soil, corresponding to 50 g on a oven-dry basis, in a plastic beaker and thoroughly mixed for 2 min using the hand mixer. Soils should be handled with care in order to avoid disaggregation and the formation of a soil paste. As quickly as possible, the samples should be transferred into the fibreglass tubes and fixed at each side by sample holders, and the measurement of the respiration rate should be started. The flow rate is adjusted for each sample to about 200 ± 200 ml/min.

Calibration

The CO_2 concentration difference has to be calibrated at least once a week, following the manual. The flow meter should be calibrated by the manufacturer every second year. An internal control using a bubble meter should be performed at least once every 6 months.

Static incubation in airtight jars (Jäggi, 1976)

Blanks

Prepare five blanks per batch without soil, handled like the soil-containing vessels, in order to be able to correct for CO_2 trapping in the alkali trap during handling and titration.

Preparation of alkali traps

Label incubation vessels and soil vessels. Add 20 ± 0.1 ml of 0.025 M sodium hydroxide solution into the incubation vessel. For one batch the same solution has to be used (to have the same relation to the blanks). The flask containing the sodium hydroxide solution has to be equipped with a CO_2 trap (soda lime) to avoid CO_2 adsorption from ambient air flowing into the flask.

First incubation time (2 h)

Glucose (60 mg) is added to the pre-incubated soil (corresponding to 20 g on an oven-dry matter basis). Soil and glucose are mixed thoroughly with a spatula and the mixture is used to fill the soil vessels. The soil vessels are transferred into a climate chamber or an incubator at 22°C for 2 h, as a starting time for glucose consumption.

Main incubation time (4 h)

After exactly 2 h the soil vessels are transferred into incubation vessels containing 20 ml 0.025 M sodium hydroxide solution, which are carefully closed and incubated for another 4 h. At least four replicates are recommended.

Titration

After 4 h, the soil vessels are removed. The sodium hydroxide solution is titrated immediately. Before titration, CO_3^{2-} is precipitated with 1 ml 0.5 M barium chloride solution. Subsequently, four drops of the phenolphthalein solution, as indicator, are added per incubation vessel and the vessel is immediately titrated with 0.025 M HCl until decoloration of the indicator. The consumed volume of the 0.025 M HCl solution is read (in ml).

Calculation

Heinemeyer soil biomass analyser

$$C_{mic} = 30\ F_s * (C_s - C_b)/SW$$

where: C_{mic} = carbon in microbial biomass (mg C/kg dry soil); F_s = flow rate of the air passing the sample (l/h); C_s = CO_2-concentration of CO_2-enriched air, coming from the sample tube (ml/l); C_b = CO_2-concentration of air coming from an empty reference tube (blank) (ml/l); SW = dry matter sample weight (kg); and 30 = constant (mg C_{mic} h/ml CO_2).

Static incubation in airtight jars (Jäggi, 1976)

The HCl readings of the five blanks are averaged (blank = BL). The amount of microbial biomass carbon (C_{mic}) for each sample (SA) is calculated as follows:

$$C_{mic} = 30(\mathrm{BL} - \mathrm{SA})\frac{k \times 22 \times 1000}{1.8295 \times \mathrm{SW} \times 4}$$

where: C_{mic} = carbon in microbial biomass (mg C_{mic}/kg dry soil); BL = mean of the HCl readings of the five blanks (ml HCl); SA = HCl readings of the samples (ml HCl); k = concentration of the HCl solution; 22 = factor (1 ml 1 M HCl corresponds to 22 mg CO_2); 1000 = conversion factor g soil into kg soil; 1.8295 = density of CO_2 at 22°C (mg/ml) (density of CO_2 at 0°C and 1013.2 hPa is 1.9768 mg/ml); SW = sample weight (g soil dry matter); and 4 = conversion factor 4 h to 1 h.

Discussion

Technical advantages of SIR

The SIR, especially when using the Heinemeyer device, has some advantages over the fumigation methods. First, it has a very low determination limit of about 5–10 mg C_{mic}/kg of soil. Thus, the method is also suitable for the assessment of subsoil samples. The standard deviation between replicates is also very low and was estimated by Höper and Kleefisch (2001), for routine analysis within the soil monitoring programme of Lower Saxony, Germany, to be in the median 2.2% and, in 90% of the cases, below 5.2% (Table 6.1).

Table 6.1. Sources of variability for substrate-induced respiration (SIR) measurements with the Heinemeyer soil biomass analyser (Heinemeyer *et al.*, 1989) in the Soil Monitoring Programme of Lower Saxony, Germany, 1996–2000 (Höper and Kleefisch, 2001).

Sources of variability	Explanation	Coefficient of variation (median)	90% quantile
Analytical	Between 3 analytical replicates	2.2	5.2
Repeated sampling	Between 5 bulk samples of 16 sample cores each of the same plot	7.4	n.d.
Spatial	Between 4 subplots of 250 m^2 in the same field	9.6 12.8[a]	17.4 37.5[a]
Temporal (interannual)	Between samples of the same plot and the same annual period over 3–5 years	14.1 15.3[a]	26.1 40.9[a]

[a]Arable versus grassland soils.

Use of SIR for monitoring of soil quality

Substrate-induced respiration has been used in several soil monitoring programmes (Beck *et al.*, 1995; Kandeler *et al.*, 1999; Höper and Kleefisch, 2001; Machulla *et al.*, 2001; Oberholzer and Höper, 2001; Rampazzo and Mentler, 2001). In the Soil Monitoring Programme of Lower Saxony (Germany), sources of variability were examined using a defined sampling strategy (16 cores of 4 cm diameter, 0–20 cm depth in arable land; 0–10 cm and 10–20 cm depth in grassland; sampling period between mid-February and mid-March). Variability due to space and year was rather low, and could be attributed partially to the fact that a baseline variation had to be considered, i.e. the variation between repeated samplings of the same area taken at the same time (Table 6.1). Due to the low spatial and interannual variation, it was already possible to detect significant changes of microbial biomass on some monitoring plots after 5 years. Between 1996 and 2000 a significant decrease in microbial biomass was found in some acid soils and in soils with a low organic matter input (Höper and Kleefisch, 2001).

For the evaluation of measurements, a reference system was established based on the prediction of microbial biomass from abiotic soil properties. To evaluate a given soil, the measured value was compared to the value calculated on the basis of a multiple regression equation with organic C or N content, pH and clay or sand content of the soil (Höper and Kleefisch, 2001; Oberholzer and Höper, 2001). It was possible to establish a common reference system for soils from Lower Saxony, northern Germany and Switzerland (Oberholzer and Höper, 2001). As a whole range of different land uses, climatic conditions and other soil conditions (e.g. water regime) were integrated, the standard error was rather high, ranging from 24% to 32% of the predicted value.

Use of SIR for ecotoxicological risk assessment

The SIR has been broadly used for ecotoxicological risk assessment of soils contaminated with heavy metals or organic contaminants (e.g. pesticides or TNT), for example by Beck (1981), Wilke (1988), Harden *et al.* (1993), Fließbach *et al.* (1994), Kandeler *et al.* (1996), Barajas *et al.* (1999), Chander *et al.* (2001) and Frische and Höper (2003). In contaminated soils, SIR biomass has been a more sensitive parameter than microbial biomass estimated by fumigation–extraction, and both methods were less correlated than in the above example (Chander *et al.*, 2001). Probably the glucose responsive and more active part of the microflora, determining the SIR biomass, is more sensitive to pollution than the total microbial biomass, as measured biochemically.

The use of substrate-induced respiration curves for ecotoxicological soil assessment is of growing interest (Palmborg and Nordgren, 1993; Johansson *et al.*, 1998) and has recently been standardized (ISO standard 17155). The shape of the respiration curve is changing with increasing contaminant content; especially, the time lag before the start of exponential growth and the time until maximal respiration rate increase and maximum growth rate decreases (Winkel and Wilke, 1997; Johansson *et al.*, 1998; Wilke and Winkel, 2000).

Drawbacks and limits of SIR

The SIR is a physiological method, based on the measurement of potential activity of microorganisms in the soil. It is calibrated against chloroform fumigation methods for measuring soil microbial biomass carbon in a large variety of soils. Nevertheless, differences between SIR and chloroform fumigation methods have been observed under specific situations, such as contaminated grassland soils or peatlands (Brake *et al.*, 1999; Chander *et al.*, 2001). Under these conditions, SIR, as a potential activity parameter, is obviously affected by soil conditions in a different way than microbial biomass estimated by chloroform fumigation methods, as a mass parameter.

Finally, SIR is a black-box method, not differentiating between different groups of microorganisms (e.g. bacteria versus fungi). Nevertheless, also as a black-box method, the benefit of SIR in monitoring of soil quality and ecotoxicological soil assessment has been shown in a large number of publications.

6.4 Enumeration and Biovolume Determination of Microbial Cells

MANFRED BÖLTER,[1] JAAP BLOEM,[2] KLAUS MEINERS[1] AND ROLF MÖLLER[1]

[1]Institute for Polar Ecology, University of Kiel, Olshausenstraße 40, D-24098 Kiel, Germany; [2]Department of Soil Sciences, Alterra, PO Box 47, NL-6700 AA Wageningen, The Netherlands

Introduction

Number and biomass of microorganisms are basic properties for soil ecological studies. They can be related to parameters describing microbial activity, soil health and other intrinsic soil descriptors (Kepner and Pratt, 1994). This often includes detailed analyses of the microbial communities, focusing on differentiation between individual organisms, growth forms, size classes, metabolic signatures or other specific properties.

The search for tools to count bacteria by microscopy in soil has a long tradition. The introduction of fluorescence stains and epifluorescence microscopy has found wide acceptance for aquatic and terrestrial ecological studies (Trolldenier, 1973; Zimmermann, 1975, 1977; Hobbie *et al.*, 1977). It has resulted in different applications and techniques for individual disciplines in marine microbiology, limnology and soil science (Bloem and Vos, 2004).

However, a basic restriction of total microscopic counts is the lack of discrimination between active and inactive cells. Another problem is posed by the small cell size, which is mostly less than 0.5 µm in diameter in natural samples, a fact that cannot be attributed only to dwarf or resting cells – although most of them are suspected to be non-culturable (Bakken and Olsen, 1987) and their growth rates can be very low (Bååth, 1994).

New techniques and image analysis have made microscopy an appropriate tool to get an independent insight into the soil communities. It has become a basic step for soil ecology and provides baselines for other indirect measures of biomass and activity (Bloem *et al.*, 1995a; Liu *et al.*, 2001; Bölter *et al.*, 2002). The following shall focus on basic ways of cell enumeration and related methodologies to evaluate microbial biovolume, and on microbial biomass calculated thereof. An example of results in contaminated soil is given.

Principles of Microbiological Counting

The list of appropriate dyes and staining procedures is long, and the use of fluorescent dyes has increased for various new applications. The introduction of molecular probes into microbial ecology is a main field for the application of fluorescent dyes. Although most procedures focus on bacteria, several attempts have also been made to develop techniques for studies on fungi and yeasts (e.g. Pringle *et al.*, 1989; Deere *et al.*, 1998). Some of the dyes are very sensitive and can be used to enumerate viruses (Fuhrman, 1999; Marie *et al.*, 1999). In most recent studies using fluorescent microscopy, the following dyes have been used.

DAPI (4',6-diamidino-2-phenylindole-dihydrochloride)

This is a blue fluorescent stain, which binds to double-stranded DNA (dsDNA) and RNA, but in different modes. It has become a popular stain in aquatic research for enumerations of bacteria and protozoa, and is widely used as a counterstain or parallel stain in fluorescence *in situ* hybridization (FISH) protocols. The fluorescence of DAPI, when bound to dsDNA, can best be excitated at 358 nm (UV); the maximum emission is at 461 nm (Haughland, 1999). When bound to RNA, the emission maximum shifts to 500 nm. Hence, excitation can best be performed by using xenon or mercury arc lamps. DAPI is also used as a stain in laser scanning microscopy and in flow cytometry.

Acridine orange

Acridine orange (AO) has strong affinity to acidic organelles (Haughland, 1999). It can be used to stain eukaryotic and prokaryotic cells. The green and red fluorescence has been used to discriminate between live and dead cells (Strugger, 1949; Bank, 1988), viable and non-viable bacterial spores (Sharma and Prasad, 1992) or to monitor physiological activity (McFeters *et al.*, 1991). This has been attributed to the different fluorescence colours when bound to DNA and RNA (Traganos *et al.*, 1977). The dye shows bright green fluorescence when it is fixed to double-stranded nucleic acid; a red light is emitted when bound to single-stranded nucleic acid. More recently, this statement has been modified: green fluorescent bacteria are in a stationary phase and actively growing bacteria show red fluorescence (Haugland, 1999).

DTAF (5-(4,6-dichlorotriazin-2-yl) aminofluorescein)

This has been preferred for automatic image analysis of bacteria in soil smears because of the low background staining (Bloem *et al.*, 1995a,b; Paul *et al.*, 1999). It binds covalently to proteins.

Differential fluorescent stain

Differential fluorescent stain (DFS) is a nucleic acid stain, which has yielded good results for bacteria, as well as fungal hyphae, in soil. This is a mixture of europium(III) thenoyltrifluoroacetonate (europium chelate) and the disodium salt of 4,4'-bis(4-anilino-6-bis(2-hydroxyethyl)amino-S-triazin-2-ylamino)2,2'-stilbene disulphonic acid, which is also called Fluorescent Brightener (FB), Calcofluor White, Tinopal, or Fluostain I. In the DFS mixture, europium chelate stains DNA and RNA red, and FB stains cellulose and polysaccharides (cell walls) blue. Blue cells are assumed to be inactive or dead. Red cells are assumed to be active because in these cells the fluorescence is dominated by the europium chelate, indicating a higher nucleic acid content and a higher growth rate. The discrimination between active and inactive bacterial cells and fungal hyphae is an advantage of DFS (Morris *et al.*, 1997). A disadvantage is that the europium cannot be detected with a confocal laser scanning microscope for automatic image analysis (Bloem *et al.*, 1995b). Europium is a phosphorescent dye, which starts light emission microseconds after excitation. This is too slow for a confocal laser scanning microscope, but it is no problem for conventional epifluorescence microscopy. The stained cells can be identified and discriminated from other particles and thus serve as a base for their enumeration, measurement or identification.

Materials and Apparatus

Sampling and storage

To obtain samples for microbiological analyses, clean, or sometimes sterile, sampling devices are needed. Samples should be stored frozen, cool or air-dried until analysis. It should be kept in mind that any storage may influence the original community, e.g. freezing may reduce bacterial number. Pre-incubation after freezing restores bacterial number, but not necessarily to the original density (Bloem *et al.*, Chapter 3, this volume). Details on storage problems can be found in the literature for various purposes – the best strategy, however, is direct observation without long times between sampling and analyses (Ross *et al.*, 1980, West *et al.*, 1986, 1987; Turley and Hughes, 1994; Stenberg *et al.*, 1998).

Sample preparation

Various methods have been employed to separate bacteria from particles (e.g. Lindahl and Bakken, 1995). Methods using detergents or ultrasonic treatments are widely used. They can provide good results in coarse materials or organic substrates, but may cause severe problems when used in samples with high loads of silt or clay. Fine particles, just in the size of small rods or cocci, often occur after ultrasonic treatments. Before attempting such procedures, normal hand shaking can be employed as a first attempt. Bloem *et al.* (1995b) have compared different methods for preparing soil suspensions and found no positive effects of detergents and deflocculants. Too much mixing and sonication results in loss of cells. Bloem *et al.* (1995b) and Paul *et al.* (1999) recommended homogenization of soil suspensions (20 g in 190 ml) for 1 min at maximum speed (20,000 rev/min) in a (Waring) blender, for counting both bacteria and fungi. Soil samples with high contents of coarse material can be allowed to settle for a short time (approximately 1 min) before subsamples are used. Soil suspensions fixed with formaldehyde (3.7% final concentration) can be stored for 1 week at 2°C. It is safest to make the preparations as soon as possible after fixation and store the stained slides.

Staining and filtration

Preparation of filters

Identification and discrimination of small objects need a well-defined background. The use of black (polycarbonate) membrane filters is recommended. A pore size of 0.2 μm should be used for counting bacteria. The staining can take place on the filter, or in solution and then be filtered by low vacuum (0.2 kP/cm^2). The filter is then mounted on a microscope slide and the stained bacteria are viewed with an epifluorescence microscope. Unbound dyes can interfere with the filter material and cause background staining, so a washing step may be necessary.

Preparation of soil smears

Bacteria can also be counted directly on a glass slide in a soil smear (Babiuk and Paul, 1970). A smear is prepared by drying 10 μl of a homogenized soil suspension on a printed microscopic slide. Bloem *et al.* (1995b) reported less background staining and less fading of fluorochromes in smears than on filters. Soil films in smears are completely flat, which is a great advantage for automatic image analysis.

Microscopy

The correct optical presentation of objects below the micrometer range may raise several problems, due to the limited depth of focus of normal lenses. The use of confocal laser scanning microscopy is thus a good tool because images of different focal planes are combined to one image with extended focus. In addition, halo effects are avoided. Illumination and halo effects may lead to overestimation of sizes. Kato (1996) makes the point for careful consideration of halo effects and low visibility of cell protoplasm with

DAPI, which may result in underestimation of biovolumes. DAPI has been reported to yield about 40% lower estimates of cell volumes than AO (Suzuki *et al.*, 1993).

Discrimination between small bacteria and unspecific particles needs to be performed carefully. Human subjectivity is another important factor when comparing and analysing data from various laboratories (Domsch *et al.*, 1979; Nagata *et al.*, 1989), beside problems with accurate size measurement (Suzuki *et al.*, 1993), which has important effects on biovolume calculations.

Equipment

Epifluorescence microscope fitted with filters for excitation of cells with blue light (wavelength *c.* 470 nm, for AO and DTAF) and UV (*c.* 365 nm, for DAPI and DFS), and equipped with a 100× oil-immersion lens for bacteria and a 40× (50×) lens(es) for fungi. For direct counting and sizing, an eyepiece graticule can be used (May, 1965), e.g. G12 New Porton Grid (Graticules Ltd, Tonbridge, Kent, UK). For digital image analysis, videocameras can be connected to personal computers equipped with image analysis software.

Chemicals and Solutions

General supplies

- Particle-free water
- Prestained polycarbonate filters (diameter: 25 mm; pore size: 0.2 µm); it is also possible to use normal polycarbonate filters stained in a solution of 2 mg/l Irgalan-Black in 2% acetic acid fixed with 0.2% formaline. After staining for about 12 h the filters are washed with particle-free distilled water until no black stain is visible in the washing water
- Cellulose acetate filters (diameter: 25 mm; pore size: 0.4–0.6 µm) must be used as backing filters to support homogeneous distribution of bacteria on the polycarbonate filter
- Clean microscope slides and cover slips
- Non-fluorescent immersion oil (e.g. Cargill Type A, Cargill Ltd, Cedar Grove, New Jersey, USA)
- Tips for micro-pipettes (200 µl and 1–10 ml)
- Box for storage of microscope slides
- For soil smears: printed microscope slides (e.g. Cel-Line (Erie Scientific, Portsmouth, New Hampshire, USA) or Bellco (Vineland, New Jersey, USA)) with a hole of 12 mm diameter in the centre. Wipe the slides with 70% ethanol and, finally, with a little undiluted liquid soap to promote even spreading of soil suspension

Solutions

Filter all the following solutions through a 0.2 μm membrane before use.

- Particle-free formalin (37%) (two times 0.2 μm filtered)
- For bacteria on filters:
  – acridine orange (AO) (e.g. Sigma Chemical Co., St Louis, Missouri, USA), 1 mg/ml (can be stored at 4°C for some days). If it is not fixed with formalin, the solution needs to be filtered (0.2 μm) before its next use. Contamination occurs frequently!
  – as an alternative to AO, the dye DAPI (e.g. Sigma Chemical Co.) can be used; prepare stock solution: 1 mg/ml (can be stored frozen in the dark for several weeks); and working solution: 10 μg/ml
- For soil smears stained with DTAF (bacteria) or DFS (bacteria and fungi):
  – buffer solution consisting of 0.05 M Na_2HPO_4 (7.8 g/l) and 0.85% NaCl, adjusted to pH 9
  – stain solution consisting of 2 mg DTAF dissolved in 10 ml of the buffer (should not be stored for longer than a day)
  – DFS solution is prepared by dissolving 3.5 g/l europium chelate (Kodak, Eastman Fine Chemicals, Rochester, New York, USA) and 50 mg/l fluorescent brightener, $C_{40}H_{42}N_{12}O_{10}S_2$ Na_2 (FW 960.9, Fluostain I, Sigma Chemical Co.), in 96% ethanol. Fluorescent brightener $C_{40}H_{42}N_{12}O_{10}S_2$ without Na_2 needs addition of NaOH (1 drop/ml) to get it into solution (Serita Frey, personal communication). After a few minutes, when the powder has dissolved completely, dilute to 50% ethanol with an equal volume particle-free water. In order to avoid high counts in blanks (preparations without soil added) due to precipitation of europium chelate, it is better to prepare the DFS solution 1 day before use and to filter the stain through a 0.2 μm pore-size membrane immediately before use. With all stains, blanks should be checked regularly.

Procedure

The protocol for the method is simple: the objects are stained with an appropriate fluorochrome, filtered on to a polycarbonate membrane and counted by epifluorescence microscopy. The individual steps are easy and can even be performed in field labs.

Preparation of the soil suspension

Weigh approximately 1 g fresh soil into a clean glass vial, add 10 ml of particle-free distilled water fixed with particle-free buffered formalin (final concentration: 1%) and shake vigorously for 1 min.

High dilution of the soil is also important to minimize masking of bacteria by soil particles. Masking is likely to occur when more than 1 mg of soil is added per cm^2 on the microscope slide (Bloem *et al.*, 1995b). Coarse particles are removed by settling for 1 min before subsamples are taken for the filtration (see below). Many laboratories homogenize the soil suspensions to disperse bacteria, e.g by using a blender for 1 min at maximum speed (Bloem *et al.*, 1995b, Paul *et al.*, 1999).

Staining

Equip the filtration unit with a cellulose-acetate backing filter (0.4–0.6 μm pore size) to spread the vacuum evenly. Put the black-stained polycarbonate filter (shiny side up) on top of it. The polycarbonate-filter must fit close and flat to the backing filter, without any air bubbles or folds.

1. Staining with AO

Pipette 5 ml of particle-free water into the funnel, add 100 μl of the soil suspension, add 500 μl of the AO solution, mix the water/sample/dye solution carefully and stain for about 3 min.

2. Staining with DAPI

Pipette 5 ml of particle-free water into the funnel, add 100 μl of the soil suspension, mix carefully and filter the samples down at low vacuum (maximum: –150 mbar) until approximately 1 ml remains. Stop filtration by releasing the vacuum from the filtration unit, add 700 μl of the DAPI working solution and stain for 8 min.

After staining (methods 1 or 2), suck the solution softly with low vacuum (maximum: –200 mbar), until all the water has gone. Transfer the dry polycarbonate filter from the vacuum device, and mount it (bacteria on top) on a microscope slide, using a thin smear of non-fluorescent immersion oil (e.g. Cargill Type A). Wynn-Williams (1985) recommends Citifluor® as a photofading retardant, which also accentuates different colour contrasts between organisms. Take care that no air is under the filter. Add a small drop of the immersion oil on top of the filter and mount a cover slip. Press the cover slip down carefully until the oil moves out from the edges of the cover slip and the filter. Lateral movements of the cover slip must be avoided; they can result in unequal distribution of the bacteria. To avoid cross-contamination, wash the funnel of the filtration unit thoroughly with particle-free distilled water between samples. The cellulose-acetate backing filter can be used for several filtrations.

3. Staining bacteria in soil smears with DTAF

Smear 10 µl soil suspension evenly in the hole on a glass slide. The water-repellent coating keeps the suspension in a well-defined area. Allow the smears to air-dry completely; this fixes the organisms to the slide. The slides are placed flat on paper tissue in a plastic tray. Flood the spots of dried soil with drops of stain for 30 min at room temperature. To prevent drying, the tissue is moistened before, and the trays are covered during staining. Rinse the slides three times for 20 min with buffer and finally for a few seconds with water, by putting them in slide holders and passing them through four baths. After air-drying, mount a cover slip with a small drop of immersion oil. The edges of the cover slip can be sealed with nail varnish. The slides can be stored at 2°C for at least a year (Bloem and Vos, 2004).

4. Staining fungal hyphae in soil smears with DFS

Prepare smear as above. After air-drying, stain the slides for 1 h in a bath with DFS solution. Flooding with drops of DFS is also possible if drying is effectively prevented during staining. Evaporation of ethanol may lead to precipitation of europium chelate, resulting in fluorescent spots, which may be confused with bacteria. Rinse the slides three times for 5 min in baths with 50% ethanol. After air-drying, mount a cover slip with a small drop of immersion oil. The slides can be stored at 2°C for at least a year (Morris *et al.*, 1997; Bloem and Vos, 2004).

Examination of the slide

Check the slide for one focal plane. A large drop of immersion oil can result in floating cells, i.e. a slide with two focal plains. Check for equal distribution of cells on the filter. Bacterial number per counting area should range between 20 and 50 cells. If the number of cells is too low or too high, different aliquots of the sample dilution (10–500 µl) or different sample inputs (0.1–3 g) can be used. Filters can be frozen immediately and stored at –20°C in the dark for later analysis.

Counting and calculations

Counting proceeds by use of an epifluorescence microscope at randomly chosen fields following a cross-pattern on the slide. To minimize subjectivity, ensure that the filter is not observed while the field of view is changed. A minimum of 400 cells or a minimum of 20 counting squares can be recommended as a general rule. The optimum number of fields depends on the average number of bacteria per field. If the bacteria are randomly distributed, and there are at least 25 cells per field, ten fields are usually sufficient. When there are only a few cells per field, it is better to increase the number of counted fields. With a free computer program, the random distribution of bacteria on the slide and the optimum number of fields can be checked during counting (Bloem *et al.*, 1992).

Total numbers of bacteria are calculated from the mean count of bacteria per counting area, the effective filtration area, the dilution factor of the soil suspension, the amount of soil used and the filtration volume as follows:

$$\mathrm{TBC} = (D \times B \times M)/W$$

where: TBC = total bacterial count per g soil (cells/g); D = dilution caused by suspension and subsequent subsampling of the soil; B = mean count of bacteria per counting area; M = microscope factor (filtration area/area of counting field); and W = weight of oven-dry soil sample (g).

Sizing of the bacteria can be performed with different methods, e.g. with eyepiece graticules or digital image analysis systems. Fungi are counted at about 400-fold magnification. Check for unstained (brown) hyphae using transmitted light instead of epifluorescence. In one or two transects over the filter (about 50–100 fields), hyphal lengths are estimated by counting the number of intersections of hyphae with (all) the lines of the counting grid (Bloem *et al.*, 1995b, Paul *et al.*, 1999). Usually, many fields are needed because most fields contain no hyphae.

The hyphal length, H (μm), is calculated as:

$$\mathrm{H} = (n \times \pi \times a)/(2 \times l)$$

where: n = number of intersections per grid; a = grid area (μm^2); l = total length of lines in the counting grid (μm).

The total length of fungal hyphae F (per mg soil) is calculated as:

$$F = H \times 10^{-6} \times (A/B) \times (1/S)$$

where: H = hyphal length (μm/grid); 10^{-6} = conversion of μm to m; A = area of the slide covered by sample; B = area of the grid; and S = amount of soil on the slide.

Calculation of biovolume

Several methods have been used for the estimation of bacterial biovolume, including electronic sizing, flow cytometry and different microscopic techniques (Bratbak, 1985). The latter are scanning electron microscopy (SEM), transmission electron microscopy (TEM), confocal laser scanning microscopy, normal light microscopy and widely used epifluorescence microscopic techniques (Bratbak and Dundas, 1984; Bratbak, 1993; Bloem *et al.*, 1995a). The microscopic estimation of cell volumes is based on measurement of the linear dimensions (length, width, perimeter) of individual bacterial cells. These parameters may be obtained with an eyepiece graticule, using photomicrographs or videocamera-equipped microscopes and digital image analysis (Bloem *et al.*, 1995a; Posch *et al.*, 1997).

Linear dimensions are converted to cell volumes using stereometric formulas. A widely used formula, applicable for cocci and rods as well as filaments, has been given by Krambeck *et al.* (1981):

$V = (\pi/4) \times W^2 \times (L - W/3)$

where: V = volume (μm^3); L = length (μm); and W = width (μm); for cocci, length = width. This equation can also be used for fungal hyphae when the diameter has been estimated.

Other calculations distinguish between distinct bacterial morphotypes and use various geometrical bodies to approximate cell shapes. Additional formulas are based on multiple measurement features derived by computer-assisted microscopy (Posch *et al.*, 1997; Blackburn *et al.*, 1998). Assuming a linear relationship between length and width for bacteria, an approximation for increasing cell width with cell length can be applied (Zimmermann, 1975; Bölter *et al.*, 1993). Special demands for volume calculations need to be taken into consideration when cyanobacteria, fungi or algae are under inspection, and special conversion factors at species level or for morphological groups need to be taken into consideration (Bölter, 1997).

Calculation of bacterial biomass

The bacterial biomass is calculated by multiplication of the product of cell number and average cell volume with a conversion factor. The easiest conversion can be performed by the assumption that 80% of the biovolume consists of water, and the remaining dry weight (20%) is considered to be 50% carbon. This rough estimate has found several refinements due to specific habitats, populations, or other considerations. Problems arise since microbial communities are natural consortia of bacteria and fungi. Van Veen and Paul (1979) describe significantly different ratios for bacteria, yeasts and filamentous fungi, depending on actual water content. The spans for specific weights of bacteria and fungi range from 0.11 g/cm^3 to 1.4 g/cm^3.

Three different models have been used to convert bacterial numbers and cell volumes into biomass values (Norland, 1993).

- The constant ratio model assumes a constant, size-independent carbon:biovolume ratio and neglects the potential condensation of carbon in smaller cells.
- The constant biomass model assumes a constant carbon content of cells with different biovolumes, e.g. a standard carbon unit per cell.
- The allometric model assumes that the dry-weight:volume ratio depends on cell volume, and that smaller bacteria have a higher dry-weight:volume ratio than larger ones. This model considers that the cell quota of different cell constituents may vary with cell size. The allometric models are expressed as power functions:

$m = C \times V^a$

where: m = carbon content; C = conversion factor (carbon per unit volume) V = cell volume; and a = scaling factor.

All models neglect the fact that cell quotas of different major constituents, including water, vary not only with cell size but also with bacterial species, and depend on the physiological condition of the organisms (Bratbak, 1985; Fagerbakke *et al.*, 1996; Troussellier *et al.*, 1997). Most of the experimentally derived conversion factors are based on single species or distinct assemblages (mainly planktonic bacteria). Transferring these factors to other environments is problematic.

As a rough estimate, fungal biomass can be calculated from the biovolume using a specific carbon content of 1.3×10^{-13} g C/μm^3. Because smaller cells tend to have a higher density, a higher specific carbon content of 3.1×10^{-13} g C/μm^3 is used for bacteria (Bloem *et al.*, 1995b).

Methodological Remarks

Clean containers and working solutions are a must for reliable results of bacterial numbers. Controls (blank filters) should be performed routinely to check for bacterial contamination of staining solutions and particle-free water used in the protocol.

A potential source of error is the subjectivity of the method, e.g. differences between the operators' judgement of what is a bacterial cell and what is detritus, and particle or size classifications. It is important to compare individual results and discuss criteria of cell identification in order to get consistency between researchers. Determination of bacterial counts of relatively small sampling sets should be performed by one person.

The effective filter area is not the total area of the filter, but the area of the filter through which the soil suspension has been filtered. The effective filter area varies with different filtration units.

Error propagation and statistical considerations

Many working steps are involved from first sample preparation to final results on counts and biomass. Each of these carries a specific error that influences the final result. Therefore, one of the most important questions is: how exact is the final result?

Errors are unavoidable, but they should be reduced to a minimum. Typical errors during biomass calculation (Table 6.2) are due to:

- *Sample collection*: this error source is not predictable. It is necessary to keep in mind that samples and subsamples have to be representative for the site of analysis.
- *Sample preparation*: weighing with lab balances (sample mass, water content) may cause an error of approximately 0.1% to maximally 1%. Dilutions and aliquots, handled with pipettes, can have an error of approximately 1%. Effective filtration or smear area can be estimated with a precision of *c.* 1% (blunt diameter).
- *Image analysis*: calibration of the length measurement is part of image analysis; its resolution is restricted to pixel size. The resulting counting field area provides another error of approximately 2%.

Table 6.2. Error Propagation – example.

	Typical value	Typical error	Relative error (%)
Sample preparation			
Sample weight	1 g	0.001 g	0.1
Water content	10 %	0.1 %	1.0
Dilution	10 ml	0.1 ml	1.0
Aliquot	100 µl	1 µl	1.0
Diameter filter	20 mm	0.1 mm	0.5
Filter area	2.9 cm²	3.0 mm²	1.0
Calibration			
Pixel length	0.1 µm	1 nm	1.0
Counting field	2340 µm²	48 µm²	2.1
Counting			
Objects per counting field due to Poisson distribution	e.g. 356/26 = 13.7	0.2	1.4
Error propagation			
Mean cell volume			+7.15 –6.68 %
Total bacterial number			+6.64 % –6.75 %
Bacterial biovolume			+13.79 % –13.90 %

Counting and calculations

Not all objects on a filter or smear can be counted. Thus, statistical methods need to be used to assess the variation and statistical significance of the counts. The first question to be answered is how many organisms need to be counted? It is recommended to analyse at least 20 counting fields, each of which should contain at least 20 objects, up to a maximum of 50. The optimal number of objects per counting field must be obtained by diluting the original sample.

All counts per counting field must fit a Poisson distribution in order to be summarized by a significant mean value. The test for Poisson distribution can be performed by the chi-square test. The error of the mean is calculated by square root of (B/n), where B is the mean count of bacteria per counting area, and n is the total count. This mean value for the counting fields can be used to calculate a total count valid for the entire sample. It is obvious that the error due to the counting procedure is much greater than other errors, e.g. the effective filter area. Hence, much effort must be put into its minimization by aiming for great accuracy in sample preparation and data evaluation. Counts of bacteria are often used for determinations of biomass, using a uniform mean cell volume or a mean carbon content per cell, or via the determination of a sample-specific biovolume. The first approach is a rough estimate, neglecting individual properties of the communities, and should be used only when details of the community cannot be obtained.

The other way offers a more precise data set and provides data for comparisons between individual communities and samples. The classical way for this procedure is the use of size classes of objects (Bölter *et al.*, 1993). The boundaries of such classes can be preset classes according to empirical knowledge about ecologically reasonable categories, which may provide a best fit to the community from known data or the size classes, or they can be calculated in order to obtain optimal fits, for example:

$$b = (x_{max} - x_{min})/[1 + 3.32\log(n)]$$

where: b = width of size class; x_{max} = maximal x value; x_{min} = minimal x value; and n = number of observations.

The use of preset size classes is recommended since comparisons between numbers of objects of individual size classes can be performed. Descriptions of communities can be performed using histograms. Contents of size classes can be compared by appropriate statistical tests. The number of organisms per size class can be aggregated if individual classes do not contain enough objects to perform such tests. It is necessary to mention that descriptors of size histograms, e.g. mean values or medians, need special attention in the case of open classes (e.g. all cells < 0.25 μm). For such classes, only medians are allowed as descriptors (Sachs, 1984; Lozan and Kausch, 1998). Extremes can be omitted by the '4 sigma-rule' (Sachs, 1984).

Comparisons between total numbers or numbers per size class become possible by various measures, their skewness and kurtosis or other statistical properties. Often, bimodal distributions or log-normal distributions can be observed. The latter are most typical for data obtained in natural environments. They are characterized by high standard deviations or high variation coefficients (Sachs, 1984). It is necessary to discriminate between rods and cocci for size classifications.

Before further use of calculated means, etc., the distribution has to be checked at least for symmetry. Normal distribution can be assessed by the David test (two-sided test, 5%) (Lozan and Kausch, 1998). As stated above, non-normal distributions are most typical for data obtained in natural environments. In this case, data have to be transformed into a normal distribution first. A logarithmic transformation is mostly successful. All relevant calculations and tests can be performed on a log-normal distribution as on a normal distribution. For further use in biomass calculations, results must be retransformed.

Estimating a maximum error of the final result

To gain a statement of the precision of the final results of mean cell volume, total bacterial number or bacterial biovolume, all individual errors have to be considered together as they appear in the equations. Following error propagation, the absolute values sum up to a maximum overall error of the final result (Table 6.2). This can be used for worst-case scenarios. An easy way to evaluate maximum error is to add up the percentages of individual errors.

Discussion

The use of direct counts and estimates of bacterial volume, biomass or other parameters derived thereof is an important tool to understand the microbial habitat and microbial processes. As an example, results from a site polluted with heavy metals are given in Fig. 6.3. At the site of a galvanizing company, soil samples were taken at the most polluted spot around a former basin, and at distances of 10 m and 50 m (unpolluted control). Nickel (Ni) and chromium (Cr) contents were 2800/430, 930/1300 and < 5.0/< 10 mg/kg dry soil, respectively. In the sandy soil under grass, the following characteristics were determined: pH-KCl 6.0, 5.7 and 5.9; organic matter 3.9%, 4.5% and 2.9% (w/w); and clay 8.0%, 2.5% and 3.7%. Bacteria were measured by image analysis in DTAF-stained soil smears (Bloem *et al.*, 1995a,b). In the most polluted soil, bacterial number was 50% lower than in the unpolluted control (Fig. 6.3a). The average cell volume was 30% smaller (Fig 6.3b). Thus, the bacterial biomass (calculated from number and volume) was only 35% of the level in the unpolluted soil (Fig. 6.3c). The results of the microscopic measurements were confirmed by independent measures of bacterial growth rate by thymidine incorporation (see Section 7.5, this volume) – the bacterial growth rate was greatly reduced (Fig. 6.3d). Similar levels of heavy metals in clay soil showed no significant ecological effects. The bioavailability of contaminants is reduced by clay, organic matter and a higher pH. Thus, actual ecological effects of contamination can be demonstrated by measuring soil microorganisms.

This view into the 'home' of the organisms under evaluation, and the use of proper descriptors, provides insights in the community structure and thus allows better and more relevant interpretations of results from biochemical or physiological approaches. Shifts in populations can be followed and their inherent changes become visible. This holds especially true for speculations on the 'active' biomass or 'active' community involved in metabolic processes. New techniques of differential staining procedures can be applied, associated with enormous progress in digital image analysis. Nevertheless, care must be taken while interpreting individual local aspects and projecting data from microscopic views into the full scale of the environment. This problem, mostly neglected, holds true not only for this method, but also needs to be respected in all calculations in environmental research.

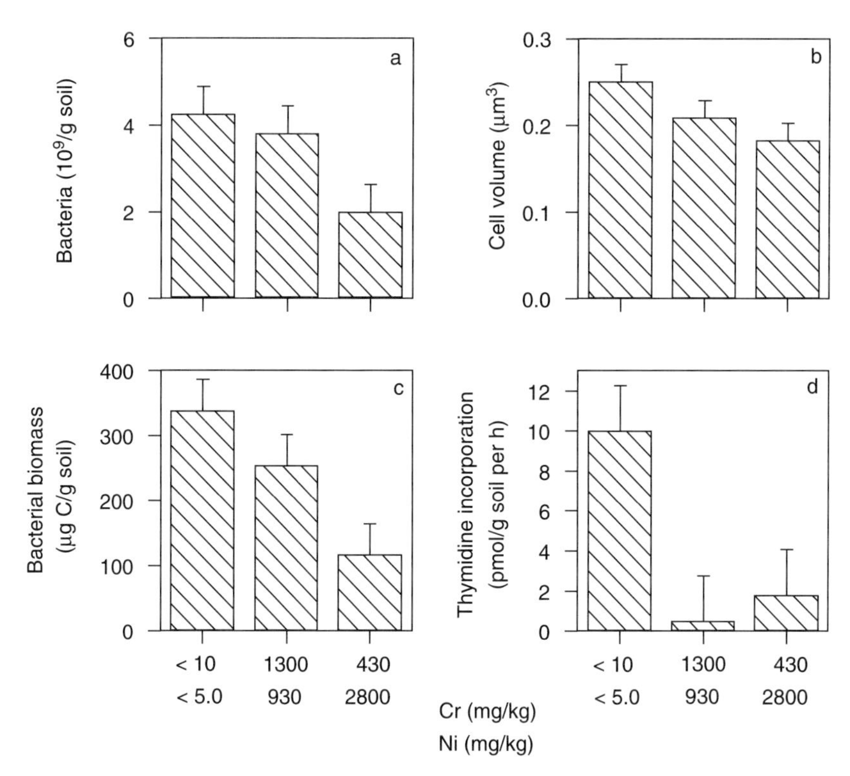

Fig. 6.3. Reduced bacterial number (a), mean cell volume (b), biomass (c) and growth rate (d) in heavy metal (chromium and nickel) contaminated soil. Error bars indicate the least significant difference at $P = 0.05$.

Normative References for Chapter 6

http://www.iso.org (accessed 27 October 2004)

ISO 14240–1 (1997) Soil quality – Determination of soil microbial biomass – Part 1: Substrate-induced respiration method.

ISO 14240–2 (1997) Soil quality – Determination of soil microbial biomass – Part 2: Fumigation–extraction method.

ISO 17155 (2002) Soil quality – Determination of abundance and activity of soil microflora using respiration curves.

References for Chapter 6

Alef, K. and Nannipieri, P. (1995) *Methods in Applied Soil Microbiology and Biochemistry*. Academic Press, London.

Amato, M. and Ladd, J.N. (1988) Assay for microbial biomass based on ninhydrin-reactive nitrogen in extracts of fumigated soils. *Soil Biology and Biochemistry* 20, 107–114.

Anderson, J.P.E. and Domsch, K.H. (1978) A physiological method for the quantitative measurement of microbial biomass in soils. *Soil Biology and Biochemistry* 10, 215–221.

Anderson, T.H. and Joergensen, R.G. (1997) Relationship between SIR and FE estimates of microbial biomass C in deciduous forest soils at different pH. *Soil Biology and Biochemistry* 29, 1033–1042.

Anonymous. (1998) Bestimmung der mikrobiellen Biomasse (Substratinduzierte Respiration; Isermeyer-Ansatz), Methode B-BM-IS. In: Agroscope FAL (ed.) *Referenzmethoden der Eidgenössischen landwirtschaftlichen Forschungsanstalten*, Vol. 2: *Bodenuntersuchungen zur Standort-Charakter-isierung*. Agroscope FAL Reckenholz, Eidgenössische Forschungsanstalt für Agrarökologie und Landbau, Zürich-Reckenholz.

Bååth, E. (1994) Thymidine and leucine incorporation in soil bacteria with different cell size. *Microbial Ecology* 27, 267–278.

Babiuk, L.A. and Paul, E.A. (1970) The use of fluorescein isothiocyanate in the determination of the bacterial biomass of grassland soil. *Canadian Journal of Microbiology* 16, 57–62.

Bai, Q.Y., Zelles, L., Scheunert, I. and Korte, F. (1988) A simple effective procedure for the determination of adenosine triphosphate in soils. *Chemosphere* 17, 2461–2470.

Bakken, L.R. and Olsen, R.A. (1987) The relationship between cell size and viability of soil bacteria. *Microbial Ecology* 13, 103–114.

Bank, H.L. (1988) Rapid assessment of islet viability with acridine orange and propidium iodide. *In Vitro Cellular and Developmental Biology* 24, 266–273.

Barajas, A.M., Grace, C., Ansorena, J., Dendooven, L. and Brookes, P.C. (1999) Soil microbial biomass and organic C in a gradient of zinc concentrations around a spoil tip mine. *Soil Biology and Biochemistry* 31, 867–876.

Beck, T. (1981) Untersuchungen über die toxische Wirkung der in Siedlungsabfällen häufigen Schwermetalle auf die Bodenmikroflora. *Zeitschrift für Pflanzenernährung und Bodenkunde* 144, 613–627.

Beck, T. (1984) Mikrobiologische und biochemische Charakterisierung landwirtschaftlich genutzter Böden. I. Mitteilung: die Ermittlung einer bodenmikrobiologischen Kennzahl. *Zeitschrift für Pflanzenernährung und Bodenkunde* 147, 456–466.

Beck, T., Öhlinger, R. and Baumgarten, A. (1993) Bestimmung der Biomasse mittels Substrat-Induzierter Respiration (SIR). In: Schinner, F., Öhlinger, R., Kandeler, E. and Margesin, R. (eds) *Bodenbiologische Arbeitsmethoden*. Springer Verlag, Berlin, pp. 68–72.

Beck, T., Capriel, P., Borchert, H. and Brandhuber, R. (1995) The microbial biomass in agricultural soils, 2nd communication. The relationships between microbial biomass and chemical and physical soil properties. *Agribiological Research* 48, 74–82.

Blackburn, N., Hagström, A., Wikner, J., Cuadros-Hansson, R. and Bjørnsen, P.K. (1998) Rapid determination of bacterial abundance, biovolume, morphology, and growth by neural network-based image analysis. *Applied and Environmental Microbiology* 64, 3246–3255.

Bloem, J. and Vos, A. (2004) Fluorescent staining of microbes for total direct counts. In: Kowalchuk, G.A., De Bruijn, F.J., Head, I.M., Akkermans, A.D.L. and van Elsas, J.D. (eds) *Molecular Microbial Ecology Manual,* 2nd edn. Kluwer Academic Publishers, Dordrecht, The Netherlands, pp. 861–874.

Bloem, J., Van Mullem, D.K. and Bolhuis, P.R. (1992) Microscopic counting and calculation of species abundances and statistics in real time with an MS-DOS personal computer, applied to bacteria in soil smears. *Journal of Microbiological Methods* 16, 203–213.

Bloem, J., Veninga, M. and Shepherd, J. (1995a) Fully automatic determination of soil bacterial numbers, cell volumes and frequencies of dividing cells by confocal laser scanning microscopy and image analysis. *Applied and Environmental Microbiology* 61, 926–936.

Bloem, J., Bolhuis, P.R., Veninga, M.R. and Wieringa, J. (1995b) Microscopic methods for counting bacteria and fungi in soil. In: Alef, K. and Nannipieri, P. (eds) *Methods in Applied Soil Microbiology and Biochemistry*. Academic Press, London, pp. 162–173.

Bölter, M. (1997) Microbial communities in

soils and on plants from King George Island (Arctowski Station, Maritime Antarctica). In: Battaglia, B., Valencia, J. and Walton, D.W.H. (eds) *Antarctic Communities: Species, Structure and Survival*. Cambridge University Press, Cambridge, pp. 162–169.

Bölter, M., Möller, R. and Dzomla, W. (1993) Determination of bacterial biovolume with epifluorescence microscopy: comparison of size distributions from image analysis and size classifications. *Micron* 24, 31–40.

Bölter, M., Bloem, J., Meiners, K. and Möller, R. (2002) Enumeration and biovolume determination of microbial cells – a methodological review and recommendations for applications in ecological research. *Biology and Fertility of Soils* 36, 249–259.

Borneman, J. and Triplett, E.W. (1997) Rapid and direct method for extraction of RNA from soil. *Soil Biology and Biochemistry* 29, 1621–1624.

Brake, M., Höper, H. and Joergensen, R.G. (1999) Land use-induced changes in activity and biomass of microorganisms in raised bog peats at different depths. *Soil Biology and Biochemistry* 31, 1489–1497.

Bratbak, G. (1985) Bacterial biovolume and biomass estimations. *Applied and Environmental Microbiology* 49, 1488–1493.

Bratbak, G. (1993) Microscope methods for measuring bacterial biovolume: epifluorescence microscopy, scanning electron microscopy, and transmission electron microscopy. In: Kemp, P.F., Sherr, B.F., Sherr, E.B. and Cole, J.J. (eds) *Handbook of Methods in Aquatic Microbial Ecology*. Lewis Publishers, Boca Raton, Florida, pp. 309–317.

Bratbak, G. and Dundas, I. (1984) Bacterial dry matter content and biomass estimations. *Applied and Environmental Microbiology* 48, 755–757.

Brookes, P.C. and McGrath, S.P. (1984) The effects of metal toxicity on the soil microbial biomass. *Journal of Soil Science* 35, 341–346.

Brookes, P.C., Landman, A., Pruden, G. and Jenkinson, D.S. (1985) Chloroform fumigation and the release of soil nitrogen: a rapid direct extraction method for measuring microbial biomass nitrogen in soil. *Soil Biology and Biochemistry* 17, 837–842.

Bundt, M., Widmer, F., Pesaro, M., Zeyer, J. and Blaser, P. (2001) Preferential flow paths: biological 'hot spots' in soils. *Soil Biology and Biochemistry* 33, 729–738.

Bürgmann, H., Pesaro, M., Widmer, F. and Zeyer, J. (2001) A strategy for optimizing quality and quantity of DNA extracted from soil. *Journal of Microbiological Methods* 45, 7–20.

Chander, K., Dyckmans, J., Höper, H. and Jörgensen, R.G. (2001) Long-term effects on soil microbial properties of heavy metals from industrial exhaust deposition. *Journal of Plant Nutrition and Soil Science* 164, 657–663.

Contin, M., Todd, A. and Brookes, P.C. (2001) The ATP concentration in the soil microbial biomass. *Soil Biology and Biochemistry* 33, 701–704.

Curci, M., Pizzigallo, M.D.R., Crecchio, C., Mininni, R. and Ruggiero, P. (1997) Effects of conventional tillage on biochemical properties of soils. *Biology and Fertility of Soils* 25, 1–6.

Deere, D., Shen, J., Vesey, G., Bell, P., Bissinger, P. and Veal, D. (1998) Flow cytometry and cell sorting for yeast viability assessment and cell selection. *Yeast* 30, 147–160.

Dilly, O. (1999) Nitrogen and phosphorus requirement of the microbiota in soils of the Bornhoeved Lake district. *Plant and Soil* 212, 175–183.

Domsch, K.H., Beck, T., Anderson, J.P.E., Söderström, B., Parkinson, D. and Trolldenier, G. (1979) A comparison of methods for soil microbial population and biomass studies. *Zeitschrift für Pflanzenernährung und Bodenkunde* 142, 520–522.

Engberding, D. (1909) Vergleichende Untersuchungen über die Bakterienzahl im Ackerboden in ihrer Abhängigkeit von äusseren Einflüssen. *Centralblatt für Bakteriologie Abt II* 23, 569–642.

Fagerbakke, K.M., Heldal, M. and Norland, S. (1996) Content of carbon, nitrogen, oxygen, sulfur and phosphorus in native aquatic and cultured bacteria. *Aquatic Microbial Ecology* 10, 15–27.

Fließbach, A., Martens, R. and Reber, H.H. (1994) Soil microbial biomass and activity in soils treated with heavy metal contaminated sewage sludge. *Soil Biology and Biochemistry* 26, 1201–1205.

Frische, T. and Höper, H. (2003) Soil microbial parameters and luminescent bacteria assays as indicators for *in-situ* bioremediation of TNT-contaminated soils. *Chemosphere* 50, 415–427.

Fuhrman, J.A. (1999) Marine viruses and their biogeochemical and ecological effects. *Nature* 399, 541–548.

Harden, T., Joergensen, R.G., Meyer, B. and Wolters, V. (1993) Soil microbial biomass estimated by fumigation–extraction and substrate-induced respiration in two pesticide-treated soils. *Soil Biology and Biochemistry* 25, 679–683.

Haughland, R.P. (ed.) (1999) *Molecular Probes: Handbook of Fluorescent Probes and Research Chemicals*. Molecular Probes Inc., 7th edn, Eugene, Oregon (CD-rom edn). Available at: http://www.probes.com (accessed 27 October 2004).

Heinemeyer, O., Insam, H., Kaiser, E.-A. and Walenzik, G. (1989) Soil microbial biomass and respiration measurements: an automated technique based on infrared gas analysis. *Plant and Soil* 116, 191–195.

Hobbie, J.E., Daley, R.J. and Jasper, S. (1977) Use of nucleopore filters for counting bacteria by fluorescence microscopy. *Applied and Environmental Microbiology* 33, 1225–1228.

Höper, H. and Kleefisch, B. (2001) Untersuchung bodenbiologischer Parameter im Rahmen der Boden-Dauerbeobachtung in Niedersachsen. Bodenbiologische Referenzwerte und Zeitreihen. *Arbeitshefte Boden* 2001/4, 1–94.

Jäggi, W. (1976) Die Bestimmung der CO_2-Bildung als Mass der bodenbiologischen Aktivität. *Schweizerische landwirtschaftliche Forschung* 15, 371–380.

Jenkinson, D.S. (1988) Determination of microbial biomass carbon and nitrogen in soil. In: Wilson J.R. (ed.) *Advances in Nitrogen Cycling in Agricultural Systems*. CAB International, Wallingford, UK, pp. 368–386.

Jenkinson, D.S. and Ladd, J.N. (1981) Microbial biomass in soil: measurement and turnover. In: Paul, E.A. and Ladd, J.N. (eds) *Soil Biochemistry*, Vol. 5. Marcel Decker, New York, pp. 415–471.

Jenkinson, D.S. and Powlson, D.S. (1976a) The effects of biocidal treatments on metabolism in soil – V. A method for measuring soil biomass. *Soil Biology and Biochemistry* 8, 209–213.

Jenkinson, D.S. and Powlson, D.S. (1976b) The effects of biocidal treatments on metabolism in soil – I. Fumigation with chloroform. *Soil Biology and Biochemistry* 8, 167–177.

Joergensen, R.G. (1996a) The fumigation–extraction method to estimate soil microbial biomass: calibration of the k_{EC} value. *Soil Biology and Biochemistry* 28, 25–31.

Joergensen, R.G. (1996b) Quantification of the microbial biomass by determining ninhydrin-reactive N. *Soil Biology and Biochemistry* 28, 301–306.

Joergensen, R.G. and Brookes, P.C. (1990) Ninhydrin-reactive nitrogen measurements of microbial biomass in 0.5 M K_2SO_4 soil extracts. *Soil Biology and Biochemistry* 22, 1023–1027.

Joergensen, R.G. and Scheu, S. (1999) Response of soil microorganisms to the addition of carbon, nitrogen and phosphorus in a forest Rendzina. *Soil Biology and Biochemistry* 31, 859–866.

Johansson, M., Pell, M. and Stenström, J. (1998) Kinetics of substrate-induced respiration (SIR) and denitrification: applications to a soil amended with silver. *Ambio* 27, 40–44.

Kaiser, E.-A., Mueller, T., Joergensen, R.G., Insam, H. and Heinemeyer, O. (1992) Evaluation of methods to estimate the soil microbial biomass and the relationship with soil texture and organic matter. *Soil Biology and Biochemistry*, 24, 675–683.

Kalembasa, S.J. and Jenkinson, D.S. (1973) A comparative study of titrimetric and gravimetric methods for the determination of organic carbon in soil. *Journal of the Science of Food and Agriculture* 24, 1085–1090.

Kandeler, E., Kampichler, C. and Horak, O. (1996) Influence of heavy metals on the functional diversity of soil microbial communities. *Biology and Fertility of Soils* 23, 299–306.

Kandeler, E., Tscherko, D. and Spiegel, H. (1999) Long-term monitoring of microbial biomass, N mineralisation and enzyme activities of a Chernozem under different tillage management. *Biology and Fertility of Soils* 28, 343–351.

Kato, K. (1996) Image analysis of bacterial cell size and diversity. In: Colwell, R.R., Simidu, U. and Ohwada, K. (eds) *Microbial Diversity in Time and Space.* Plenum Press, New York, pp. 141–147.

Kepner, R.L. and Pratt, J.R. (1994) Use of fluorochromes for direct enumeration of total bacteria in environmental samples: past and present. *Microbiology and Molecular Biology Reviews* 58, 603–615.

Krambeck, C., Krambeck, H.-J. and Overbeck, J. (1981) Microcomputer-assisted biomass determination of plankton bacteria on scanning electron micrographs. *Ambio* 42, 142–149.

Lin, Q. and Brookes, P.C. (1996) Comparison of methods to measure microbial biomass in unamended, ryegrass-amended and fumigated soils. *Soil Biology and Biochemistry* 28, 933–939.

Lindahl, V. and Bakken, L.R. (1995) Evaluation of methods for extraction of bacteria from soil. *FEMS Microbiology Ecology* 16, 135–142.

Liu, J., Dazzo, F.B., Glagoleva, O., Yu, B. and Jain, A.K. (2001) CMEIAS: a computer-aided system for the image analysis of bacterial morphotypes in microbial communities. *Microbial Ecology* 41, 173–194.

Lozan, J.L. and Kausch, H. (1998) *Angewandte Statistik für Naturwissenschaftler*, 2nd edn. Parey, Berlin.

Machulla, G., Barth, N., Heilmann, H. and Pälchen, W. (2001) Bodenmikrobiologische Untersuchungen an landwirtschaftlich genutzten Boden-Dauerbeobachtungsflächen (BDF II) in Sachsen. *Mitteilungen der Deutschen Bodenkundlichen Gesellschaft* 96, 359–360.

Macrae, A., Lucon, C.M.M., Rimmer, D.L. and O'Donnell, A.G. (2001) Sampling DNA from the rhizosphere of *Brassica napus* to investigate rhizobacterial community structure. *Plant and Soil* 233, 223–230.

Mäder, P., Fließbach, A., Dubois, D., Gunst, L., Fried, P. and Niggli, U. (2002) Soil fertility and biodiversity in organic farming systems. *Science* 296, 1694–1697.

Marie, D., Brussard, C.P.D., Thyrhaug, R., Bratbak, G. and Vaulot, D. (1999) Enumeration of marine viruses in culture and natural samples by flow cytometry. *Applied and Environmental Microbiology* 65, 45–52.

Martens, R. (2001) Estimation of ATP in soil: extraction methods and calculation of extraction efficiency. *Soil Biology and Biochemistry* 33, 973–982.

May, K.R. (1965) A new graticule for particle counting and sizing. *Journal of Scientific Instruments* 42, 500–501.

McFeters, G.A., Singh, A., Byun, S., Williams, S. and Callis, P.R. (1991) Acridine orange staining reaction as an index of physiological activity in *Escherichia coli. Journal of Microbiological Methods* 13, 87–97.

Moore, S. (1968) Amino acid analysis: aqueous dimethyl sulfoxide as solvent for the ninhydrin reaction. *Journal of Biological Chemistry* 243, 6281–6283.

Moore, S. and Stein, W.H. (1948) Photometric ninhydrin method for use in the chromatography of amino acids. *Journal of Biological Chemistry* 176, 367–388.

Morris, S.J., Zink, T., Conners, K. and Allen, M.F. (1997) Comparison between fluorescein diacetate and differential fluorescent staining procedures for determining fungal biomass in soils. *Applied Soil Ecology* 6, 161–167.

Nagata, T., Someya, T., Konda, T., Yamamoto, M., Morikawa, K., Fukui, M., Kuroda, N., Takahashi, K., Oh, S.-W., Mori, M., Araki, S. and Kato, K. (1989) Intercalibration of the acridine orange direct count method of aquatic bacteria.

Bulletin of the Japanese Society of Microbial Ecology 4, 89–99.
Norland, S. (1993) The relationship between biomass and volume of bacteria. In: Kemp, P.F., Sherr, B.F., Sherr, E.B. and Cole, J.J. (eds) *Handbook of Methods in Aquatic Microbial Ecology*. Lewis Publishers, Boca Raton, Florida, pp. 303–307.
Oberholzer, H.R. and Höper, H. (2001) Reference systems for the microbiological evaluation of soils. *VDLUFA-Schriftenreihe* 55/II, 19–34.
Palmborg, C. and Nordgren, A. (1993) Soil respiration curves, a method to test the abundance, activity and vitality of the microflora in forest soils. In: Torstensson, L. (ed.) *MATS Guidelines: Soil Biological Variables in Environmental Hazard Assessment*. Swedish Environmental Protection Agency, Uppsala, Sweden, pp. 149–153.
Paul, E.A., Harris, D., Klug, M. and Ruess, R. (1999) The determination of microbial biomass. In: Robertson, G.P., Coleman, D.C., Bledsoe, C.S. and Sollins, P. (eds) *Standard Soil Methods for Long-term Ecological Research*. Oxford University Press, New York, pp. 291–317.
Posch, T., Pernthaler, J., Alfreider, A. and Psenner, R. (1997) Cell-specific respiratory activity of aquatic bacteria studied with the tetrazolium reduction method, cytoclear slides and image analysis. *Applied and Environmental Microbiology* 63, 867–873.
Powlson, D.S. and Jenkinson, D.S. (1976) The effects of biocidal treatments on metabolism in soil. II. Gamma irradiation, autoclaving, air-drying and fumigation. *Soil Biology and Biochemistry* 8, 179–188.
Powlson, D.S., Brookes, P.C. and Christensen, B.T. (1987) Measurement of soil microbial biomass provides an early indication of changes in total soil organic matter due to straw incorporation. *Soil Biology and Biochemistry* 19, 159–164.
Pringle, J.R., Preston, R.A., Adams, A.E., Stearns, T., Drubin, D.G., Haarer, B.K. and Jones, E.W. (1989) Fluorescence microscopy methods for yeasts. *Methods in Cell Biology* 31, 357–435.
Rampazzo, N. and Mentler, A. (2001) Influence of different agricultural land use on soil properties along the Austrian–Hungarian border. *Bodenkultur* 52(2), 89–115.
Ross, D.J., Tate, K.R., Cairns, A. and Meyrick, K.F. (1980) Influence of storage on soil microbial biomass estimated by three biochemical procedures. *Soil Biology and Biochemistry* 12, 369–374.
Sachs, L. (1984) *Angewandte Statistik*, 6th edn. Springer-Verlag, Berlin.
Sharma, D.K. and Prasad, D.N. (1992) Rapid identification of viable bacterial spores using a fluorescence method. *Biotechnic and Histochemistry* 67, 27–29.
Stenberg, B., Johansson, M., Pell, M., Sjödahl-Svensson, K., Stenström, J. and Torstensson, L. (1998) Microbial biomass and activities in soils as affected by frozen and cold strorage. *Soil Biology and Biochemistry* 30, 393–402.
Strugger, S. (1949) *Fluoreszenzmikroskopie und Mikrobiologie*. Verlag Schaper, Hanover, Germany.
Suzuki, M.T., Sherr, E.B. and Sherr, B.F. (1993) DAPI direct counting underestimates bacterial abundances and average cell size compared to AO direct counting. *Limnology and Oceanography* 38, 1566–1570.
Taylor, J.P., Wilson, B., Mills, M.S. and Burns, R.G. (2002) Comparison of microbial numbers and enzymatic activities in surface soils and subsoils using various techniques. *Soil Biology and Biochemistry* 34, 387–401.
Torsvik, V., Sørheim, R. and Goksøyr, J. (1996) Total bacterial diversity in soil and sediment communities – a review. *Journal of Industrial Microbiology* 17, 170–178.
Traganos, F., Darzynkiewicz, Z., Sharpless, T. and Melamed, M.R. (1977) Simultaneous staining of ribonucleic and deoxyribonucleic acids in unfixed cells using acridine orange in a flow cytofluorometric system. *Journal of Histochemistry and Cytochemistry* 25, 46–56.
Trolldenier, G. (1973) The use of fluorescence microscopy for counting soil microorganisms. In: Rosswall, T. (ed.) *Modern Methods in the Study of Microbial Ecology*. *Bull Ecol Res Comm (Stockholm)* 17, 53–59.

Troussellier, M., Bouvy, M., Courties, C. and Dupuy, C. (1997) Variation of carbon content among bacterial species under starvation condition. *Aquatic Microbial Ecology* 13, 113–119.

Turley, C.M. and Hughes, D.J. (1994) The effect of storage temperature on the enumeration of epifluorescence-detectable bacterial cells in preserved sea-water samples. *Journal of the Marine Biological Association of the United Kingdom* 74, 259–262.

Van Veen, J.A. and Paul, E.A. (1979) Conversion of biovolume measurements of soil organisms, grown under various moisture tensions, to biomass and their nutrient content. *Applied and Environmental Microbiology* 37, 686–692.

Vance, E.D., Brookes, P.C. and Jenkinson, D.S. (1987) An extraction method for measuring soil microbial C. *Soil Biology and Biochemistry* 19, 703–707.

West, A.W., Ross, D.J. and Cowling, J.C. (1986) Changes in microbial C, N, P, and ATP contents, numbers and respiration on storage of soil. *Soil Biology and Biochemistry* 18, 141–148.

West, A.W., Sparling, G.P. and Grant, W.D. (1987) Relationships between mycelial and bacterial populations in stored, air-dried and glucose-amended arable and grassland soils. *Soil Biology and Biochemistry* 19, 599–605.

Wilke, B.M. (1988) Long term effects of inorganic pollutants on the microbial activity of a sandy Cambisol. *Zeitschrift für Pflanzenernährung und Bodenkunde* 151, 131–136.

Wilke, B.M. and Winkel, B. (2000) Einsatz mikrobiologischer Methoden in der Bewertung sanierter Böden und in der Ökotoxikologie. *VDLUFA-Schriftenreihe* 55/II, 35–46.

Winkel, B. and Wilke, B.M. (1997) Wirkung von TNT (2,4,6-Trinitrotoluol) auf Bodenatmung und Nitrifikation. *Mitteilungen der Deutschen Bodenkundlichen Gesellschaft* 85, 631–634.

Wu, J., Joergensen, R.G., Pommerening, B., Chaussod, R. and Brookes, P.C. (1990) Measurement of soil microbial biomass C – an automated procedure. *Soil Biology and Biochemistry* 22, 1167–1169.

Wynn-Williams, D.D. (1985) Photofading retardant for epifluorescence microscopy in soil micro-ecological studies. *Soil Biology and Biochemistry* 17, 739–746.

Zelles, L. (1999) Fatty acid patterns of phospholipids and lipopolysaccharides in the characterisation of microbial communities in soil: a review. *Biology and Fertility of Soils* 29, 111–129.

Zimmermann, R. (1975) Entwicklung und Anwendung von fluoreszenz- und rasterelektronen-mikroskopischen Methoden zur Ermittlung der Bakterienmenge in Wasserproben. PhD thesis, Universität Kiel, Kiel, Germany.

Zimmermann, R. (1977) Estimation of bacterial number and biomass by epifluorescence microscopy and scanning electron microscopy. In: Rheinheimer, G. (ed.) *Microbial Ecology of a Brackish Water Ecosystem*. Springer, Heidelberg, Germany, pp. 103–120.

7 Soil Microbial Activity

7.1 Estimating Soil Microbial Activity

OLIVER DILLY

Lehrstuhl für Bodenökologie, Technische Universität München, D-85764 Neuherberg and Ökologie-Zentrum, Universität Kiel, Schauenburgerstraße 112, D-24118 Kiel, Germany; Present address: *Lehrstuhl für Bodenschutz und Rekultivierung, Brandenburgische Technische Universität, Postfach 101344, D-03013 Cottbus, Germany*

Why and How to Estimate Soil Microbial Activity

Microbial communities in soil consist of a great diversity of species exploring their habitats by adjusting population abundance and activity rates to environmental factors. Soil microbial activities lead to the liberation of nutrients available for plants, and are of crucial importance in biogeochemical cycling. Furthermore, microorganisms degrade pollutants and xenobiotics, and are important in stabilizing soil structure and conserving organic matter for sustainable agriculture and environmental quality. Microbial activities are regulated by nutritional conditions, temperature and water availability. Other important factors affecting microbial activities are proton concentrations and oxygen supply.

To estimate soil microbial activity, two groups of microbiological approaches can be distinguished. First, experiments in the field that often require long periods of incubation (e.g. Hatch *et al.*, 1991; Alves *et al.*, 1993) before significant changes of product concentrations are detected, e.g. 4–8 weeks for the estimation of net N mineralization. In this case, variations of soil conditions during the experiment are inevitable, e.g. aeration and site-specific temperature, and may influence the results (Madsen, 1996). Secondly, short-term laboratory procedures, which are usually carried out with sieved samples at standardized temperature, water content and pH value. Short-term designs of 2–5 h minimize changes in community structure during the experiments (Brock and Madigan, 1991). Such microbial activity measurements include enzymatic assays that catalyse substrate-specific transformations and may be helpful to ascertain effects of soil management, land use and specific environmental conditions (Burns, 1978).

Laboratory methods have the advantage of standardizing environmental factors and, thus, allowing the comparison of soils from different geographical locations and environmental conditions, and also data from different laboratories. In contrast, approaches in the field are considered advantageous for integrating site-specific environmental factors, such as temperature, water and oxygen availability and the microbial interactions with plants and animals. Besides net N mineralization measurements, decomposition experiments with litterbags are frequently used. Litterbags of approximately 20 × 20 cm length and width, with 2–5 mm mesh size are filled with site-specific litter (e.g. 10 g dry material) or cellulose (e.g. filter paper) with or without additional N, distributed in the investigated soil and sampled throughout the year to determine remaining mass, physical, chemical and biological characteristics (Dilly and Munch, 1996). Litterbag studies with smaller mesh size are used when estimating the participation of meso- and macrofauna. However, suppressive and stimulating effects of the fauna on the soil microbiota are not considered in such studies (Mamilov *et al.*, 2001). Finally, soil respiration determined in the field suffers from separating the activity of microorganisms and other organisms, such as animals and plants, which vary significantly in different systems and throughout the season (Dilly *et al.*, 2000).

The group of methods on soil microbial activities embraces biochemical procedures revealing information on metabolic processes of microbial communities. They are frequently used to gain information on 'functional groups'. However, laboratory results refer to microbial capabilities, as they are determined under optimal conditions of one or more factors, such as temperature, water availability and/or substrate. These activities have common units: 1/h, 1/day or 1/year.

Here, six methods have been selected for the estimatation of soil microbial activity. Two methods (Section 7.2 'Soil Respiration' and Section 7.3 'Soil Nitrogen Mineralization') refer to C and N cycling, respectively, and no substrate is added. Section 7.4 'Nitrification in Soil' and Section 7.5 'Thymidine and Leucine Incorporation to Assess Bacterial Growth Rate' follow the transformation after addition of substrate and tracer, respectively. 'Nitrification in soil' can also be estimated without substrate (ammonium) addition. Occasionally, assays without substrate addition are identified as 'actual activity' and those with substrate addition 'potential activity'. However, this classification is critical and confusing, since actual activity should refer to activity under the natural environmental conditions and the response to, for example, changing temperature (Q_{10} values) and site-specific water supply. The *in situ* method described in Section 7.6 'N_2O Emissions and Denitrification from Soil' considers specific pathways of the nitrogen cycle and estimates levels of one of the most important radiatively active trace gases in the atmosphere, contributing to at least 5% of observed global warming (Myhre *et al.*, 1998). Enzyme activities in soil are responsible for the flux of carbon, nitrogen and other essential elements in biogeochemical cycles. Measuring 'Enzyme Activity Profiles' (Section 7.7), and understanding the factors that regulate enzyme expression and the rates of substrate turnover, are the first stages in characterizing soil metabolic potential, fertility and quality. The highly abundant and diverse microorganisms in soil have high metabolic potentials. Generally, soil microorganisms are growth limited and, thus, may poorly exploit their capabilities. Combining measurements with reference to both carbon and nitrogen cycling may give information concerning the microbial adjustment to nutritional conditions. Microbial activities related to microbial biomass are used for evaluating environmental conditions, for example the metabolic quotient, which is the ratio between CO_2 production and microbial C content (Anderson and Domsch, 1993). Finally, soil microbial activities of C and N cycles should be related to soil C and N stocks, providing information concerning transformation intensity in labile pools by looking at substrate transformation and product formation.

7.2 Soil Respiration

MIKAEL PELL, JOHN STENSTRÖM AND ULF GRANHALL

Department of Microbiology, Swedish University of Agricultural Sciences, Box 7025, SE-750 07, Uppsala, Sweden

Introduction: Definition of, and Objectives for, Measuring Soil Respiration

Respiration is probably the process most closely associated with life. It is the aerobic or anaerobic energy-yielding process whereby reduced organic or inorganic compounds in the cell serve as primary electron donors and imported oxidized compounds serve as terminal electron acceptors. During the respiration process, the energy-containing compound falls down a redox ladder, commonly consisting of glycolysis, the citric acid cycle (CAC) and, finally, the electron transport chain.

In a less strict sense, respiration can be defined as the uptake of oxygen while, at the same time, carbon dioxide is released. However, in the soil ecosystem CO_2 is also formed by other processes, such as fermentation and abiotic processes, e.g. CO_2 release from carbonate. In addition, several types of anaerobic respiration can take place where, for example, NO_3^- or SO_4^{2-} are used by microorganisms as electron acceptors; hence, O_2 is then not consumed as in aerobic respiration. Thus, when CO_2 or O_2 are used as indices of respiration, they actually represent carbon mineralization or aerobic respiration, respectively.

Basal respiration (BAS) is the steady rate of respiration in soil, which originates from the turnover of organic matter (predominantly native carbon). The rate of BAS reflects both the amount and the quality of the carbon source. BAS may therefore constitute an integrated index of the potential of the soil biota to degrade both indigenous and antropogenically introduced organic substances under given environmental conditions. In the following, respiration, and measurements thereof, refer to BAS unless otherwise stated.

Soil respiration is a key process for carbon flux to the atmosphere. Soil water content, oxygen concentration and the bioavailability of carbon are the main factors that regulate soil respiration. Perhaps the most important regulator is water, since it will dissolve organic carbon as well as oxygen and, by diffusion, control the access rate of these substances to the cell. Hence, water facilitates the availability of organic carbon and energy, while, at the same time, it restricts access to oxygen. Moreover, water will delay the exchange of CO_2 between the soil surface and the atmosphere. The diffusion constraints often obstruct the interpretation of the results in terms of enzyme kinetics. The optimal water content in soil for respiration is thought to be 50–70% of the soil's water-holding capacity (Orchard and Cook, 1983).

Soil respiration is attributed to a wide range of microorganisms, such as fungi, bacteria, protozoa and algae. Moreover, the soil fauna contributes significantly. Generally, the microbial contribution to the total release of CO_2 (excluding root respiration) is thought to be about 90%, compared to 10% released by the fauna (Paustian *et al.*, 1990). Although fungal biomass often dominates microbial biomass (Hansson *et al.*, 1990; Bardgett and McAlister, 1999) the relation fungi:bacteria with respect to respiration may vary considerably, due to, for example, type of ecosystem or soil management (Persson *et al.*, 1980). To complete the picture, plant roots also contribute between 12% and 30% to the total release of CO_2 through respiration in the field (Buyanovsky *et al.*, 1987; Steen, 1990).

When the carbon mineralization capacity is estimated by means of soil respiration, it is important to consider that the soil may act as a sink for CO_2. Both chemolithotrophic bacteria and phototrophic bacteria and plants fix CO_2 into their biomass. Also, different volumes of O_2 are needed for the mineralization of specific amounts of various carbon sources, i.e. the respiration quotient (RQ) is seldom the often-assumed quotient of 1. Dilly (2001) reported that RQ values for BAS in various soil ecosystems are frequently < 1. Moreover, if the oxygen content of the soil is lowered, mineralization can occur through anaerobic respiration or fermentation, meaning that CO_2 is released without O_2 being consumed.

Principle of Measurements

Methods can be divided into those intended for measuring respiration: (i) in the field; and (ii) in the laboratory.

Measurement of soil respiration in the field is usually accomplished by covering a specific soil surface with a gas-tight chamber. During incubation for a specific time, under ambient climate conditions, changes in gas composition (CO_2 or O_2) are monitored. Alternatively, soil probes can be pushed into the soil and gases withdrawn from a desired depth. Field measurements are the only way to assess the general microbial activity under natural conditions. Hence, field methods give the sum of respiration of all organisms (including roots) under conditions that can seldom be controlled by the investigator and therefore often result in large spatial and temporal variations in gas fluxes.

One way to simplify and standardize the work is to sample a large number of intact soil cores and bring them to the laboratory for incubation under constant temperature and/or moisture content. Establishment of a temperature–response curve can partly transform the results into those encountered under field conditions. However, the pore volume of the soil core is probably reduced by the sampling procedure, and the incubation vessel will generate 'wall effects', resulting in altered gas fluxes in a core incubation device.

Laboratory-based techniques, although usually having less resemblance to natural soil conditions, are easier to handle. Besides allowing the measurement of the basal soil respiration under standardized conditions, such techniques also permit well-designed and controlled experiments to be performed, addressing specific questions (Torstensson and Stenström, 1986). Substrates containing inorganic N and P can be added to eliminate limiting factors other than the carbon source, thereby enabling measurement of the enzymatic capacity to mineralize the intrinsic organic material or some added organic test compound.

In all laboratory-based techniques, proper sample treatment and storage are essential for accurate results. Besides the recommendations given in ISO 10381–6 (1993), the soil samples should be transported from the field to the laboratory as quickly as possible (within hours), or be stored in refrigerated containers. At the laboratory, the moist samples should be sieved through a screen with 2–5 mm mesh width. Sometimes the soil must be partially dried, at a constant temperature of +2°C to +4°C, before sieving is possible. Soils from northern countries (Scandinavia), at least, can be stored at –20°C for up to 12 months without affecting the activity (Stenberg *et al.*, 1998a). Before performing an analysis of respiration, the soil should be preincubated for at least 1 week, to allow the initial carbon flush to diminish. When handling a soil, it is important to know that all kinds of soil disturbances, such as agitation and cycles of drying–wetting and freezing–thawing, will result in bursts of CO_2.

Whether measurements of soil respiration are based on the analyses of consumption of O_2 or the production of CO_2, two methods could be used (Table 7.1).

1. Static methods, where the gases are collected within a closed incubation system containing the soil, or an incubation chamber placed over the soil surface.
2. Dynamic systems, where CO_2-free air flows continuously through the incubation system and the gas composition is analysed at the outlet.

More complete lists of principles and examples of performances can be found in Stotzky (1965), Anderson (1982), Bringmark and Bringmark (1993), Zibilske (1994), Alef (1995) and Öhlinger *et al.* (1996). International standards for tests of soil quality by respiration are suggested in ISO 16072 (2001) and ISO 17155 (2001).

We have chosen to focus on two static methods, both suitable for routine measurements of large numbers of soil samples. Both methods can be used in agricultural soils (Hadas *et al.*, 1998; Yakovchenko *et al.*, 1998; Goyal *et al.*, 1999; Svensson, 1999; Stenström *et al.*, 2001) as well as the mor layer of forest soils (Palmborg and Nordgren, 1993; Bringmark and Bringmark, 2001a,b).

The first method, 'Basal respiration by titration', is simple and can be performed by most soil laboratories. The method can be modified easily for determination of biomass or degradation capacity of organic ^{14}C compounds. Several variations of this old method exist and the method

Table 7.1. Determination of CO_2 and O_2 in static and dynamic set-ups for the measurement of soil respiration.

Gas	Sample	Principle
CO_2	Absorption in alkali	Gain of weight
		Titration of remaining OH^-
		Decrease in conductivity
CO_2	Headspace concentration	IR absorption
		Gas chromatography
O_2	Headspace concentration	Gas chromatography
		Electrode
		Change in partial pressure or gas volume (only static system)

IR, infrared.

described below, based on experience from our lab, is a modification of methods described by Bringmark and Bringmark (1993), Zibilske (1994) and Öhlinger *et al.* (1996).

The second method, 'Basal respiration, substrate-responsive soil microbial biomass, and its active and dormant part', requires special equipment, but is ideal for detailed studies of kinetics and soil microbial subpopulations. This method quantifies:

1. The basal respiration rate (BAS);
2. The substrate-induced respiration rate (SIR); and
3. The distribution between active (r) and dormant (K) microorganisms in the substrate-responsive biomass.

Computerized equipment that allows frequent and automated measurement and storage of data on CO_2 production is needed. The method is based on experience from our laboratory and has been used to study the kinetics of the reversible r $\leftrightarrow$ K transition (Stenström *et al.*, 2001) and how the biomass and the r/K distribution are affected by the concentration and contact time of silver in the soil (Johansson *et al.*, 1998), and by the soil water content and the antibiotic cycloheximide (Stenström *et al.*, 1998).

Basal Respiration by Titration

Principles of the method

Carbon dioxide produced from soil is trapped in a sodium hydroxide solution according to the formula:

$$CO_2 + 2NaOH \rightarrow Na_2CO_3 + H_2O$$

As long as the alkaline solution contains a large excess of OH^-, the chemical reaction is forced to the right as CO_2 is dissolved. At the end of the incubation the non-consumed OH^- is titrated with an acid, e.g. hydrochloric acid (HCl).

Materials and apparatus

- Standard laboratory equipment
- The incubation vessel should have a wide opening and hold at least 500 ml. The vessel should be sealable in a rapid and reliable manner. Glass jars for food preserving with a rubber gasket and metal mount are suitable
- Small plastic cups, one (50 ml) with a perforated lid to hold the soil sample and the other (e.g. a scintillation vial) to hold the NaOH solution
- Burette, magnetic stirrer and magnetic stirring bar (4×14 mm)

Auto-titration equipment will assist the analysis; for example, in our lab this includes:

- Autoburette ABU 80
- Titrator TTT 60
- Standard pH meter PHM82
- Combined pH electrode E16M306

(supplied by Radiometer, Copenhagen, Denmark).

Chemicals and solutions

- Freshly prepared sodium hydroxide (NaOH) solutions (1 M and 0.1 M)
- Diluted hydrochloric acid (HCl) solution (0.05 M)
- Barium chloride ($BaCl_2$) solution (0.05 M)
- Autoburette or phenolphthalein indicator solution (0.1 g/100 ml 60% (v/v) ethanol)

All chemicals should be prepared using CO_2-free distilled water. Boil the water and, after some cooling, close the flasks with stoppers. Concentrates for preparation of 0.1 M NaOH and 0.05 M HCl can be bought from commercial manufacturers. The concentrations of NaOH should be adjusted according to the rate of BAS and duration of incubation.

Procedure

- 40 g of soil with a water-holding capacity (WHC) of 50–70% is weighed out into a plastic cup. Close the perforated lid and place the sample in a humid place at constant room temperature (see below).
- Pre-incubate for 10 days to let the initial flush of CO_2 cease. During the pre-incubation, loss of water from the soil must be compensated for.
- Alternatively, fill a scintillation vial with 2 ml 1 M NaOH and place the opened sample cup and scintillation vial in the incubation vessel on a filter paper moistened with CO_2-free water. Close the vessel and pre-incubate for 10 days at constant room temperature of 22°C.

- At the start of the basal respiration measurement, open the incubation vessel and replace the scintillation vial with a new vial with 2 ml of fresh 0.1 M NaOH. Make sure that the filter paper is moistened. Close the vessel and incubate for 24 h at 22°C.
- Remove the absorption cup and add 4 ml 0.05 M $BaCl_2$ to precipitate carbonate as $BaCO_3$. Titrate the remaining OH^- to a pH of 8.30 with 0.1 M HCl, using an autoburette; or add three to four drops of phenolphthalein and titrate with 0.05 M HCl to the endpoint of the indicator.
- If respiration measurements are to be continued, repeat the BAS measurement and titration steps above until the study period is finished. At least triplicate samples should be measured. Three incubation vessels without soil should be used as blanks.
- Dispensers for dosing NaOH will increase the start-up speed. A high degree of neutralization of the OH^- solution by CO_2 will result in less reliable results and should therefore be avoided. Continuous incubation for several days can be achieved if additional vials with NaOH are placed in the incubation vial.
- After establishment of the basal respiration, SIR could be determined after the addition of a SIR substrate containing glucose (Alvarez and Alvarez, 2000; Section 6.3, this volume).
- The degradation capacity of a ^{14}C-labelled substance can be determined by amending the soil with this substance (Torstensson and Stenström, 1986). A second scintillation vial with NaOH is then placed into the incubation vessel. After incubation, this vial is analysed for labelled $[^{14}C]CO_2$ using a liquid scintillation counter. Soil without added substance is used as a control.

Calculations

The basal respiration rate (BAS) in units of μg CO_2-C/g DW per hour can be calculated from the formula:

$$\text{BAS} = \frac{M_C \times (V_b - V_s) \times 0.05}{S_{dw} \times t \times 2} \times 10^3 \tag{7.1}$$

where M_C is the molar weight of carbon (Mw = 12.01); V_b and V_s the volume, in ml, of 0.05 M HCl consumed in the titration of the blanks (mean of three replicates) and the sample, respectively; S_{dw} is gram dry weight of soil; and t is the incubation time in hours. Since two OH^- are consumed per CO_2 precipitated, a factor of 2 must be included in the formula. If other concentrations of HCl are used, the formula must be adjusted.

Discussion

Due to the many and diverse groups of respiring microorganisms in soil, even a small change in basal activity must be considered as a serious effect.

Soil respiration is therefore a major component of the soil quality concept (cf. ISO TC 190) and is included in monitoring programmes to assess soil quality (Torstensson *et al.*, 1998), as well as responses to various natural or anthropogenic-induced influences (Brookes, 1995; Franzluebbers *et al.*, 1995; Pankhurst *et al.*, 1995; Yakovchenko *et al.*, 1996; Stenberg *et al.*, 1998b; Benedetti, 2001; Svensson and Pell, 2001).

With regard to soil quality, standard microbial respiration (CO_2 evolved) is included as one of only two indicators suggested for assessing 'soil life and soil biodiversity' (Benedetti, 2001). It is also included in the ISO TC 190 'Soil Quality' parameters necessary to consider regarding soil restoration. Differences between temperate and tropical soils concerning the energy content of their organic matter pool have also been studied by COST members using basal respiration (Grisi *et al.*, 1998). The energetic efficiency of the microbial communities using respiration quotients has been discussed largely by Dilly and co-workers (Dilly, 2001).

Besides being a generally accepted measure of total soil organism activity, basal respiration may, with certain modifications, give additional information regarding environmental impact. For example, small increases in basal respiration that follow small additions of easily metabolizable C compounds can be used to study the resulting priming effects on overall C-mineralization (Anderson and Domsch, 1978; Dalenberg and Jager, 1981, 1989; Falchini *et al.*, 2001). Secondly, microbial-specific respiration (an indicator of stress) has been used particularly for monitoring metal-contaminated soils (Brookes and McGrath, 1984; Chander and Brookes, 1991; Bringmark and Bringmark, 2001a,b; Pasqual *et al.*, 2001).

Basal Respiration, Substrate-responsive Microbial Biomass and its Active and Dormant Part

Principles of the method

Carbon dioxide produced from soil in a closed vessel is trapped in a potassium hydroxide solution. This results in a proportional decrease in its electrical conductivity, which is measured with platinum electrodes in each incubation vessel.

Apparatus and materials

The Respicond respirometer (Nordgren Innovations AB, Umeå, Sweden) is fully computerized and suitable for the measurement of 96 samples at 30-min intervals. Since the conductometric measurements are very temperature sensitive, the respirometer should be installed in a room at constant temperature (22 ± 0.02°C).

Chemicals and solutions

- Potassium hydroxide (KOH) solution (0.2 M), prepared using CO_2-free (boiled) water
- *C:N:P substrate*: a mixture of 7.5 g glucose, 1.13 g $NH_4(SO_4)_2$ and 0.35 g KH_2PO_4 is thoroughly ground and mixed in a mortar and thereafter mixed into 10 g of talcum powder. This substrate can be stored at room temperature

Procedure

- 10 ml of the 0.2 M KOH solution is added to the conductometric cup of the incubation vessel.
- Portions of 20 g (dry weight) of soil, with moisture content of 60% WHC, are weighed into the 250 ml incubation vessels. The vessels are closed and installed in the water bath.
- The programme is started with measurements of accumulated CO_2 twice every hour for each vessel.
- Include at least three vessels without soil in the measurement.
- When the initial flush of CO_2 has ceased, continue the measurement for two more days.
- Replace the KOH solution in the cup of each vessel with 10 ml of a new solution if more than 10% of the KOH has been consumed in any vessel (i.e. if more than 4.4 mg of CO_2 has been produced).
- Thoroughly mix 0.19 g of the substrate into each soil sample. Record the time at which the substrate is added to each vessel. Continue the measurements until the maximum respiration rate has been obtained in all soil samples.

Calculations

Export data on CO_2 accumulation and rate of CO_2 production to a spreadsheet program.

Transform the CO_2 data (in mg) to CO_2-C (in μg)/g dry weight (DW) of soil per hour by multiplying each value by $0.2728 \times 1000/\text{DW} = 13.64$.

Due to the disturbance in temperature obtained during addition of the substrate, some initial erratic data have to be removed from the data set (Fig. 7.1). Data belonging to this disturbed period are identified by plotting the rate data for the empty vessels against t.

Calculate the basal respiration rate (BAS), i.e. the slope by linear regression of accumulated CO_2 data against t for the 48 h before substrate addition.

Upon addition of a saturating amount of glucose to a soil sample, active and dormant microorganisms in the soil behave differently. The active

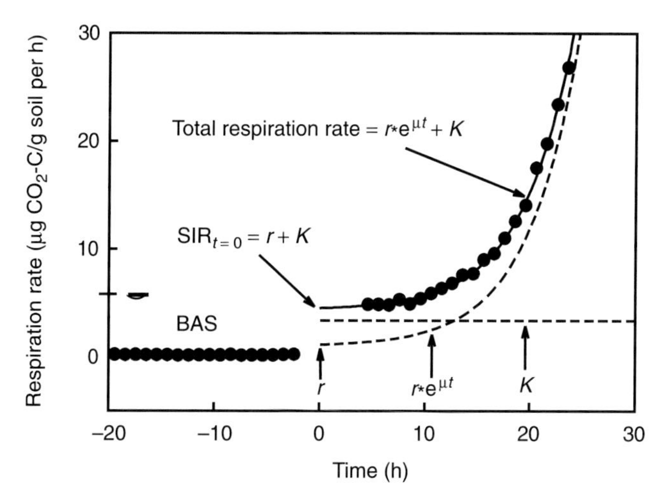

Fig. 7.1. Respiration curve divided into the two phases, basal respiration (BAS) obtained before substrate addition, and the total respiration rate after substrate addition. Substrate-induced respiration (SIR) is the sum of respiration rate of active (r) and dormant (K) soil microorganisms triggered by substrate addition at $t = 0$.

organisms (r) immediately start to grow exponentially, while the dormant ones (K) respond by initiating a constant production rate of CO_2 (Stenström *et al.*, 2001). Thus,

$$\frac{dp}{dt} = re^{\mu t} + K \tag{7.2}$$

where dp/dt is the total rate of CO_2 formation and μ is the specific growth rate of active organisms. The accumulated amount of CO_2 is obtained by integrating Eq. (7.2):

$$p = p_0 + \frac{r}{\mu}\left(e^{\mu t} - 1\right) + Kt \tag{7.3}$$

where p_0 is a fitting parameter that accounts for the initial conductivity of the KOH solution of each jar at the time of addition of the substrate. The SIR rate (Fig. 7.1) is defined as the respiration rate obtained instantaneously on the addition of the substrate. Thus,

$$\mathrm{SIR} = r + K \tag{7.4}$$

which is obtained by solving Eq. (7.2) for $t = 0$. The value of SIR obtained can be converted to biomass as discussed by Höper (Section 6.3, this volume). Numerical parameter values of Eq. (7.3) are obtained by non-linear regression of accumulated data that belong to the period during which growth is exponential and K is constant. This period is identified as the time from substrate addition until data plotted according to $\ln(dp/dt)$ against t start to fall below the straight line so obtained.

Discussion

This method refers directly to the objectives of the COST Action 831, as follows:

- to improve the effectiveness of microbial and molecular methods so to improve monitoring, conservation and remediation of soil;
- to use new microbial parameters as better indicators of environmental impact;
- to help early detection of any fertility decline of natural ecosystems by the setting up of efficient and rapid methods of soil pollution diagnosis.

The arguments for use of this method are:

- Standardized (basal) respiration measurements are relatively insensitive means for monitoring environmental impact or soil health (Brookes, 1995).
- Microbial and respiratory quotients may be more useful to determine trends with time and to compare soils than measurements of basal or substrate-induced respiration only; but if the quality of the carbon substrates differs greatly, or if soils are of different types, the interpretation of such quotients becomes very difficult (Sparling, 1997; Granhall, 1999).
- The method described here, being more sensitive and accurate, could help in the early detection of soil pollution and its following remediation, as it describes the basic respiration rate, the total glucose-responsive biomass, and the sizes of the active (r) and the dormant (K) microbial populations.

7.3 Soil Nitrogen Mineralization

Stefano Canali and Anna Benedetti

Consiglio per la ricerca e la sperimentazione in Agricoltura, Istituto Sperimentale per la Nutrizione delle Piante, Via della Navicella, 2, 00184 Rome, Italy

Introduction

The process of nitrogen (N) mineralization in the soil can be defined as the conversion of organic N into mineral forms available to plants, which takes place through the biochemical transformation mediated by microorganisms (Stevenson, 1985).

The first step (ammonification) involves the conversion of organic N into NH_4^+ and is performed exclusively by heterotrophic microorganisms, able to operate in both aerobic and anaerobic conditions. The second step (conversion of NH_4^+ into NO_3^-), defined as nitrification, occurs mainly through the activity of two groups of autotrophic aerobic bacteria: *Nitrosomonas* (from NH_4^+ into NO_2^-) and *Nitrobacter* (from NO_2^- into NO_3^-). In cultivated, well-aerated soils, nitrate is the predominant available mineral form of the element. On the other hand, if low O_2 concentrations are present in the soil (i.e. waterlogged conditions), NH_4^+ accumulates (Stevenson, 1985).

Nearly always, the mineralization process is accompanied by the immobilization of N, due to the activities of the soil living biomass and, since the two gross processes take place simultaneously, the increase of mineral N concentration at a defined time indicates the net mineralization (Powlson and Barraclough, 1993).

The possible use of soil N mineralization as an index of soil quality is relevant because of the relation of this process with the capacity of the soil to supply N for crop growth, and also because of the risk of water and atmospheric pollution. Thus, N mineralization is often included in *minimum data sets* set up to evaluate the capacity of a soil to operate within the boundaries of the ecosystem: to promote biological productivity, to maintain environmental quality and to safeguard the health of plants and animals (Doran and Parkin, 1994).

The mineralization process is driven by factors able to influence the microbial activity in the soil. The most relevant factors are soil temperature, moisture, pH, O_2 concentration, energy and other nutritive elements available for the accumulation of produced mineral N in the system. If, during the N mineralization measurement, these factors are not controlled, the values may vary within the range and the measurements are performed under 'current conditions'. On the other hand, if one (or more) factor that drives the microbial mineralization process is controlled and set at its optimal value, the measurements are carried out under 'potential conditions'– these are used in laboratory incubation methods.

According to the considerations outlined above, methods for the evaluation of N mineralization in the soil can be divided into those for the evaluation of the gross or the net mineralization activity, and these procedures are classified into two groups: methods for measuring the actual mineralization (field condition) and methods for the evaluation of potential mineralization. The aim of this section is to review the most widespread methods used to measure N mineralization in agricultural soils, describing procedures and discussing the advantages and limitations of their application.

Methods for measuring the actual mineralization activity (field conditions)

Net mineralization measurements

Net nitrogen mineralization can be evaluated by the application of the N balance equation to a defined soil–plant system (Powlson and Barraclough, 1993):

$$M = \Delta NH_4^+ + \Delta NO_3^- + \Delta Plant + loss$$

where ΔNH_4^+ and ΔNO_3^- represent the differences of ammonia and nitrate concentrations at the end and at the beginning of a defined period of time. 'ΔPlant' is the amount of nitrogen uptake by the crop and 'loss' is the N leaving the system via leaching and gaseous emission (ammonia volatilization, denitrification and nitrogen oxide evolution during the nitrification) in the same period.

Although direct measurement of N uptake by annual herbaceous crops is relatively easy, great difficulties arise in the case of perennial tree crops (e.g. fruit crops). In any case, it is probably true that the most important limitation of this procedure is that the measurement of losses under field conditions is an extremely difficult task and, for this reason, the nitrogen balance is often applied without taking into account all the losses in the system in question, which means that the results obtained are not accurate.

An indirect approach to the evaluation of the net rate of N mineralization by applying the N balance consists of the use of some soil contents that prevent the N uptake on the part of the plant and reduce the losses in the system (Hart *et al.*, 1994). These methods employ closed-top solid cylinders or buried bags, carried out in intact soil cores to reduce soil disturbances.

The main advantage of this procedure is that, even when it is performed in a simplified way (no account for crop uptake and losses), it supplies a rough indication of the actual net mineralization, which is useful for the evaluation of the impact of the process on field conditions in the medium or long term.

Gross mineralization measurements: the isotopic dilution technique

Techniques using ^{15}N label and employing the pool dilution principle are based on the measurement of the change in ^{15}N abundance of a labelled ammonium soil pool receiving nitrogen at a natural abundance through mineralization (Powlson and Barraclough, 1993). The measured changes are then described according to a specific set of equations (Kirkham and Bartholomew, 1954; Barraclough *et al.*, 1985; Nishio *et al.*, 1985; Mary *et al.*, 1998) capable of quantifying the size of the soil N pools and the rate of the gross processes (i.e. mineralization, immobilization, nitrification). This method has proven very useful in understanding how soil and crop management affect the N turnover, *in situ* or in intact soil cores (Recous *et al.*, 2001).

In order to obtain useful quantitative estimates, the application of the technique requires four basic assumptions.

- Ammonium-consuming processes (plant uptake and nitrification) do not discriminate between ^{14}N and ^{15}N.
- Added label mixes with native soil ammonium, such that labelled and unlabelled N are used in proportion to the relative amounts present in the system.
- Over the experimental period, all rate processes can be described by a zero-order equation.
- Labelled nitrogen immobilized over the experimental period is not re-mineralized.

These assumptions have been verified exclusively for a short-term period, and the procedure supplies reliable information only for instantaneous rates of nitrogen mineralization.

The limited popularity of this technique is probably due to the (relatively) high costs of the equipment required to perform ^{15}N analyses, and because of the unwieldy formulation of the equations used to describe the processes quantitatively.

Methods for measuring the potential mineralization activity (laboratory incubation methods)

These methods involve the laboratory incubation of the soil for a defined period of time and in conditions that promote the N mineralization from organic sources. The N mineral produced is then measured.

The methods are generally classified on the basis of the length of the incubation period and the incubation conditions. In this section we shall discuss short-term static incubation methods (aerobic and anaerobic) and a long-term aerobic dynamic method.

All laboratory incubation procedures considered measure the net nitrogen mineralization and, since at least the temperature, which is one of the most relevant factors affecting the process, is set to an optimal value, the N mineralization activity is measured under potential conditions. Consequently, the results obtained must be interpreted as a relative indication of the process rate.

Short-term static incubation procedures

The static aerobic incubation procedure was described first by Bremner (1965). The soil sample is mixed with a defined amount of washed sand and the mixture is moistened with water and incubated at 30°C for 15 days. At the end of the incubation, NH_4^+-N, NO_3^--N and NO_2^--N are extracted by means of a concentrated saline solution (KCl) and then evaluated. According to this procedure, the two main factors affecting the mineralization (water contents of the system and the temperature) are set at defined optimal values and the presence of sand should allow adequate O_2 availability.

This is a widespread and well-known method and, in accordance with its general principles, many minor modifications have been proposed by different authors in order to reduce the incubation time and to facilitate laboratory procedures.

Waring and Bremner (1964) and then Keeney (1982) proposed the short-term static anaerobic incubation procedure. The rate of the N mineralization process is determined by measuring the quantity of NH_4^+-N produced during 7 days of waterlogged incubation at a temperature of 40°C.

In the above-mentioned incubation conditions, mineralization of organic N is performed only by microorganisms that are able to operate under anaerobic conditions. This method has been used to study nitrogen transformation in paddy soils and, according to this procedure, Sahrawat and Ponnamperuma (1978) and Sahrawat (1983) measured the net mineralization process in tropical rice soils.

This method can also be applied to agricultural aerated soils. In fact, even in this type of soil, waterlogged conditions can occur for a limited period of time (rainy periods) and anaerobic conditions can be found at any time inside soil aggregates. Indeed, Keeney and Nelson (1982), Stanford (1982) and Meisinger (1984) found satisfactory relationships between the results obtained through the anaerobic and the aerobic procedures. Our hypothesis is that even in the Mediterranean soil of southern Italy, characterized by extreme conditions in which waterlogged situations hardly ever occur, the anaerobic and the aerobic procedures for measuring the net N mineralization process supply substantially the same indicative results.

This method, as compared to the aerobic static procedure, has several advantages that make it an attractive solution when a rapid procedure is needed to supply a relative assessment of N mineralization. These advantages include: (i) simplicity and ease of adaptation in laboratory routines; (ii) a short incubation time (7 days); (iii) a low impact of sample pretreatment on test results; (iv) the elimination of problems related to optimum water contents and water loss during the incubation; and (v) the requirement of only a few machines and reagents (Bremner, 1965; Keeney, 1982; Lober and Reeder, 1993).

The long-term dynamic incubation procedure

This procedure was initially proposed by Stanford and Smith (1972). It is based on aerobic long-term incubation at 35°C, and, within the incubation period, the mineral nitrogen produced (NH_4^+-N, NO_3^--N and NO_2^--N) is leached at predetermined intervals (after 2, 4, 8, 16, 22 and 30 weeks). N mineralization is described by fitting the cumulative experimental data of mineralized N according to a first-order kinetic model.

Soil samples are incubated in a combined filtration–incubation container. Full-potential conditions are imposed by mixing soil with washed quartz sand (1:1 p/p) to allow for an adequate O_2 availability in the system. Water is added to reach the optimum pH value. Mineral nitrogen is removed by leaching to reduce the re-immobilization of mineralized nitrogen and/or to avoid feedback effects on the mineralization process. In order to prevent any limiting effects due to the absence (or reduced concentration) of other elements, after each leaching a nutritive solution, minus N, is applied to the soil.

The long-term incubation should allow problems linked to the influence of pre-treatment of the samples in the N mineralization process to be overcome; for this purpose, Stanford and Smith (1972) suggested that the results obtained during the first 2 weeks of incubation should be ignored. Benedetti *et al.* (1994, 1996) suggested that the results obtained during the first 2–3 weeks of incubation give an indication of the size of the labile mineralizable N pool and/or the biomass nitrogen pool of soils.

Furthermore, after the first weeks of incubation, depending on the specific characteristics of the different soil types, the system generally reaches a steady-state condition. When this new balance is observed, it is possible to obtain indications regarding basal N mineralization (MN_{Bas}). Thus, it is possible to use basal N mineralization to calculate the N mineralization coefficient (MN_{Bas}/N_{tot}) and the N mineralization quotient ($Q_N = MN_{Bas}/N_{bio}$).

When the first-order kinetic model is applied in order to describe the experimental results of potentially mineralizable nitrogen (N_0), and the rate constant (k) is calculated (Stanford and Smith, 1972), the first parameter (N_0) is considered to be a nitrogen availability index and, together with the k constant (the dimension of which is $1/t$), it is capable of describing the long-term N fertility status of the soils.

Benedetti and Sebastiani (1996) determined N_0 and k by the Stanford and Smith method for a number of Italian soils. These authors also confirmed that significantly less reliable estimates of potentially mineralizable N were obtained by using data for up to only 22 weeks of incubation.

Despite the advantages described above, this method is characterized by several disadvantages which are concerned with the substantial time and apparatus requirements (Bundy and Meisinger, 1994). For this reason, this method is usually used only when reliable and long-term N mineralization information (i.e. basal N mineralization) is required.

Conclusions

Actual net and gross N mineralization rate (field conditions) should be determined when the aim of the study is to evaluate the absolute rate of the process (i.e. the capability of soils or organic fertilizers to supply N to the crop).

Methods to assess the potential mineralization (laboratory methods) may be used when performing a comparison of different soil samples or applying several methods for evaluating the same soil.

The anaerobic static incubation procedure is strongly recommended when a quick, work-saving, inexpensive procedure (i.e. field survey approach on a large, regional scale) is needed. We do not yet have confirmation about the potential correlation between the results obtained with the aerobic and the anaerobic static procedure for all climatic and pedological situations.

The aerobic long-term dynamic procedure (Stanford and Smith method) can be applied whenever reliable and exhaustive quantitative information about net N mineralization in potential conditions is necessary.

Anaerobic N Mineralization

Principle of the method

This method is based on incubation of waterlogged soil for 7 days at 40°C. At the end of the incubation, accumulated ammonium is measured.

Materials and apparatus

- Incubator at 40°C
- Shaking apparatus
- Filter paper (N-free)
- Glass or plastic tubes (50 ml) with rubber stoppers and screw caps
- Glass Erlenmeyer flasks (300 ml)
- Plastic bottles (100 ml)

Chemicals and solution

- *4 M KCl solution*: dissolve 298 g of KCl in 750 ml of distilled water in a 1000 ml glass flask; bring up to volume with distilled water

Procedure

- Put 40 ml of distilled water into the tube and add 16 g of soil (three replicates).
- Close the tube by the rubber stopper and the cap, and then shake manually until the soil is completely suspended.
- Incubate for 7 days at 40°C. During the incubation, re-suspend the soil daily by manual shaking.
- After the incubation, transfer the soil–water suspension into the 300 ml Erlenmeyer flask.
- Wash the incubation tube with 10 ml of the 4 N KCl solution and transfer the obtained suspension into the Erlenmeyer flask; repeat the same procedure three times, in order to use 40 ml of the 4 N KCl solution in all. No soil particles should remain in the incubation tube.
- Shake the Erlenmeyer flask containing the soil suspension for 1 h and then filter the supernatant through a filter paper into the plastic 100 ml bottle.
- Use the clear filtrate for ammonium determination, according to the chosen method.
- Samples that are not to be incubated are prepared by putting 16 g of the same soil + 40 ml of distilled water + 40 ml of the 4 N KCl solution into the 300 ml Erlenmeyer flask. Shake and filter according to the procedure reported above.

Calculation

- Mineralized nitrogen during 7 days of incubation is calculated by subtracting the ammonium measured (μg NH_4^+-N/g soil) in the sample that was not incubated from that measured in the incubated sample.
- In order to verify that anaerobic conditions occurred during incubation, the presence of nitrate and nitrite should be assessed – only traces of NO_3^--N and NO_2^--N should be found.

Long-term Aerobic N Mineralization

Principle of the method

The method is based on the aerobic incubation of soil (mixed with quartz sand) in optimal conditions of moisture and temperature for soil microbial activity. Mineralized nitrogen is removed periodically by leaching, and then determined at fixed times for 30 weeks.

Materials and apparatus

- Ceramic Buchner funnel (outer diameter 55 mm)
- Thermostatic system

Chemicals and solutions

- Quartz sand (granulometry: 0.2–0.8 mm)
- Glass-fibre diskettes (diameter chosen according to the Buchner funnel used)
- *Leaching solution*: 0.01 M calcium chloride ($CaCl_2$)
- *Nutritive* N-minus *solution*: 0.002 M $CaSO_4.2H_2O$; 0.002 M $MgSO_4$; 0.005 M $Ca(HPO_4)_2. H_2O$; 0.0025 M K_2SO_4

Procedure

- Mix 10 g of soil (air-dried and sieved at 2 mm) with quartz sand in the ratio 1:1, and put the mixture in a ceramic Buchner funnel (outer diameter 55 mm), on the bottom of which is placed a glass-fibre diskette (three replicates).
- Remove the mineral nitrogen (present in the soil at the beginning of the experiment) by leaching the system 'soil + quartz sand' with 180 ml of 0.01 M $CaSO_4$, followed by the addition of 20 ml of the nutritive *N-minus* solution. Remove excess water with a vacuum system (60 cmHg). Incubate at 30°C. Gaseous exchange through the opened portion of the funnels guarantees the maintenance of aerobic conditions.
- After 2 weeks, remove the mineralized nitrogen by leaching again with 100 ml of 0.01 M $CaSO_4$ and then add 20 ml of the nutritive *'N-minus'* solution, followed by the vacuum application as described above. Incubate again, performing a series of leachings after 2, 4, 8, 12, 16, 22 and 30 weeks. In order to avoid soil dryness during the incubation period, add distilled water to the mixture of soil + quartz sand, until reaching the optimal value of soil moisture.
- Determine NH_4^+-N, NO_3^--N and NO_2^--N in the clear leached solution according to the chosen method.

Calculation

The total mineralized nitrogen is calculated as the sum of NH_4^+-N, NO_3^--N and NO_2^--N content after each leaching.

N_0 (potentially mineralized nitrogen) could be calculated by fitting the cumulative experimental data by the first-order equation:

$$N_t = N_0 (1 - e^{-kt})$$

where: N_t = mineralized nitrogen (cumulated value, in mg N/kg soil); N_0 = potentially mineralizable nitrogen (mg N/kg soil); k = kinetic constant (in 1/week); and t = time (in weeks).

7.4 Nitrification in Soil

Annette Bollmann

University of Aarhus, Department of Microbial Ecology, Ny Munkegade Bygning 540, 8000 Aarhus C, Denmark; Present address: *Northeastern University, Department of Biology, 360 Huntington Avenue, 134 Mugar Life Science Building, Boston, MA 02115, USA*

State of the Art

Ammonium (NH_4^+) plays a central role in the terrestrial nitrogen cycle. It is produced by mineralization of organic matter or by nitrogen (N_2) fixation by nitrogen-fixing bacteria. Consumption of NH_4^+ occurs through assimilation by bacteria and plants, and conversion to nitrite and nitrate by nitrification. Nitrification is the oxidation of ammonia (NH_3) to nitrate (NO_3^-) via nitrite (NO_2^-), carried out by two separate groups of specialized bacteria. The ammonia-oxidizing bacteria are obligate aerobic, chemolithoautotrophic bacteria with a pH optimum around 7–8, which convert ammonia into nitrite. They belong to the β-subgroup of the *Proteobacteria*. The nitrite-oxidizing bacteria, catalysing the conversion of nitrite to nitrate, are physiologically and phylogenetically more diverse than the ammonia-oxidizing bacteria. Some are capable of mixotrophic or heterotrophic growth and dissimilatory nitrate reduction, and they belong to different classes of the *Proteobacteria*. These bacteria live in soils, sediments, fresh and marine water, and wastewater treatment systems (Prosser, 1989; Koops and Pommerening-Röser, 2001; Kowalchuk and Stephen, 2001).

There are different ways to determine nitrification: measurement of gross nitrification rates by ^{15}N methods, measurement of net nitrification rates by determining the nitrate accumulation, and measurement of the potential nitrification by measuring rates under optimal conditions. The available methods are reviewed by Schinner *et al.* (1992), Mosier and Schimel (1993), Hart *et al.* (1994) and Schmidt and Belser (1994).

Gross nitrification

^{15}N tracer studies

The substrate pool (NH_4^+) is labelled to follow the fate of the label in the system (Myrold and Tiedje, 1986). This method has been used to determine gross nitrification, but some problems have become evident. The *in situ* ammonium concentration is changed by the addition of the labelled substrate and/or by the *in situ* production of ammonium by other processes. In addition, the product of nitrification (nitrate/nitrite) can be removed by assimilation or dissimilation processes.

^{15}N dilution method

The product pool (NO_3^-) is labelled and, by following the dilution of the ^{15}N NO_3^- pool, the gross nitrification rate can be calculated (Kirkham and Bartholomew, 1954). Good detailed descriptions of this highly recommended method to determine gross nitrification rates are available in Davidson *et al.* (1991), Mosier and Schimel (1993) and Hart *et al.* (1994).

Net nitrification

The measurement of net nitrification in the laboratory and in the field is based on the determination of the accumulation of NO_3^- over time (Beck, 1979; Hart and Binkley, 1985; Davidson *et al.*, 1990). Measuring the accumulation of NO_3^- does not give a good indication about the real size of the nitrification process, because it does not take into account the immobilized NO_3^- fraction, the denitrified NO_3^- fraction and the NO_3^- fraction converted back to NH_4^+. Additionally, the long incubation periods of up to 3 weeks can cause changes in the populations of the ammonia- and nitrite-oxidizing bacteria.

Nitrification potential

The nitrification potential is a method to determine the maximum nitrifying activity under optimal conditions. This method will be described in detail.

Description of the ammonia oxidizer population

Molecular fingerprinting methods, such as denaturing gradient gel electrophoresis (DGGE) and restriction fragment analysis, are available to describe communities of ammonia-oxidizing bacteria, based on the 16S rRNA gene or *amo*A gene (Kowalchuk *et al.*, 1997; Rotthauwe *et al.*, 1997; Stephen *et al.*, 1999; Nicolaisen and Ramsing, 2002).

Principle of the Method

The nitrification potential is a method to measure the maximal nitrifying activity under optimized conditions (Belser, 1979; Hart *et al.*, 1994). By keeping the basic requirements of the ammonia-oxidizing bacteria (NH_4^+, O_2, pH and water) within the optimal range, effects of toxic substances, fertilization or management processes can be determined. Changes in the populations of the ammonia- or nitrite-oxidizing bacteria will not have a significant influence, because potential rates can be determined in short-term experiments (within 24 h). This method, or very similar methods, has been used to determine the nitrification potential in paddy soils (Bodelier *et al.*, 2000), in agricultural soils (Mader, 1994; Stoyan *et al.*, 2000) and in sediments (Bodelier *et al.*, 1996).

Materials and Apparatus

- 250 ml Erlenmeyer flasks, closed with aluminium foil to reduce evaporation
- Shaker
- Eppendorf pipettes and pipette tips with cut-off tips, to prevent blocking of the tip with soil or sediment for sampling
- 1.5 ml Eppendorf cups and Eppendorf cup centrifuge with cooling
- HPLC (high-pressure liquid chromatograph), IC (ion chromatograph) or autoanalyser to determine nitrite and nitrate
- In case no HPLC, IC or autoanalyser is available: photometer and other equipment to determine nitrite and nitrate photometrically (Keeney and Nelson, 1982)

Chemicals and Solutions

- *0.1 M potassium dihydrogen phosphate solution*: dissolve 13.61 g KH_2PO_4 in 1 l of water
- *0.1 M dipotassium hydrogen phosphate solution*: dissolve 17.42 g K_2HPO_4 in 1 l water
- *Ammonium stock solution*: 1 M NH_{4+}: dissolve 66.07 g ammonium sulphate $(NH_4)_2SO_4$ in 1 l of water
- Sodium hydroxide solution and sulphuric acid to adjust the pH value
- *Reagents for photometric* NO_2^- *determination*:
  – sulphanilamide solution: add 150 ml of ortho-phosphoric acid carefully to 700 ml water, add 10 g sulphanilamide, stir and warm up a little to dissolve the sulphanilamide, add 0.5 g α-naphthyl ethylene diamine dihydrochloride and top up to 1 l (store in a dark bottle in the fridge)
  – calibration stock solution: 10 mM NO_2^-: dissolve 0.69 g sodium nitrite in 1 l of water
- *Reagents for photometric* NO_3^- *determination*:
  – 0.5% sodium salicylate solution (prepare fresh): dissolve 0.5 g sodium salicylate in 100 ml water
  – concentrated sulphuric acid (95–97%)
  – seignette salt solution: dissolve 300 g sodium hydroxide in 800 ml water, then add 6 g potassium sodium tartrate and fill up to an end volume of 1 l

– calibration stock solution: 10 mM NO_3^-: dissolve 0.849 g sodium nitrate in 1 l of water

Procedure

Nitrification potential

- Prepare a working solution (1 mM PO_4^{3-}, 5 mM NH_4^+, pH 7.5): add per l, 2 ml of the KH_2PO_4, 8 ml of the K_2HPO_4 and 5 ml NH_4^+ stock solution and adjust the pH value to 7.5.
- Put 10 g soil into a 250 ml Erlenmeyer flask (at least three replicates).
- Determine the gravimetric water content of the soil.
- Add 100 ml of the working solutions to the flasks.
- Put on the shaker at 200 rpm.
- Incubate for 26 h and take samples directly after the start and after 1, 3, 6, 22 and 26 h.
- Take a 1.5 ml sample into an Eppendorf cup, by pipetting 2 × 0.75 ml with a cut-off 1 ml pipette tip on an Eppendorf pipette; shake samples directly before sampling, because the sample should have the same soil:solution ratio as the slurry.
- Put the sample immediately on ice.
- Spin down for 10 min at 12,000 × *g* at 4°C.
- Transfer the supernatant to a new Eppendorf cup.
- Store the samples overnight at 4°C or for longer storage at –20°C.
- Dilute the samples and determine nitrite and nitrate.

Determination of nitrite

- Put 2 ml of sample (concentration lower than 50 μM NO_2^-) or calibration solution (0–50 μM NO_2^-) into a test tube.
- Add 0.5 ml sulphanilamide solution and mix well.
- Incubate in the dark for 15 min.
- Measure the extinction at 540 nm against water.

Determination of nitrate

- Put 1 ml of sample (concentration lower than 1 mM NO_3^-) or calibration solution (0–1 mM NO_3^-) into a test tube.
- Add 0.5 ml sodium salicylate solution and mix well.
- Dry the mixture in an oven at 110°C overnight.
- The next steps should be performed carefully in a fumehood:

– add 0.5 ml concentrated sulphuric acid to dissolve the dried solution in the test tube;
– add 4 ml water and mix well;
– add 3 ml seignette salt-solution and mix well;
– incubate for 10 min;
– measure the extinction at 430 nm against water.

Calculations

Calibration

- Plot the extinctions against the NO_2^- or NO_3^- concentrations (μM).
- Determine the slope and the intercept with the *Y*-axis via linear regression.

Determination of the NO_2^- or NO_3^- concentrations in the sample

Determine the NO_2^- and the NO_3^- concentrations (μM) from the calibration curve:

$$NO_2^- \text{or } NO_3^- \text{conc}\,(\mu\text{M}) = \frac{(E_{\text{xs}} - E_{\text{xb}}) - \text{int}}{\text{slo}}$$

where: E_{xs} = extinction of the sample; E_{xb} = extinction of the blank; int = intercept with the *Y*-axis calculated from the calibration curve; slo = slope of the calibration curve.

Transform the concentrations from (μM) to (μg N/g DW)

$$NO_2^- \text{or } NO_3^- \text{conc}\,(\mu\text{g N / g DW}) = NO_2^- \text{or } NO_3^- \text{conc}\,(\mu\text{M}) \times \frac{\left(\frac{100 + (10 - \text{DW})}{1000}\right) \times 14}{\text{DW}}$$

where: DW = weight of the oven-dry soil (g); 14 = conversion factor (1 M nitrogen = 14 g nitrogen); and

$$\left(\frac{100 + (10 - \text{DW})}{1000}\right) = \text{conversion factor for the soil solution.}$$

Determination of the potential nitrification rate

- Sum up the NO_2^- and NO_3^- concentrations (= NO_x).
- Plot the NO_x concentrations (μg N/g DW) against the time (h) and determine the potential nitrification rate by calculating the slope via linear regression (μg N/g DW/h).

Note: The potential nitrification rate is the maximum nitrification rate of a sample under optimal conditions without growth of the cells during the incubation period. Under certain circumstances ammonia-oxidizing bacteria can have a lag phase until they become fully active, indicating that they reach maximal activity after several hours of incubation. In other cases they are very active – the ammonium can be completely consumed after several hours of incubation and the NO_x concentration no longer increases. Therefore, it is very important to take several samples during the time course of the incubation, to calculate the potential nitrification rate from the maximal increase of NO_x concentration against time and to cross-check the NO_x concentrations with the added ammonium concentrations.

Discussion

Measurement of the potential nitrification rate gives a good insight into the quality of a soil. Nitrification is a process carried out by highly specialized bacteria, which are very sensitive to changes in the environment, environmental stress, and toxic substances. Therefore, nitrification is a useful microbiological parameter for describing soil quality.

Optimal initial ammonium concentration

There is a strong relationship between the ammonia oxidation (potential nitrification) rate and the initial NH_4^+ concentration. This relationship can differ a lot between different soils (Stark and Firestone, 1996). So it would be useful to conduct a pre-experiment to determine the relationship between NH_4^+ concentration and ammonium oxidation rate for the particular soil. Then the nitrification potential at different initial NH_4^+ concentrations (usually between 0.1 mM NH_4^+ and 5 mM NH_4^+) can be determined and the following experiments can be done with an initial NH_4^+ concentration at which the ammonia oxidation rate is maximal.

Addition of chlorate (ClO_3^-) to inhibit nitrite oxidation

ClO_3^- inhibition of nitrite oxidation is a widely used method to determine potential ammonia oxidation rates (Belser and Mays, 1980; Berg and Rosswall, 1985). ClO_3^- acts as a competitive inhibitor of nitrite oxidation. It is most effective at low nitrite concentrations (Belser and Mays, 1980). The use of ClO_3^- is critical because *Nitrobacter winogradskii* (a nitrite oxidizer) is able to convert ClO_3^- to chlorite (ClO_2^-). ClO_2^- inhibits *Nitrosomonas europaea* (Hynes and Knowles, 1983) and other ammonia-oxidizing bacteria. Therefore, the use of the chlorate inhibition method is questionable.

7.5 Thymidine and Leucine Incorporation to Assess Bacterial Growth Rate

Jaap Bloem and Popko R. Bolhuis

Department of Soil Sciences, Alterra, PO Box 47, NL-6700 AA Wageningen, The Netherlands

Introduction

Besides biomass and respiration, bacterial growth rate is a key parameter involved in microbial functioning in soil food webs and nutrient cycling (Bloem *et al.*, 1997). Moreover, growth or reproduction is very sensitive to contamination, and is thus a useful indicator of stress. Growth rate cannot easily be calculated from increases in biomass or cell number, because often increases are balanced by losses. Losses may be caused by bacterivores such as protozoa and nematodes, or by viruses. Since about 1980, measurement of [^{3}H]thymidine incorporation during short incubations (about 1 h) has become the method of choice to determine bacterial growth rate in water and sediments (Fuhrman and Azam, 1982; Moriarty, 1986). Thymidine is incorporated into bacterial DNA and thus reflects DNA synthesis or cell division. Since about 1990, the method has also been used for bacteria in soil (Bååth, 1990; Michel and Bloem, 1993). Thymidine incorporation has been found to be more sensitive to contamination than biomass and respiration, in both water and soil (Jones *et al.*, 1984; Bååth, 1992). A plausible explanation for a reduced growth rate in contaminated environments is that microorganisms under stress (e.g. heavy metals or pH) divert energy from growth to cell maintenance (Kilham, 1985; Giller *et al.*, 1998). Physiological processes for detoxification require additional energy. Thus, less energy is available for synthesis of new biomass (growth). This may explain why bacterial growth rate appears to be one of the most sensitive indicators of heavy metal stress in contaminated soils (Bloem and Breure, 2003).

Leucine incorporation into proteins has been introduced as an alternative to the commonly used thymidine method, but can also serve as an independent check (Kirchman *et al.*, 1985). Using [^{3}H]thymidine and [^{14}C]leucine in a dual-label approach, both parameters can be measured in a single assay. Here we describe a protocol for simultaneous measurement of thymidine and leucine incorporation into soil bacteria. Differences between these protocols for thymidine incorporation and those published in previous handbooks (Bååth, 1995; Christensen and Christensen, 1995) are discussed. We describe measurement of thymidine incorporation in a soil slurry. Bacteria may also be first extracted from the soil and further treated like a water sample. This relatively simple, but more indirect, method is especially useful for determining soil bacterial community tolerance to heavy metals and pH, for example (Bååth, 1992). A thorough review of thymidine incorporation has been given by Robarts and Zohary (1993).

Principle of the Method

Thymidine is a precursor of thymine, one of the four bases in the DNA molecule. Leucine is an amino acid which is incorporated in proteins. When sufficient thymidine is added, *de novo* synthesis is inhibited, and labelled thymidine is incorporated. Bacterial growth rate is reflected by the incorporation rate of [^{3}H]thymidine and [^{14}C]leucine into bacterial macromolecules during a short incubation of 1 h. If the incubation period is short enough, growth rate is not affected by incubation. Using a dual-label approach (^{3}H and ^{14}C) both parameters are measured in a single assay.

Thymidine versus leucine

Thymidine incorporation is more proportional to growth rate than leucine incorporation because bacterial DNA content (usually 2–5 fg/cell) is more constant than protein content. Fungi do not incorporate thymidine because they lack the key enzyme thymidine kinase. Growth rates of fungi can be estimated by [^{14}C]acetate incorporation into ergosterol (Bååth, 2001). Only bacteria incorporate thymidine, but not all bacteria are able to do so (e.g. pseudomonads, some anaerobes, nitrifiers and sulphate reducers cannot incorporate thymidine). All bacteria incorporate leucine, but leucine can also be incorporated by other organisms. However, the usual concentration of about 2 μM is probably too low, and the incubation time of 1 h is probably too short, to label cells bigger than most bacteria. Bigger cells have a less suitable surface to volume ratio. Incorporation rates of leucine into proteins are an order of magnitude higher than those of thymidine into DNA, thus lower growth rates can be measured with greater precision. Since both methods have advantages and limitations, their use is complementary.

Materials and Apparatus

- Centrifuge
- Polypropylene centrifuge tubes (13 ml) with screw cap (e.g. Falcon) for incubation of samples
- Incubator at 30°C
- Ice
- Cellulose nitrate membrane filters, 0.2 μm pore size: filters are prewashed with unlabelled thymidine and leucine to minimize adsorption of unincorporated labelled thymidine and leucine

- Filtration manifold to handle larger number of samples (e.g. Millipore, for 12 filters)
- Liquid scintillation counter to measure ^3H and ^{14}C

Chemicals and Solutions

Radioisotopes

- Methyl-[^{3}H]thymidine (TRK 300, 1 mCi/ml, 25 Ci/mmol = 925 GBq/mmol, 40 μM) and L-[U-^{14}C]leucine (CFB 67, 0.05 mCi/ml, 0.31 Ci/mmol = 11.5 GBq/mmol) (Amersham Ltd, Amersham, UK)
- The isotopes are stored at 4°C before use; we store them no longer than 3 months
- Per sample (tube) we use: 1.5 μl [^{14}C]leucine, 2.0 μl [^{3}H]thymidine and 16.5 μl unlabelled thymidine (2.35 mg/l). This corresponds to 2 μM and 2.78 kBq (= 0.075 μCi) [^{14}C]leucine and 2 μM and 74 kBq (= 2 μCi) [^{3}H]thymidine per tube

PJ mineral solution

Soil is suspended in a mineral solution (Prescott and James, 1955), made up as follows:

- Make up three stock solutions, each with 100 ml of distilled (or Milli Q) water:
  – stock solution A: 0.433 g $CaCl_2.2H_2O$ and 0.162 g KCl
  – stock solution B: 0.512 g K_2HPO_4
  – stock solution C: 0.280 g $MgSO_4.7H_2O$
- The final dilution consists of 1 ml of each stock solution and 997 ml of distilled water

Solution for extraction of macromolecules

5 ml per tube of:

- 0.3 N NaOH (12 g/l; solubilizes DNA, hydrolyses RNA)
- 25 mM EDTA (9.3 g/l; breaks up aggregates)
- 0.1% SDS (1 g/l; sodium dodecyl sulphate, lyses membranes)

Both SDS and EDTA inhibit nucleases and breakdown of DNA

- 1 N HCl (82.6 ml/l; 1.3 ml/tube)
- 29% TCA (trichloroacetic acid; 290 g/l; 1.3 ml/tube)
- 5% TCA (50 g/l; 15 ml/tube)

- 5 mM thymidine (1.21 g/l) and 5 mM leucine (0.65 g/l) (1 ml/tube; unlabelled)
- 0.1 N NaOH (4 g/l; 1 ml/vial)
- Ethylacetate (1 ml/vial)

Procedure

Incubation of soil suspension with labelled thymidine and leucine

- Homogenize 20 g soil in 95 ml mineral medium (PJ) by hand shaking for 30 s.
- Allow settling for 1 min to remove coarse soil particles.
- Add 100 μl soil suspension (20 mg soil) to the centrifuge tubes.
- Add 20 μl labelled [^{3}H]thymidine and [^{14}C]leucine (both 2 μM final concentration).
- Incubate for 1 h at *in situ* temperature; growing cells incorporate thymidine and leucine.
- Stop incorporation by adding 5 ml of extraction mixture (0.3 N NaOH, 25 mM EDTA and 0.1% SDS).
- Blanks are prepared by adding the extraction mixture immediately after the start of the incubation.
- At this stage the samples may be stored. We found no effect of 3 days' storage at 4°C in the dark, on thymidine and leucine incorporation. Dixon and Turley (2000) incubated marine sediment slurries on board ship, and stored them in a freezer until later analysis in the laboratory.

Extraction of labelled macromolecules

- The suspension is left at 30°C overnight (18–20 h) to extract macromolecules in the warm base solution.
- Mix and centrifuge (40 min, 5000 × *g*, 25°C) to remove soil particles.
- Aspire the supernatant with macromolecules and cool on ice for 5 min.
- Neutralize the base suspension with 1.3 ml ice-cold 1 N HCl.
- Add 1.3 ml ice-cold 29% TCA (w/v) and cool on ice for 15 min to precipitate the macromolecules.
- Collect the precipitated macromolecules on a 0.2 μm cellulose nitrate membrane filter.
- Wash filters three times with 5 ml ice-cold 5% TCA to remove unincorporated label, and transfer to glass scintillation vial.
- Add 1 ml 0.1 N NaOH to dissolve macromolecules and 1 ml ethylacetate to dissolve the filter. Shake, leave for half an hour and shake again.
- Add (15 ml) scintillation fluid. Shake vial and leave for at least 1 h to reduce chemoluminescence.
- Count radioactivity (dpm) of ^{3}H and ^{14}C.

Calculation

The number of blanks equals the number of replicates. Because the blanks have no individual relationship with the replicates, the mean value of the blanks is subtracted from each replicate. After subtraction of blanks, the counted dpm of ^{3}H and ^{14}C, respectively, are multiplied by 0.00284 to calculate pmol thymidine (1 picomol = 10^{-12} mol), and by 0.0759 to calculate pmol leucine incorporated per g soil per hour. These factors are based on the specific activity (Bq/mol, 1 Bq = 1 dps or 60 dpm) of thymidine and leucine as given above.

Using conversion factors, pmol/g per h can be converted to cells and carbon produced. Conversion factors have been established by growing bacteria at well-defined growth rates in continuous cultures (Bloem *et al.*, 1989; Ellenbroek and Cappenberg, 1991; Michel and Bloem, 1993). For thymidine, the mean conversion factor was 0.5×10^{18} cells/mol thymidine. The range was $0.2–1.1 \times 10^{18}$. In the literature, much higher factors (range $0.13–7.9 \times 10^{18}$) have been published (Bååth and Johansson, 1990; Christensen, 1991). Given the uncertainties, we usually avoid conversion factors and express thymidine incorporation as pmol/g soil/h.

Discussion

Bacterial growth rate appears to be a more sensitive indicator of contamination than biomass or respiration. We compared the effects of copper pollution on bacterial biomass (microscopic measurements), growth rate (thymidine incorporation) and respiration (CO_2 evolution) in an arable sandy soil. The soil was originally unpolluted, with a background concentration of 25 mg Cu/kg. Jars were filled with 180 g gamma-sterilized soil, amended with 360 mg lucerne meal and 40 mg wheat straw meal, and re-inoculated with soil organisms (Bouwman *et al.*, 1994). In addition, different amounts of $CuSO_4$ were added, using three microcosms per treatment. Two days after amendment, bacterial growth rate, as determined by thymidine incorporation, was already significantly reduced at a low copper addition of 10 mg/kg (Fig. 7.2). Bacterial biomass and respiration were reduced at higher copper concentrations of 100 mg/kg and 1000 mg/kg.

In this experiment, thymidine incorporation was converted to growth rate using a conversion factor of 0.5×10^{18} cells/mol (Michel and Bloem, 1993), a cell volume of 0.2 μm^3 and a specific carbon content of 3.1×10^{-13} g C/μm^3 (Bloem *et al.*, 1995). This resulted in a calculated growth rate of 50 μg C/g soil/day (Fig. 7.2). However, the respiration rate was only 10 μg C/g soil/day. Assuming a growth efficiency of, at most, 50%, the growth rate should roughly equal the respiration rate. Thus, the growth rate was probably overestimated by at least a factor of 5. Using advanced methods and avoiding conversion factors, Harris and Paul (1994) determined specific rates of DNA synthesis from the specific activities of the DNA precursor deoxythymidine triphosphate (dTTP) and purified bacterial DNA

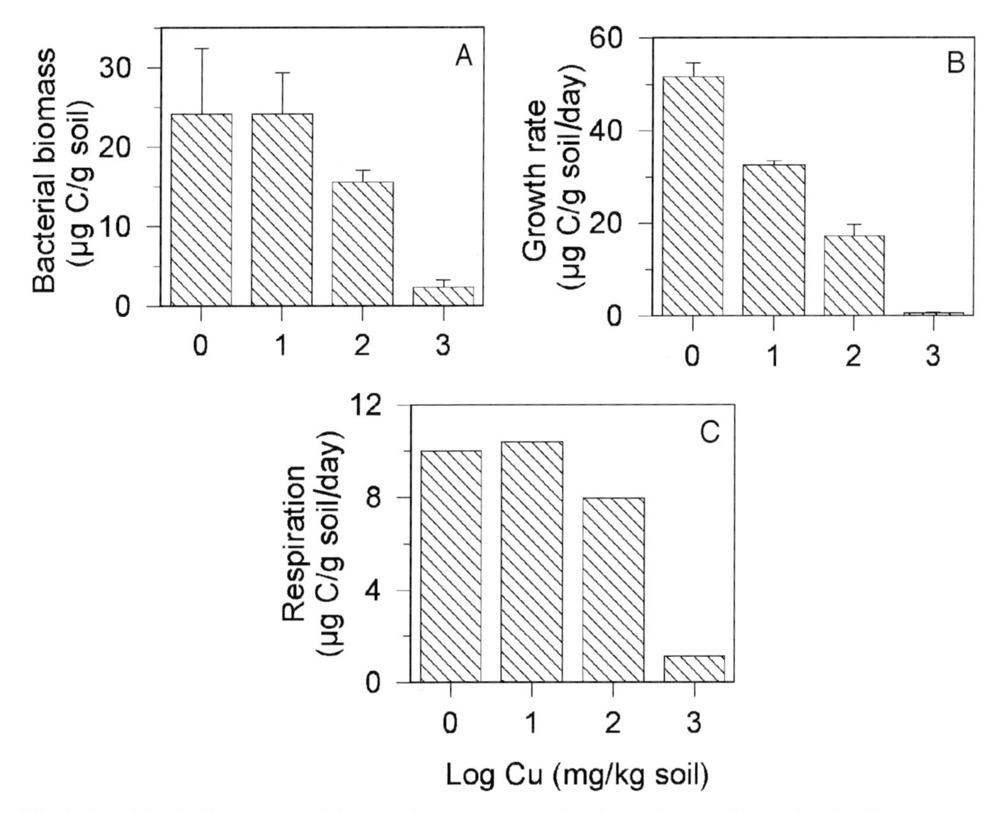

Fig. 7.2. Effects of copper pollution on biomass, growth rate and respiration rate of soil bacteria in microcosms, measured 2 days after amendment. Error bars indicate SD, $n = 3$.

after [^{3}H]thymidine incorporation. Their growth rate estimates were in agreement with respiration rates and much lower than those found in some other studies. This indicates that commonly used conversion factors tend to overestimate bacterial growth rate.

The thymidine incorporation procedure described here is based mainly on Findlay *et al.* (1984) and Thorn and Ventullo (1988), with some modifications (Michel and Bloem, 1993). The procedure can be summarized as follows:

- Incubate soil suspension with [^{3}H]thymidine and [^{14}C]leucine for 1 h.
- Extract macromolecules in warm base overnight.
- Remove soil particles by centrifugation.
- Cool supernatant on ice and precipitate macromolecules with cold acid (TCA).
- Collect macromolecules (including DNA and proteins) on a cellulose nitrate membrane filter.
- Count radioactivity (^{3}H and ^{14}C).

With this procedure, thymidine and leucine incorporation can be determined simultaneously. The procedure differs in some aspects from the procedures for thymidine incorporation described in two previous handbooks. Each procedure probably works well for the conditions it was originally tested for, but it may have shortcomings under other conditions.

Christensen and Christensen (1995):

- Used a concentration of 200 nM thymidine. For some soils this is sufficient, but in other soils at least 1 μM may be needed for maximum incorporation. Ideally, the optimum concentration should be determined by a saturation experiment for each soil, to minimize dilution by unlabelled precursors of thymine which may be present in the environment. Specific techniques to estimate isotope dilution are time consuming and not very accurate (Michel and Bloem, 1993; Robarts and Zohary, 1993). For routine use with large numbers of different samples (in monitoring networks) we use 2 μM thymidine (and leucine).
- Extracted in 0.6 N NaOH for 1 h at 60°C. This may cause degradation of DNA and lower recovery than a milder extraction in 0.3 N NaOH at 30°C for at least 12 h (Findlay *et al.*, 1984; Thorn and Ventullo, 1988).
- Added no EDTA and SDS. EDTA and SDS have been reported to increase recovery by inhibiting nucleases, and promoting cell lysis and breakdown of soil aggregates (Findlay *et al.*, 1984; Thorn and Ventullo, 1988; Bååth and Johansson, 1990).
- Used cellulose acetate filters. Cellulose nitrate is recommended to bind DNA (Robarts and Zohary, 1993).
- Did not pre-wash filters with cold (unlabelled) thymidine: this may cause higher values of blanks (Robarts and Zohary, 1993).

The procedure of Bååth (1995) can be summarized as follows:

- Incubate with [^{3}H]thymidine.
- Centrifuge to collect bacteria and soil.
- Extract macromolecules in warm base overnight at 60°C.
- Centrifuge to remove soil particles.
- Precipitate macromolecules with acid (TCA) on ice.
- Centrifuge twice with 5% TCA to wash macromolecules.
- Hydrolyse DNA in 5% TCA at 90°C for 30 min.
- Centrifuge.
- Collect supernatant ('DNA fraction') and measure radioactivity.

This procedure uses five centrifugation steps and may be more time consuming. Bååth and Johansson (1990) found little effect of extraction temperature on the recovery of added [^{14}C]DNA. In contrast, at 60°C and 90°C, we found 30% and 90% lower thymidine incorporation than at 30°C. This may indicate degradation of DNA at higher temperatures. In Bååth's protocol, the incorporation into total macromolecules is not measured, but the DNA fraction is isolated using hot acid. In principle, it is better to isolate the DNA fraction. However, DNA has been reported to hydrolyse incompletely (about 50%) in hot acid, and this procedure may lead to underestimates of DNA synthesis (Servais *et al.*, 1987; Robarts and Zohary, 1993). Using hot acid hydrolysis, we sometimes found 80% incorporation into the 'DNA fraction'. On other occasions we found as little as 0% incorporation into the 'DNA fraction' of actively growing (and incorporating) bacteria from continuous cultures (Bloem *et al.*, 1989). Robarts and Zohary (1993) criticized (mis)use of acid–base hydrolysis, and alternatively proposed to purify the DNA by rinsing the filters with 50% phenol–chloroform to remove proteins, and with 80% ice-cold ethanol to remove lipids. Using this method, we found between 10% and 100% incorporation into DNA. Since proteins are removed, this method requires separate samples for measuring leucine incorporation.

Thus, a minor fraction of the total thymidine incorporation may be into DNA. Measured incorporation into DNA underestimated actual DNA synthesis by a factor of 6–8 (Jeffrey and Paul, 1988; Ellenbroek and Cappenberg, 1991). This may be caused by a rate-limiting step in the incorporation of labelled thymidine, resulting in intracellular dilution with unlabelled thymine. Part of the (apparently) low incorporation into DNA may be caused by incomplete hydrolysis of DNA in hot acid (Robarts and Zohary, 1993). However, two alternative methods, using enzymatic degradation with DNase or the DNA synthesis inhibitor mitomycin, confirmed a low (about 35%) incorporation into DNA (Ellenbroek and Cappenberg, 1991). Even incorporation into total macromolecules (DNA, proteins, lipids) underestimated real DNA synthesis in continuous culture by at least a factor of two (Bloem *et al.*, 1989; Ellenbroek and Cappenberg, 1991).

Given the difficulties with measuring actual DNA synthesis, and the apparently consistent empirical relationship between (total) thymidine incorporation and bacterial growth rate (Bloem *et al.*, 1989; Ellenbroek and Cappenberg, 1991; Michel and Bloem, 1993), we have chosen to use incorporation of [^{3}H]thymidine and [^{14}C]leucine into total macromolecules. This is generally sufficient to indicate relative effects on bacterial growth rates.

7.6 N_2O Emissions and Denitrification from Soil

ULRIKE SEHY,[1] MICHAEL SCHLOTER,[1] HERMANN BOTHE[2] AND JEAN CHARLES MUNCH[1]

[1]GSF – National Research Centre for Environment and Health, Institute for Soil Ecology, Ingolstädter Landstraße 1, D-85758 Oberschleissheim, Germany; [2]University of Cologne, Institute of Botany, Gyrhofstraße 15, D-50923 Cologne, Germany

Introduction

The emission of nitrogenous gases from soil plays a crucial role in the global nitrogen cycle, as nitrate or ammonia are turned into gaseous products, which return to the atmosphere. Gases involved include ammonia, nitric oxide, nitrous oxide and dinitrogen. For nitric oxide, nitrous oxide and dinitrogen, two microbial processes, nitrification and denitrification, are considered to be the most important biotic sources (Granli and Bøckman, 1994). Both processes may occur simultaneously in microsites and are controlled by various soil- and weather-dependent parameters.

Denitrification

During denitrification, nitrate is reduced via nitrite, NO and N_2O to N_2. Denitrification is an energy-yielding process, in which microorganisms utilize nitrate as an alternative terminal respiratory electron acceptor under oxygen-limited conditions. The ability for denitrification is widespread amongst microorganisms. It is known that approximately 50% of all known bacteria can denitrify (or at least carry out some partial denitrification reactions). In addition, some species of fungi and archaebacteria have also been described as denitrifiers. Overall, the occurrence of denitrification in organisms of totally unrelated affiliations suggests that denitrification has been distributed evolutionarily by lateral gene transfer.

The single steps of the denitrification process are catalysed by specific reductases. The first reaction, the conversion of nitrate to nitrite, is catalysed by a molybdenum (Mo)-containing nitrate reductase. All nitrate reductases have a molybdopterin cofactor and contain FeS centres, and, in addition, some possess cytochrome *b* or *c*. Several forms of nitrate reductases exist (membrane-bound, periplasmic and also an assimilatory enzyme with similar characteristics to the dissimilatory enzymes), which make the study of this step interesting but also complex, particularly since some organisms contain more than one type of enzyme.

The next step in the denitrification pathway is the reduction of nitrite to nitric oxide, which is catalysed by nitrite reductase. Bacteria express two different forms of nitrite reductase, containing either a cytochrome (nirS) or copper (nirK) in their prosthetic group. Organisms possess one of the two enzymes. The cytochrome-containing nitrite reductase appears to be more widespread among bacteria, whereas the Cu-enzyme apparently is evolutionarily more conserved. The product of the reaction, NO, is highly reactive and toxic, and is also an important signalling molecule for plant or animal life. Intact denitrifying organisms evolve, at best, minute amounts of this gas, which is effectively utilized by the cytochrome *b, c* containing nitric oxide reductase. The conversion of NO to N_2O, catalysed by this enzyme, involves the formation of the dinitrogen bond, which is biochemically an extremely interesting, and currently poorly understood, reaction. Finally, nitrous oxide is reduced to the dinitrogen molecule by nitrous oxide reductase, which contains Cu atoms in a novel tetra-nuclear cluster at the active site.

Nitrification

While denitrification generates N, NO and N_2O, nitrification can produce only NO and N_2O. Nitrification is the conversion of inorganic or organic nitrogen from a reduced to a more oxidized state. Chemoautotrophic bacteria are largely, or solely, responsible for the nitrification in soils with a pH above 5.5 (Focht and Verstraete, 1977); at lower pH values there is evidence for acid-tolerant heterotrophic nitrifiers (Schimmel *et al.*, 1984). The heterotrophic nitrifiers (e.g. bacteria, such as strains from *Arthrobacter*, and fungi, such as *Aspergillus*) do not derive energy from the oxidation of NH_4^+. In arable soils, the production of nitrate by heterotrophs appears to be insignificant in relation to that brought about by chemoautotrophs (Paul and Clark, 1989), which is in contrast to potential nitrification levels in acid forest soils (Kilham, 1987).

Consequences of the activity of denitrifying and nitrifying microorganisms

Due to the action of denitrifying and nitrifying microorganisms, the global dinitrogen content in the atmosphere is in balance (due to the formation of the nitric oxide, nitrous oxide and dinitrogen from nitrate and ammonia). On the other hand, nitrogenous oxides, also released from soils and waters, have several impacts on the chemistry of the atmosphere and radiation processes. Nitrous oxide is next to carbon dioxide (CO_2) and methane (CH_4) in its importance as a potent greenhouse gas. Nitric acid and its chemical oxidation product, NO_2, are major constituents of acid rain, and NO and also N_2O interact with ozone in complex reactions and are major causes of the destruction of the protective ozone layer in the stratosphere.

Nitrate is the main source of nitrogen for the growth of plants in agriculture, but it can simultaneously be used by microorganisms in soils.

Denitrification is generally regarded as an anaerobic process, but there are indications that it may also take place in well-aerated soils. The conditions that favour denitrification in soils have been elucidated in detail. It is clear that any use of nitrate by bacteria means a loss of N for the growth of plants. Thus, denitrification also has a severe impact on agriculture.

In addition, products of denitrification (nitrate respiration) have manifold other, mainly adverse, effects on the atmosphere and on waters.

Measurement of NO, N_2O and N_2

While the products N_2O and NO can be easily measured as trace gases, using gas chromatography, the determination of N_2 is not straightforward because comparatively small amounts of N_2 produced during denitrification have to be distinguished from a large background of 78% N_2 in the atmosphere (Aulakh *et al.*, 1992). The methods available for measuring denitrification in the field are based on the use of the stable isotope ^{15}N, or on acetylene for blockage of the enzyme N_2O-reductase.

Principle of the Method

The acetylene inhibition method (AIM) was developed utilizing the fact that acetylene (C_2H_2) blocks the enzymatic reduction of N_2O to N_2 (Balderston *et al.*, 1976; Yoshinari and Knowles, 1977) when present in a range of 1–10% v/v (Granli and Bøckman, 1994). In soils treated with acetylene, the amount of N_2O released thus represents both N_2O and N_2 produced during denitrification.

There are basically two methods available for the application of the AIM in the field:

1. *Chamber methods*: in which enclosures are placed over the soil to separate the soil atmosphere from the ambient atmosphere (Ryden *et al.*, 1979; Burton and Beauchamp, 1984; Mosier *et al.*, 1986; Aulakh *et al.*, 1991). Denitrification measurements can be made either with open chambers, where acetylene is forced through the cover and N_2O is collected at the outlet for analysis, or with closed chambers, where N_2O is allowed to accumulate before withdrawal for analysis.

2. *Soil core methods*: in which undisturbed soil cores are taken and incubated in the laboratory or in the field (Parkin *et al.*, 1985; Ryden *et al.*, 1987; Aulakh *et al.*, 1991; Jarvis *et al.*, 2001). For denitrification measurements, soil cores are collected in the field, by driving a small cylinder into the soil, then incubated together with acetylene. Two methods are commonly used: one is to incubate soil cores statically in containers with acetylene and to collect gas samples through rubber septa (Burton and Beauchamp, 1984). In another method, soil air and acetylene are recirculated through the macropores of the soil and the outgoing air is directly connected to a gas chromatograph for quantification of N_2O (Parkin *et al.*, 1984, 1985).

The following sections will focus on the measurement of N_2O from denitrification. However, with slight modifications of the methods (no acetylene inhibition), it is possible to measure all nitrous gases (see also Fig. 7.3).

Materials and Apparatus

Chamber method

- PVC rings (diameter, 30 cm; area, 0.07 m^2; volume, 15 l)
- Rubber stopper
- Stainless-steel tubes (6 mm diameter)
- 10 ml Vacutainers™ (Becton and Dickinson, Heidelberg, Germany)
- Gas chromatograph equipped with a flame ionization detector (e.g. Shimadzu GC 14, Duisburg, Germany)
- ^{63}Ni-electron capture detector

Soil core method

- Plastic tubes of 9 cm length and 4.5 cm diameter
- Rubber stopper
- Stainless-steel tubes (6 mm diameter)
- 10 ml Vacutainers™ (Becton and Dickinson)
- Gas chromatograph equipped with a flame ionization detector (e.g. Shimadzu GC 14)
- ^{63}Ni-electron capture detector

Chemicals and solutions

- Acetone-free acetylene

Procedure

Chamber method (Mogge *et al.*, 1998, 1999)

For each experimental site, PVC rings are driven into the soil to a depth of 10 cm. During the sampling process the rings are sealed by a lid equipped with a rubber stopper. Gas samples should be taken at the beginning and at the end of the sampling period. Beforehand, a linear increase in N_2O concentration during the sampling period should be proven. *In situ* measurements of N_2O should usually be repeated weekly.

For measurement of denitrification N losses, additional soil chambers are supplied with acetone-free acetylene by diffusion, using four perforated stainless-steel tubes (6 mm diameter) for each cover. The use of flow meters, together with a timer, allows the addition of different amounts of acetylene into different soil types, to ensure equal concentrations of acetylene (0.5–1.5%) during a period of 48 h. Acetylene concentrations are determined by a gas chromatograph equipped with a flame ionization detector (Shimadzu GC 14). *In situ* measurements of denitrification N losses are usually repeated every 14 days.

Evacuated 10 ml Vacutainers™ should be used for sampling and storing gas samples from the chamber atmosphere, after a pretreatment described by Heinemeyer and Kaiser (1996). Gas samples can be analysed for N_2O by a ^{63}Ni-electron capture detector equipped with an automatic sample-injection system (Heinemeyer and Kaiser, 1996).

Soil core methods (Rudaz *et al.*, 1999)

Three to 5 months before the start of the measurements, plastic tubes must be put into the soil and placed in 450 ml jars without removing the soil core. The jars are closed with a lid containing a rubber septum to take gas samples from the headspace. The jars should be placed in the soil and covered with grass in order to avoid a change in temperature. Samples from each site are treated with acetylene (12 kPa) to estimate total N flux from denitrification. To determine N_2O fluxes, controls are not treated with acetylene. Each sample should be incubated for 6 h. Every 2 h, gas samples (7 ml) are withdrawn with a syringe and transferred in a pre-evacuated gas sample flask (9.1 ml). The procedure described above is carried out in the field. In the lab, N_2O is analysed with a gas chromatograph fitted with an electron capture detector. Figure 7.3 shows a possible set-up for the measurement of nitrous oxide emissions with and without acetylene-inhibition techniques (from Mogge *et al.*, 1998).

Calculations

From Bauernfeind (1996).

If a calibration gas mixture (with known mixing ratio) is used for calibration, the μl N_2O must be calculated from the injected volume. To set up the calibration curve, plot the injected μg N_2O under standard conditions (20°C, 293 K and 101,300 Pa).

$$\mu g\ N_2O\ \text{injected} = \frac{P \times V_n \times 10^{-9} \times MW \times 10^6}{R \times T}$$

From the calculated μg N_2O of the sample from the calibration curve, the μg N_2O-N/g dm/h can be calculated.

$$\mu g\ N_2O\text{-}N/g\ dm/h = \frac{X \times V \times 0.6363 \times 100}{IV \times t \times SW + \%dm}$$

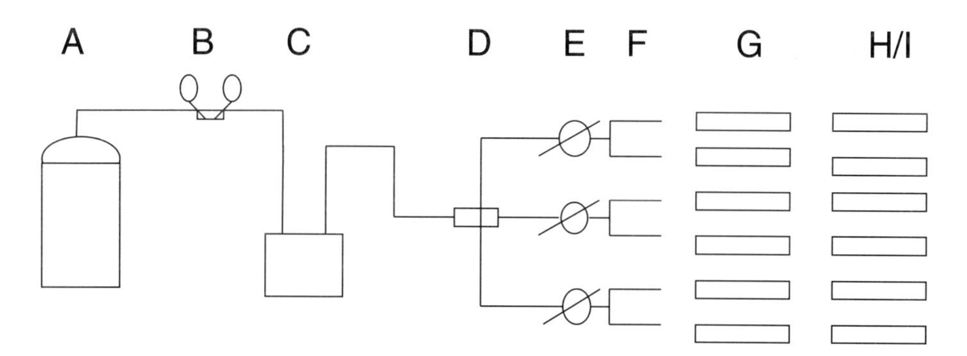

Fig. 7.3. Set-up for the measurement of nitrous oxide emissions with and without the acetylene-inhibition technique. A, gas vessel; B, pressure gauge, magnetic valve, timer; C, acetone trap (sulphuric acid, 95%); D, gas pipe; E, quickfit connector; F, flow meter; G, diffusion tubes; H, soil ring (with acetylene); I, soil ring (without acetylene).

where: P = standard atmospheric pressure (101,300 Pa); V_n = injected volume of N_2O standard (µl); 10^{-9} = conversion factor (1 µl = 10^{-9} m^3); MW = molecular weight of N_2O; 10^6 = conversion factor (1 g = 10^6 µg); R = gas constant (8.31 J/mol.K); T = standard temperature (293 K); X = µg N_2O of the injected sample volume; V = total volume of incubation flask minus soil volume (ml); 0.06363 = factor to convert N_2O to N_2O-N; IV = injected sample volume (ml); t = incubation time (h); SW = initial soil weight; and %dm = soil dry matter in percent.

Discussion

General advantages and disadvantages of the acetylene inhibition method

The principles of the described methods, as well as advantages and disadvantages, have been reviewed by Aulakh *et al.* (1992) and Tiedje *et al.* (1989). Major advantages of the AIM are the applicability in undisturbed and in fertilized ecosystems, the low cost of the sampling and the analytical equipment (especially compared to ^{15}N-based methods) and the high sensitivity (a loss of about 1 g N/ha per day can be measured; Duxbury, 1986). The main disadvantage of the acetylene inhibition method is the inhibition of nitrification by acetylene; thus, denitrification can only proceed from nitrate already present in the soil (McCarty and Bremner, 1986). This might lead to underestimation of real denitrification rates, especially in unfertilized, natural ecosystems (Tiedje *et al.*, 1989; Aulakh *et al.*, 1992). Furthermore, the dispersal of acetylene in the soil and diffusion of N_2O out of the soil can be severely hindered in heavy and/or wet soils, or in compacted zones (Klemedtsson *et al.*, 1990; Granli and Bøckman, 1994). Finally, acetylene can be biodegraded after a long time exposure of the microflora. However, this can be avoided by restricting measurements to less than a week (Terry and Duxbury, 1985) and by choosing new points for measurements.

Soil core versus chamber methods

Advantages of chamber methods are the minimal disturbance of natural conditions in the field (Klemedtsson *et al.*, 1990). With chamber methods, actual *fluxes of N gases* from the soil to the atmosphere are measured, while cores give more direct estimates of *N gas production* by biological processes (Tiedje *et al.*, 1989).

Soil core methods have been considered superior to chamber methods in very wet soils, due to better gas diffusion of acetylene into, and of N_2O out of, the smaller soil volume monitored in the former (Ryden *et al.*, 1987).

A major criticism of the soil core method is that the aeration status of the soil is influenced by the coring procedure and the incubation. This has recently been addressed by Jarvis *et al.* (2001), who designed an incubation box for square-section soil cores with minimum exposure of surfaces to air.

Other methods

The various ^{15}N methods used in denitrification research have been reviewed by Myrold (1990). The principle of the ^{15}N mass balance method is the measurement of the amount of applied ^{15}N-labelled fertilizer in different soil pools and in the plant. The amount of N unaccounted for is assumed to be lost by denitrification (Aulakh *et al.*, 1992). In the ^{15}N gas flux method, highly enriched (20–80%) ^{15}N-labelled fertilizer is applied to soil; the soil is then covered with a chamber, and the flux of N_2 and N_2O following denitrification is measured by quantifying the increase in ^{15}N-labelled gases in the chamber headspace (Stevens and Laughlin, 1998).

It has been stated that ^{15}N methods are preferable to the acetylene inhibition method in heavy-textured soils where acetylene diffusion is hindered (Granli and Bøckman, 1994). However, quantification of denitrification rates using the ^{15}N gas flux method may also suffer from hindered diffusion of N_2O out of the soil, as well as from N_2O being dissolved in soil water (Myrold, 1990). Furthermore, a possible increase in the denitrification rate and change in the N_2O/N_2 ratio might occur after addition of $^{15}N\text{-}NO_3^-$ (Granli and Bøckman, 1994). In addition, dilution of $^{15}NO_3^-$ with soil NO_3^- by various soil processes will cause an underestimate of denitrification if not corrected for. Finally, a uniform distribution of added $^{15}NO_3^-$ is difficult. However, a uniformly labelled soil nitrate pool is the prerequisite for the correct calculation of N_2 flux in the ^{15}N gas flux method (Hauck *et al.*, 1958).

Tiedje *et al.* (1989) reviewed a number of studies comparing ^{15}N and acetylene inhibition methods. They concluded that both methods gave similar estimates of N loss, but high spatial variability of the denitrification process may be partly responsible for the lack of statistical difference between the methods. Recently, Watts and Seitzinger (2000) showed that denitrification rates measured with a ^{15}N gas flux method and acetylene inhibition methods may differ substantially. Malone *et al.* (1998) combined the ^{15}N gas flux method with the acetylene inhibition technique in order to confirm the assumption that acetylene had completely blocked N_2O reductase for their particular soil and condition.

7.7 Enzyme Activity Profiles and Soil Quality

Liz J. Shaw[1] and Richard G. Burns[2]

[1]Imperial College London, Department of Agricultural Sciences, Wye Campus, Wye TN25 5AH, UK; [2]School of Land and Food Sciences, The University of Queensland, Brisbane, Queensland 4072, Australia

Introduction

Enzyme activities in soil are primarily the expression of bacteria, fungi and plant roots, and are responsible for the flux of carbon, nitrogen and other essential elements in biogeochemical cycles. Measuring enzymatic catalysis and understanding the factors that regulate enzyme expression and the rates of substrate turnover are the first stages in characterizing soil metabolic potential, fertility and quality, as well as a guide to the resilience of the soil when subjected to various natural and anthropogenic impacts. Furthermore, substrate catalysis and enzyme responses, when combined with the other soil properties described in this handbook, may provide enough information to allow the rational manipulation of soil processes for commercial and environmental benefit. Three obvious advantages of a better understanding of soil enzyme activities are the enhancement of plant nutrient solubilization, the inhibition of phytopathogens, and the stimulation of pollutant degradation. These beneficial influences are likely to be most strongly expressed in the rhizosphere (Shaw and Burns, 2003).

Soil enzyme assays are far from uniform and usually employ incubation conditions that are dissimilar to those encountered in the natural environment. As a consequence, interpretation of the data is difficult, controversial, and sometimes downright misleading. Furthermore, even when the problems associated with the design of meaningful assays appear to have been considered and resolved, there are large variations in activity in both space and time; soil is heterogeneous. For example, enzyme–substrate interactions in macro-aggregates are different from those in micro-aggregates (Ladd *et al.*, 1996), reaction rates in the rhizosphere are likely to be dissimilar to those in the non-rhizosphere (Reddy *et al.*, 1987; Kandeler *et al.*, 2002), and the surface soil will display a different range and level of activities than the sub-surface soil (Taylor *et al.*, 2002). Perhaps introducing even greater complexity is the fact that most biogeochemical processes are mediated, *in toto*, by many enzymes located in one or more locations within the soil matrix.

Most enzymes in soil are directly associated with viable cells and function within the confines of the microbial cell membrane (i.e. are truly intracellular). Others, the extracellular enzymes, are secreted and catalyse reactions at the outer surfaces of the cell wall and in the surrounding environment. In addition, many strictly intracellular enzymes (especially hydrolases) are released from leaking cells and lysing dead cells, but remain functional for a period of time. This is because these, together with the truly extracellular enzymes, become intimately associated with clays and organic (humic) colloids. Clay– and humic–enzyme complexes form a long-term persistent catalytic component of soils, but one that may be poorly correlated with microbial numbers and the biomass (Burns, 1982; Nannipieri *et al.*, 2002).

Thus, estimation of a suite of soil enzyme activities integrates both the intra- and extracellular biogeochemical activities of the soil biological system and should be a key aspect of soil quality assessment.

Numerous methods have been developed to measure the activities of the hundreds of enzymes that catalyse reactions involved in soil nutrient cycling (for an extensive review, see Tabatabai and Dick, 2002). Here we describe the methods for measuring three undeniably important, yet contrasting, enzymatic processes. The first of these is an exclusively intracellular process and depends on the reduction of tetrazolium salts. The second is the result of both intracellular and extracellular microbial activity during which fluorescein diacetate is hydrolysed. In both assays, the products of the reaction are a result of the activities of many different enzymes. For example, fluorescein diacetate can be hydrolysed by the action of lipases, proteases and esterases (Alef and Nannipieri, 1995), and tetrazolium salts are reduced by a number of enzymes integral to the intact cell and reflect the total oxidative potential of the soil microbial community (Dick, 1997). While useful to describe overall microbial activity in soil (in terms of either electron transport activity or hydrolytic capability), these activities do not yield much information regarding the rate of specific catalytic steps, such as nutrient acquisition and biogeochemical cycling, processes which define soil health (Dick, 1997). Therefore, the third type of assay is more appropriate for estimating the activity of hydrolytic enzymes involved in, for example, specific stages in C, N, P and S acquisition. Table 7.2 shows examples of hydrolases commonly assayed, and the soil health function they mediate. As can be seen, a common approach when assaying the activity of soil hydrolases is to use artificial *p*-nitrophenyl-linked substrates. These esters are hydrolysed to *p*-nitrophenol, which is easily determined by colorimetry. Thus, since the principle of the *p*-nitrophenyl linked substrate hydrolase assay is the same, one enzyme, phosphomonoesterase, is chosen as an example.

Dehydrogenase Activity

Principle of the method

The aerobic microbial oxidation of organic substrates is mediated by the membrane-bound electron transport chain that transfers electrons or

Table 7.2. Examples of commonly assayed soil hydrolases, their function and measurement method (adapted from Alef and Nannipieri (1995); Acosta-Martinez and Tabatabai (2001)).

Enzyme Commission classification	Enzyme	Function	Assay substrate	Assay product
Phosphatases				
3.1.3.1 and 3.1.3.2	Alkaline and acid phosphatase	$RNa_2PO_4 + H_2O \rightarrow R\text{-}OH + Na_2HPO_4$	*p*-Nitrophenyl phosphate[b]	*p*-Nitrophenol
3.1.4.1	Phosphodiesterase	$R_2NaHPO_4 + H_2O \rightarrow R\text{-}OH + RNaHPO_4$	*Bis*-*p*-Nitrophenyl phosphate[b]	*p*-Nitrophenol
Sulphatase				
3.1.6.1	Arylsuphatase	$ROSO_3^- + H_2O \rightarrow R\text{-}OH + RNaHPO_4$	*p*-Nitrophenyl sulphate[b]	*p*-Nitrophenol
Glycosidases				
3.2.1.20	α-Glucosidase	Glucoside-R + H_2O → glucose + R-OH	*p*-Nitrophenyl-α-D-glucopyranoside[b]	*p*-Nitrophenol
3.2.1.21	β-Glucosidase	Glucoside-R + H_2O → glucose + R-OH	*p*-Nitrophenyl-β-D-glucopyraniside[b]	*p*-Nitrophenol
3.2.1.22	α-Galactosidase	Galactoside-R + H_2O → galactose + R-OH	*p*-Nitrophenyl-α-D-galactopyraniside[b]	*p*-Nitrophenol
3.2.1.23	β-Galactosidase	Galactoside-R + H_2O → galactose + R-OH	*p*-Nitrophenyl-β-D-galactopyraniside[b]	*p*-Nitrophenol
Amidohydrolases and arylamidases				
3.5.1.1	L-Aspariginase	L-Asparagine + H_2O → L-Aspartate + NH_3	L-Asparagine[a]	NH_4-N
3.5.1.2	L-Glutaminase	L-Glutamine + H_2O → L-Glutamate + NH_3	L-Glutamine[a]	NH_4-N
3.5.1.4	Amidase	$R\text{-}CONH_2 + H_2O \rightarrow NH_3 + R\text{-}COOH$	Formamide[a]	NH_4-N
3.5.1.5	Urease	Urea + $H_2O \rightarrow CO_2 + 2NH_3$	Urea[a]	NH_4-N

[a]Natural substrate.
[b]Synthetic substrate.

hydrogen from substrates via electron carrier proteins and oxidoreductases to O_2, the final electron acceptor. Thus, dehydrogenases exist as an integral part of intact cells and represent the total oxidative activities of soil microorganisms during the initial stages of organic matter breakdown (Dick, 1997). The concept of estimating microbial activity in soil by quantifying dehydrogenase activity relies on the ability of a tetrazolium salt, iodonitrotetrazolium chloride (INT), to act as an artificial electron acceptor in the place of oxygen. Tetrazolium compounds are characterized by a heterocyclic ring structure which readily accepts hydrogen atoms (and electrons) and becomes reduced. Thus, upon incubation, water-soluble INT becomes biologically reduced to form the purple, water-insoluble, iodonitrotetrazolium formazan (INTF). This can be extracted using an organic solvent and the amount produced determined colorimetrically. The method, detailed below, and recommended as the standard assay, is an INT-based method originally described by Benefield *et al.* (1977), and adapted from Trevors (1984a,b) and vonMersi and Schinner (1991).

Materials and apparatus

- Glass McCartney bottles (28 ml) with rubber-lined screw caps, sterilized by autoclaving
- Semi-micro clear polystyrene spectrophotometer cuvettes (1 ml)
- Microcentrifuge with a rotor to fit 1.5 ml microcentrifuge tubes
- Autoclave
- Spectrophotometer set to 464 nm
- Incubator at 25°C

Chemicals and solutions

- *INT solution*: dissolve INT (Sigma-Aldrich Co. Ltd, Dorset, UK) in distilled water to give a 0.2% (w/v) solution and sterilize by passing through a 0.2 μm filter into a clean sterile container. Note that INT is sparingly soluble in water (solubility limit – 0.3% (w/v), therefore allow for sufficient time (> 2 h) for the INT to become completely dissolved
- *Extractant*: *N,N*-dimethyl formamide:ethanol (1:1, v/v)
- *INTF master standard solution* (500 μg/ml): dissolve 25.0 mg of INTF (Sigma) in ~40 ml of extractant. Decant to a 50 ml volumetric flask and bring up to the mark with extractant
- *INTF working standard solutions*: prepare INTF working solutions by pipetting the following volumes of the INTF master solution to 10 ml volumetric flasks and make up to the mark with a mixture of extractant:distilled water (5:2, v/v)

Working INTF standard concentration (μg/ml)	Volume of master solution (μl) in 10 ml
0	0
1	20
2	40
4	80
6	120
10	200
15	300
25	500

Procedure

- Soil, freshly collected from the field, should be sieved (< 2.8 mm), field moist and stored at 4°C prior to the assay. (NB. This is essential for all enzyme assays.)
- Weigh replicate soil samples (1 g) into sterile McCartney bottles and add 4 ml of INT solution (0.2%).
- Close the lids and incubate at 25°C for 48 h in the dark.
- Sterile controls (to account for any abiotic INT reduction) should consist of autoclaved soil (121°C, 20 min on three consecutive days). Spectrophotometer blanks for both autoclaved and non-autoclaved treatments should consist of soil with the INT solution replaced with distilled water. Controls and blanks should be treated like the samples.
- After the incubation period, extract soil by addition of 10 ml of the extractant and incubate in the dark with agitation for 1 h.
- After the extraction period, transfer ~1.2 ml of the extractant/soil mixture to 1.5 ml microcentrifuge tubes and remove soil by centrifugation (relative centrifugal force (RCF) = 11,600 × g for 5 min).
- Transfer 1 ml of supernatant fraction to spectrophotometer cuvettes and determine absorbance at 464 nm (OD_{464nm}) against the appropriate blank.
- At the same time, construct a calibration curve by determining OD_{464nm} values for the working standard solutions of INTF (0–25 μg/ml).

Calculation

Using the calibration curve, calculate INTF concentrations from the corresponding OD_{464nm} value.

$$\text{Dehydrogenase activity } (\mu\text{g INTF/g dry soil/48 h}) = \frac{([INTF_s] - [INTF_c]) \times 14}{edw}$$

where: $[INTF_s]$ is the INTF concentration (μg/ml) in the sample; $[INTF_c]$ is the INTF concentration in the sterile control; edw is the equivalent dry weight of 1 g of soil (determined by loss of weight of field-moist subsamples after heating at 105°C until constant weight); and 14 is the volume (ml) of solution added in the assay (INT + extractant).

Discussion

Although INT reductase activity is used widely in soil ecotoxicology research (Gong *et al.*, 1999; Welp, 1999; Moreno *et al.*, 2001), no standard method has been adopted, and, consequently, assay parameters, such as pH, incubation temperature and time, substrate concentration and extractant solvent used, vary depending on the soil type and experiment being conducted.

The 48-h incubation period in the method described above is based on the research of Trevors (1984a,b), which allows the accumulation of measurable product and the estimation of dehydrogenase activity at temperatures (i.e. 25°C) more likely to be found in the natural environment. In addition, in the method described above, soil is assayed without use of a buffer and the 'natural' dehydrogenase activity is determined (vonMersi and Schinner, 1991). A pH optimum of 7–7.5 has been determined for INT reduction (vonMersi and Schinner, 1991), with very little activity below a pH of 6.6 or above 9.5 (Trevors, 1984a). Thus, in order to obtain statistically meaningful values for the dehydrogenase actvity of either acidic or alkaline soils, it may be necessary to use a buffered system (vonMersi and Schinner, 1991; Taylor *et al.*, 2002). Other studies (Trevors *et al.*, 1982; Trevors, 1984b) have determined INT reduction in soils incubated with labile carbon sources (glucose and yeast extract) in an attempt to quantify 'potential' as well as 'natural' activity.

The dehydrogenase assay has been a useful parameter in comparable ecotoxicological studies within the same soil type. For example, Welp (1999) used INT reductase activity to examine the toxicity of total and water-soluble concentrations of metals in a loess soil. Gong *et al.* (1999) also used dehydrogenase activity in the toxicity assessment of TNT-contaminated soil. However, Obbard (2001) has shown that INTF undergoes abiotic interaction with copper, leading to decreased absorbance. Thus, in copper-contaminated soils, biotic complexation between INTF and Cu may lead to an underestimate of dehydrogenase activity.

Fluorescein Diacetate Hydrolysis

Principle of the method

Fluorescein diacetate (FDA) is a relatively non-polar compound. As a result of this it is assumed that it diffuses easily through the cell membrane, where it is hydrolysed by non-specific esterases to the fluorescent compound, fluorescein (Rotman and Papermaster, 1966). In addition, FDA can also be hydrolysed by extracellular enzymes produced by the soil microflora, such as proteases, lipases and esterases. Thus, FDA hydrolysis has been suggested as a measure of the global hydrolytic capacity of soils and a broad-spectrum indicator of soil biological activity (Bandick and Dick, 1999; Perucci *et al.*, 1999).

The method described below is based on that described by Adam and Duncan (2001). Soil is incubated with the substrate, FDA, at 25°C for 30 min. The amount of fluorescein formed is determined colorimetrically following extraction with an organic solvent mixture.

Materials and apparatus

- Glass McCartney bottles (28 ml) with rubber-lined screw caps; sterilized by autoclaving
- Semi-micro clear polystyrene spectrophotometer cuvettes (1 ml)
- Centrifuge(s) with rotor(s) to fit 1.5 ml microcentrifuge tubes and glass McCartney bottles
- Spectrophotometer set to 490 nm
- Shaking incubator at 25°C

Chemicals and solutions

- *Potassium phosphate buffer* (60 mM, pH 7.6): dissolve 8.7 g of K_2HPO_4 and 1.3 g of KH_2PO_4 in 1 l of distilled water; sterilize by autoclaving (121°C, 20 min)
- *FDA solution* (1 mg/ml): dissolve 25 mg fluorescein diacetate (3′6′-diacetyl-fluoresein, Sigma-Aldrich) in 25 ml acetone; store at –20°C
- *Extractant* (2:1 chloroform:methanol): add 666 ml of chloroform (AR grade) to a 1-l volumetric flask, make up to the mark with methanol (AR grade) and mix thoroughly
- *Fluorescein master solution* (2000 µg/ml): dissolve 113.2 mg fluorescein disodium salt in 50 ml of potassium phosphate buffer (60 mM, pH 7.6)
- *Fluorescein working standard solutions*: prepare fluorescein working solutions by pipetting the following volumes of the fluorescein master solution into 100 ml volumetric flasks and make up to the mark with potassium phosphate buffer:

Working fluorescein standard concentration (µg/ml)	Volume of master solution (µl) in 100 ml
0	0
1	50
2	100
3	150
4	200
5	250

Procedure

- Soil, freshly collected from the field, should be sieved (< 2.8 mm), field moist and stored at 4°C prior to the assay.
- Weigh 1 g soil sample into sterile McCartney bottles and add 7.5 ml potassium phosphate buffer (pH 7.6, 60 mM) and allow to equilibrate at 25°C on an end-over-end shaker.
- Start the reaction by the addition of 0.1 ml FDA solution (1000 μg/ml) and return the samples to the shaker. Incubate at 25°C for 30 min.
- Spectrophotometer blanks should consist of the soil and buffer mixture with the FDA solution replaced by 0.1 ml acetone. Incubate as the samples.
- After the 30 min have elapsed, immediately add 7.5 ml of chloroform:methanol (2:1) to samples and blanks to stop the reaction. Replace the lids and mix the contents thoroughly (10 s, vortex mixer).
- Centrifuge the tubes at a low speed (RCF = 300 × *g* for 2 min) to clarify the phases.
- Transfer 1.2 ml of the upper phase to a 1.5 ml microcentrifuge tube and centrifuge (RCF = 16,500 × *g* for 5 min) to remove suspended fines.
- Decant 1 ml of supernatant fraction to a 1 ml cuvette and measure absorbance at 490 nm on a spectrophotometer against the soil blank.
- To construct a calibration curve, pipette triplicate 7.5 ml aliquots of each concentration of fluorescein working standards into McCartney bottles, extract with chloroform:methanol and determine the OD_{490nm} of the clarified upper phase as for the samples.

Calculation

Using the calibration curve, calculate the mass of fluorescein produced in each assay from the corresponding OD_{490nm} value and divide by the equivalent dry weight of soil (determined by loss of weight of field-moist subsamples after heating at 105°C until constant weight). Fluorescein diacetate hydrolysis activity is expressed as μg fluorescein/g dry soil/0.5 h.

Discussion

The method for FDA hydrolysis described above does not use abiotic controls. We tested the FDA hydrolysis assay with four soil types and found no abiotic hydrolysis of FDA using autoclaved soil as the control (Shaw and Burns, unpublished results). Based on this evidence, we suggest that, once established, it may not be necessary to include an abiotic control, despite a single report that FDA can be hydrolysed spontaneously without microbial activity (Guilbault and Kramer, 1964).

Buffer addition can directly or indirectly alter the nutrient conditions of the microbial community and therefore its hydrolytic activity (Battin, 1997).

Consequently, to better estimate the 'natural' FDA hydrolysis capacity of a soil, a non-buffered system should be used. However, sensitivity of the assay may be problematic for acidic soils, due to the decreased fluorescence of fluorescein at low pH. Nevertheless, it may be possible for post-assay adjustment of the pH of the aqueous phase prior to measurement to improve the sensitivity of fluorescein detection. It should also be mentioned that spontaneous hydrolysis of FDA is reported at pH values greater than 8.5 (Guilbault and Kramer, 1964), and this should be considered when making decisions regarding FDA hydrolysis measurements in alkaline soils.

Dumontet *et al.* (1997) suggested that FDA hydrolysis may be a suitable tool to measure early detrimental effects of pesticides on soil microbial biomass, as it is both sensitive and non-specific. Indeed, the FDA assay has been used in ecotoxicological studies to investigate the influence of pollutants (Vischetti *et al.*, 1997; Perucci *et al.*, 1999) or field management effects (Bandick and Dick, 1999; Haynes and Tregurtha, 1999) on soil microorganisms. However, variable responses have been recorded for pollutant effects on FDA hydrolysis. For example, Vischetti *et al.* (1997) reported increases in FDA hydrolysis rates in silty clay loam soil treated with rimsulfuron (a sulphonylurea herbicide), whereas Perucci *et al.* (2000) described a decline in FDA hydrolysis activity after application of the same herbicide to a soil-swelling clay. Since Vekemans *et al.* (1989) have reported that FDA hydrolysis and microbial biomass content are closely correlated, negative pollutant effects on FDA hydrolysis activity can generally be explained in terms of reduced biomass due to the toxicity of the pollutant. Positive effects of organic pollutants on FDA activity have been interpreted as being either due to the utilization of the molecule as a source of growth by the microbial biomass, or due to the toxic effect of the pollutant, causing cell lysis and release of intracellular hydrolytic enzymes (Perucci *et al.*, 1999). Thus, the challenge is to develop methodology to distinguish between intracellular and extracellular activity. Perucci *et al.* (2000) introduced the concept of specific hydrolytic activity (qFDA), where the per cent FDA hydrolysed is expressed per unit of microbial biomass carbon. This approach may help distinguish between increases in FDA activity which resulted from the growth of microorganisms on the pollutant or from the lysis of cells, releasing endocellular hydrolases.

Phosphomonoesterase Activity

Principle of the method

Soil humic material contains significant amounts of organic phosphates, in which phosphorus is bound to carbon via ester linkages. Phosphatases (which may be of plant, microbial or animal origin) catalyse the hydrolysis of phosphate esters to inorganic phosphorus. Phosphatases are classified according to the type of substrate upon which they act. For example, phosphomonoesterases (e.g. phytase, nucleotidases, sugar phosphatases) catalyse the hydrolysis of organic phosphomonoesters, whereas phosphodiesterases (e.g. nucleases, phospholipases) catalyse the hydrolysis of organic phosphodiesters (Alef and Nannipieri, 1995) (see Table 7.2). However, it is likely that phosphatases have broad substrate specificity, such that phosphomono- and phosphodiesterases share common substrates (Pant and Warman, 2000). It has been estimated that phosphatase-labile organic P makes up a significant component of the total soil P pool and, therefore, is potentially an important source of P for plants (Hayes *et al.*, 2000). Thus, phosphatase activity is of paramount importance as a soil quality indicator.

The methodology described below involves the use of an artificial substrate, *p*-nitrophenyl phosphate (*p*-NPP). The product of phosphomonoesterase activity, *p*-nitrophenol, is a yellow chromophore under alkaline conditions and can be detected colorimetrically. The much used method of Tabatabai and Bremner (1969) is described.

Materials and apparatus

- Erlenmeyer flasks (50 ml) and stoppers
- Incubator at 37°C
- Spectrophotometer set to 400 nm
- Semi-micro clear polystyrene spectrophotometer cuvettes (1 ml)
- Filter papers (Whatman No. 12) or centrifuge
- Volumetric flasks (1000 ml and 100 ml)

Chemicals and solutions

- Toluene
- *Modified universal buffer (MUB) stock solution*: dissolve 12.1 g Tris, 11.6 g maleic acid, 14 g citric acid and 6.3 g boric acid in 500 ml 1 M NaOH; dilute the solution to 1000 ml with distilled water; store at 4°C
- *MUB* (pH 6.5): adjust 200 ml of MUB stock solution to pH 6.5 using 0.1 M HCl. Make the volume up to 1000 ml with distilled water
- p-*NPP solution* (115 mM): dissolve 2.13 g of disodium *p*-nitrophenyl phosphate hexahydrate (Sigma-Aldrich) in 50 ml MUB; store at 4°C
- Sodium hydroxide solution (0.5 M)
- $CaCl_2$ solution (0.5 M)
- p-*Nitrophenol master solution* (1000 μg/ml): dissolve 1 g *p*-nitrophenol (spectrophotometric grade; Sigma-Aldrich) in 1000 ml of distilled water; store at 4°C
- p-*Nitrophenol working solution* (10 μg/ml): dilute 1 ml of *p*-nitrophenol master solution to 100 ml with distilled water

Procedure

- Soil, freshly collected from the field, should be sieved (< 2.8 mm), field moist and stored at 4°C prior to the assay.
- Place 1 g soil in a 50 ml Erlenmeyer flask and add 4 ml of MUB, 0.25 ml of toluene and 1 ml of *p*-NPP solution.
- Swirl contents to mix, stopper and incubate at 37°C for 1 h.
- At the end of the incubation, add 1 ml 0.5 M $CaCl_2$ and 4 ml 0.5 M NaOH and swirl to mix.
- Remove soil by filtration (Whatman No. 12) or centrifugation.
- Transfer the filtrate/supernatant fraction to a spectrophotometer cuvette and determine the absorbance at 400 nm.
- For the controls, add 1 ml of *p*-NPP solution after the addition of $CaCl_2$ (1 ml) and NaOH (4 ml) but immediately before filtration/centrifugation.
- For the calibration, pipette 0, 1, 2, 3, 4 and 5 ml aliquots of the working *p*-nitrophenol standard solution into 50 ml Erlenmeyer flasks. Adjust the volume to 5 ml by addition of distilled water and proceed as described for determining *p*-nitrophenol after the incubation of the soil samples. Plot the mass of *p*-nitrophenol in each reaction (0–50 μg) against the OD_{400nm} reading.

Calculation

Using the calibration curve, calculate the mass of *p*-nitrophenol produced in each assay from the corresponding OD_{400nm} value, correct using blank readings, and divide by the equivalent dry weight of soil (determined by loss of weight of field-moist subsamples after heating at 105°C until constant weight).

Phosphomonoesterase activity is expressed as μg *p*-nitrophenol/g dry soil/h.

Discussion

Phosphomonoesterase is a generic name for a group of enzymes which catalyse the hydrolysis of esters of phosphoric acid. However, both acid (pH optimum of 4–6.5) and alkaline (pH optimum of 9–10) phosphatases have been found in soil (Speir and Ross, 1978). It is suggested that the rates of synthesis of acid and alkaline phosphatases are dependent on the soil pH and that acid phosphatase is predominant in acid soils and alkaline phosphatase is predominant in alkaline soils (Juma and Tabatabai, 1978). The assay described above is buffered to pH 6.5, since this was the maximum activity recorded in the two soils tested by Tabatabai and Bremner (1969). However, the MUB may be adjusted to different pH values by titration with either 0.1 M NaOH or 0.1 HCl (Alef and Nannipieri, 1995). In addition, 'natural' activity may be determined using a non-buffered system, substituting distilled water for the MUB (Tabatabai and Bremner, 1969).

Apart from varying the pH of the assay and other slight modifications, for example using less soil in determinations of phosphatase activity in rhizosphere soil (Tarafdar and Jungk, 1987) or its application to measuring the activity associated with roots (Asmar and Gissel-Nielsen, 1997), the *p*-NPP method has become standard for most studies. However, in soils with high organic matter content, the method is hampered by large amounts of interfering organic materials (Trasar-Cepeda and Gilsotres, 1987; Freeman *et al.*, 1995). Schneider *et al.* (2000) were able to get over this problem for high organic matter forest soils by increasing the $CaCl_2$ concentration (to 2 M) and decreasing the NaOH concentration (to 0.2 M). In peat-accumulating wetland soil samples, however, Freeman and co-workers employed the use of methylumbelliferyl (MUF)-linked phosphate coupled to high-performance liquid chromatography to separate the interferences from the compound of interest (Freeman, 1997). This method has the potential to quantify both the MUF-substrate and free-MUF product (Freeman and Nevison, 1999).

Acid and alkaline phosophomonoesterases have been used successfully, in combination with other enzyme activities, to detect heavy metal contamination in Mediterranean soil (Belen Hinojosa *et al.*, 2004).

Interpretation of Assay Data and Conclusions

Definitions of a high-quality soil relate mainly to its ability to produce healthy and abundant crops, but also include the soil's capacity to function as a mature and sustainable ecosystem, capable of degrading organic inputs (Trasar-Cepeda *et al.*, 2000). Although soil physical and chemical properties are important for soil function, it has been suggested that soil biochemical properties are the most useful indicators of soil quality, as they intimately reflect soil nutrient cycles and the ability of soil to break down organic matter (including organic xenobiotics) (Trasar-Cepeda *et al.*, 2000).

However, interpretation of enzyme assay data can be problematic, in particular, dissecting what underlying mechanism(s) is responsible for a measured change in the activity of a particular enzyme. Potential mechanisms relate to microbial growth/death, enzyme de-repression/repression, and enzyme inhibition/activation. For example, an increase in the measured activity of a particular soil enzyme in response to disturbance may be interpreted in terms of five different non-mutually exclusive processes: (i) enzyme induction; (ii) enzyme activation; (iii) a shift in community structure to a species which produces greater quantities of the enzyme; (iv) microbial growth; and (v) cell lysis and release of intracellular enzymes.

Consequently, in order to aid interpretation of enzyme measurements, concurrent measurement of other parameters, such as those describing the relative contribution of intracellular and extracellular enzymes (reviewed in detail by Nannipieri *et al.*, 2002), microbial biomass (Vischetti *et al.*, 1997;

Perucci *et al.*, 1999) and microbial diversity (Waldrop *et al.*, 2000), are extremely valuable.

The existence of methodological artefacts should also be taken into account when interpreting assay data. For example, physical disturbance associated with the collection of the soil sample and sample pretreatment (e.g. air-drying and sieving) will disrupt soil structure at the aggregate and sub-aggregate scale. Consequently, the accessibility of substrates, previously physically inaccessible due to their location within aggregates and small pores, will be increased, thereby altering the environmental parameters that determine the total amount or activity present (Tate, 2002). Thus, before the soil sample is actually used in an enzyme assay, it should be recognized that, compared to the *in situ* activity expressed in the field, sample collection and storage will undoubtedly have resulted in changed activity profiles (Shaw and Burns, 1996).

Artefacts arising from the actual assay methodology itself are numerous and have been reviewed by Speir and Ross (2002). Briefly, the following points should be recognized when interpreting assay data:

- Most enzymes are assayed by using artificial substrates that either may not serve as a substrate for all enzymes catalysing the reaction in nature, or may be more easily transformed than the natural substrate.
- Assays are usually conducted under artificial conditions; the system is buffered to the optimal pH of the enzyme, at enzyme-saturating substrate concentrations, and at temperatures higher than normally encountered in soil.
- Most assay protocols involve the use of water-saturated systems, and often agitation of the samples to negate diffusional constraints. Thus, the interaction between the enzyme and substrate is maximized.

In this section, we have described simple colorimetric methods for the assay of three soil enzyme activities: dehydrogenase, fluorescein diacetate hydrolase and phosphomonoesterase. We highlight studies where the methodology has been employed in soil quality and ecotoxicological assessment. When discussing problems associated with assay interpretation, it should be remembered that there is a methodological revolution under way in which microscopy (e.g. atomic force and confocal microscopy), genomic and post-genomic (stable isotope probing, reporter genes, soil proteomics) techniques are being combined with the more established biochemical methods to solve the long-established problems and anomalies associated with soil enzymology.

Normative References (ISO, DIN) for Chapter 7

Basal respiration by titration

ISO 10381-6 (1993) Soil quality – Sampling – Guidance on the collection, handling and storage of soil for the assessment of aerobic microbial processes in the laboratory.

ISO 11465 (1993) Soil quality – Determination of dry matter and water content on mass basis – Gravimetric method.
ISO 11274 (1998) Soil quality – Determination of the water retention characteristics.
ISO 16072 (2001) Soil quality – Laboratory methods for determination of microbial soil respiration.
ISO 17155 (2001) Soil quality – Determination of abundance and activity of soil microflora using respiration curves.

Basal respiration, substrate-responsive microbial biomass, and its active and dormant part

For handling the soils see above.
ISO 17155 (2001) Soil quality – Determination of abundance and activity of soil microflora using respiration curves.
ISO 16072 (2001) Soil quality – Laboratory methods for determination of microbial soil respiration.

Nitrification

The ISO and DIN norm describes a method very similar to the measurement of net nitrification. As already stated, underestimations of the nitrification rates are possible when using those methods.

Dehydrogenase activity

ISO/DIN 23753–2 (draft standard). Soil quality – Determination of dehydrogenase activity in soil – Part 2 method using iodotetrazolium chloride (INT).

Fluorescein diacetate hydrolysis

None available.

Phosphomonoesterase activity

None available.

References for Chapter 7

Acosta-Martinez, V. and Tabatabai, M.A. (2001) Arylamidase activity in soils: effect of trace elements and relationships to soil properties and activities of amidohydrolases. *Soil Biology and Biochemistry* 33, 17–23.

Adam, G. and Duncan, H. (2001) Development of a sensitive and rapid method for the measurement of total microbial activity using fluorescein diacetate (FDA) in a range of soils. *Soil Biology and Biochemistry* 33, 943–951.

Alef, K. (1995) Soil respiration. In: Alef, K. and Nannipieri, P. (eds) *Methods in Applied Soil Microbiology and Biochemistry*. Academic Press, London, pp. 214–218.

Alef, K. and Nannipieri, P. (eds) (1995) *Methods in Applied Soil Microbiology and Biochemistry*. Academic Press, London.

Alvarez, C.R. and Alvarez, R. (2000) Short-term effects of tillage systems on active microbial biomass. *Biology and Fertility of Soils* 31, 157–161.

Alves, B.L.R., Urquiaga, S., Cadisch, G., Souto, C.M. and Boddy, R.M. (1993) *In situ* estimation of soil nitrogen mineralization. In: Mulongoy, K. and Merckx, R. (eds) *Soil Organic Matter Dynamics and Sustainability of Tropical Agriculture*. Wiley-Sayce, Leuven, Belgium, pp. 173–180.

Anderson, J.P.E. (1982) Soil respiration. In: Page, A.L. *et al.* (eds) *Methods of Soil Analysis, Part 2. Chemical and Microbiological Properties*, 2nd edn. Agron. Monogr. 9 ASA and SSSA, Madison, Wisconsin, pp. 831–871.

Anderson, J.P.E. and Domsch, K.H (1978) A physiological method for the quantitative measurements of microbial biomass in soils. *Soil Biology and Biochemistry* 10, 215–221.

Anderson, T.H. and Domsch, K.H. (1993) The metabolic quotient for CO_2 (qCO_2) as a specific activity parameter to assess the effects of environmental conditions, such as pH, on the microbial biomass of forest soils. *Soil Biology and Biochemistry* 25, 393–395.

Asmar, F. and Gissel-Nielsen, G. (1997) Extracellular phosphomono- and phosphodiesterase associated with and released by the roots of barley genotypes: a non-destructive method for the measurement of the extracellullar enzymes of roots. *Biology and Fertility of Soils* 25, 117–122.

Aulakh, M.S., Doran, J.W. and Mosier, A.R. (1991) Field evaluation of four methods for measuring denitrification. *Soil Science Society American Journal* 55, 1332–1338.

Aulakh, M.S., Doran, J.W. and Mosier, A.R. (1992) Soil denitrification – significance, measurement and effects of management. *Advances in Soil Sciences* 18, 1–57.

Bååth, E. (1990) Thymidine incorporation into soil bacteria. *Soil Biology and Biochemistry* 22, 803–810.

Bååth, E. (1992) Measurement of heavy metal tolerance of soil bacteria using thymidine incorporation into bacteria after homogenization-centrifugation. *Soil Biology and Biochemistry* 24, 1167–1172.

Bååth, E. (1995) Incorporation of thymidine into DNA of soil bacteria. In: Akkermans, A.D.L., van Elsas, J.D. and DeBruijn, F.J. (eds) *Molecular Microbial Ecology Manual*. Kluwer Academic Publishers, Dordrecht, The Netherlands, pp. 2.8.2, 1–9.

Bååth, E. (2001) Estimation of fungal growth rates in soil using ^{14}C-acetate incorporation into ergosterol. *Soil Biology and Biochemistry* 33, 2011–2018.

Bååth, E. and Johansson, T. (1990) Measurement of bacterial growth rates on the rhizoplane using ^{3}H-thymidine incorporation into DNA. *Plant and Soil* 126, 133–139.

Balderston, W.L., Sherr, B. and Payne, W.J. (1976) Blockage by acetylene of nitrous oxide reduction in *Pseudomonas perfectomarinus*. *Applied and Environmental Microbiology* 31, 504–508.

Bandick, A.K. and Dick, R.P. (1999) Field management effects on soil enzyme activities. *Soil Biology and Biochemistry* 31, 1471–1479.

Bardgett, R.D. and McAlister, E. (1999) The measurement of soil fungal:bacterial biomass ratios as an indicator of ecosystem self-regulation in temperate meadow grasslands. *Biology and Fertility of Soils* 29, 282–290.

Barraclough, D., Geens, E.L., Davies, G.P. and Maggs, J.M. (1985) Fate of fertiliser nitrogen. III. The use of single and double labelled ^{15}N ammonium nitrate to study nitrogen uptake by ryegrass. *Journal of Soil Science* 36, 593–603.

Battin, T.J. (1997) Assessment of fluorescein diacetate hydrolysis as a measure of total esterase activity in natural stream sediment biofilms. *The Science of the Total Environment* 198, 51–60.

Bauernfeind, G. (1996) Actual and potential denitrification rates by acetylene-inhibition technique. In: Schinner, F., Öhlinger, R., Kandeler, E. and Margesin, R. (eds) *Methods in Soil Biology*. Springer Verlag, Berlin, pp. 151–155.

Beck, T.H. (1979) Die Nitrifikation in Böden. *Zeitschrift für Pflanzenernährung und Bodenkunde* 142, 344–364.

Belen Hinojosa, M., Carreira, J.A., Garcia-Ruiz, R. and Dick, R.P. (2004) Soil moisture pre-treatment effects on enzyme activities as indicators of heavy metal-contaminated and reclaimed soils. *Soil Biology and Biochemistry* 36, 1559–1568.

Belser, L.W. (1979) Population ecology of nitrifying bacteria. *Annual Review of Microbiology* 33, 309–333.

Belser, L.W. and Mays, E.L. (1980) Specific inhibition of nitrite oxidation by chlorate and its use in assessing nitrification in soils and sediments. *Applied and Environmental Microbiology* 39, 505–510.

Benedetti, A. (2001) Defining soil quality: introduction to Round Table. In: Benedetti, A. *et al.* (eds) *COST Action 831 – Joint Meeting of Working Groups Biotechnology of Soil Monitoring, Conservation and Remediation.* Round Table: Defining Soil Quality, Rome, 10–11 December 1998. EUR 19225 – COST Action 831 – Report of 1997–1998. European Communities, Luxembourg, pp. 266–269.

Benedetti, A. and Sebastiani, G. (1996) Determination of potentially mineralizable nitrogen in agricultural soil. *Biology and Fertility of Soils* 21, 114–120.

Benedetti, A., Canali, S. and Alianiello, F. (1994) Mineralisation dynamics of the organic nitrogen: soil management effects. *Proceedings of the N-immobilisation workshop.* November, Macaulay Land Use Research Institute, Aberdeen, UK.

Benedetti, A., Vittori Antisari, L., Canali, S., Giacchini, P. and Sequi, P. (1996) Relationship between the fixed ammonium and the mineralization of the organic nitrogen in soil. In: Van Cleemput, O. *et al.* (eds) *Progress in Nitrogen Cycling Studies.* Kluwer Academic Publishers, Dordrecht, The Netherlands, pp. 23–26.

Benefield, C.B., Howard, P.J.A. and Howard, D.M. (1977) The estimation of dehydrogenase activity in soil. *Soil Biology and Biochemistry* 9, 67–70.

Berg, P. and Rosswall, T. (1985) Ammonium oxidizer numbers, potential and actual oxidation rates in two Swedish arable soils. *Biology and Fertility of Soils* 1, 131–140.

Bloem, J. and Breure, A.M. (2003) Microbial indicators. In: Markert, B., Breure, A.M. and Zechmeister, H. (eds) *Bioindicators/Biomonitors – Principles, Assessment, Concepts.* Elsevier, Amsterdam, pp. 259–282.

Bloem, J., Ellenbroek, F., Bär-Gilissen, M.J.B. and Cappenberg, T.E. (1989) Protozoan grazing and bacterial production in stratified Lake Vechten, estimated with fluorescently labelled bacteria and thymidine incorporation. *Applied and Environmental Microbiology* 55, 1787–1795.

Bloem, J., Bolhuis, P.R., Veninga, M.R. and Wieringa, J. (1995) Microscopic methods for counting bacteria and fungi in soil. In: Alef, K. and Nannipieri, P. (eds) *Methods in Applied Soil Microbiology and Biochemistry*. Academic Press, London, pp. 162–173.

Bloem, J., de Ruiter, P.C. and Bouwman, L.A. (1997) Food webs and nutrient cycling in agro-ecosystems. In: van Elsas, J.D., Trevors, J.T. and Wellington, E. (eds) *Modern Soil Microbiology*. Marcel Dekker, New York, pp. 245–278.

Bodelier, P.L.E., Libochant, J.A., Blom, C.W.P.M. and Laanbroek, H.J. (1996) Dynamics of nitrification and denitrification in root-oxygenated sediments and adaptation of ammonia-oxidizing bacteria to low-oxygen or anoxic habitats. *Applied and Environmental Microbiology* 62, 4100–4107.

Bodelier, P.L.E., Hahn, A.P., Arth, I.R. and Frenzel, P. (2000) Effects of ammonium-based fertilisation on microbial processes involved in methane. *Biogeochemistry* 51, 225–257.

Bouwman, L.A., Bloem, J., van den Boogert, P.H.J.F., Bremer, F., Hoenderboom, G.H.J. and de Ruiter, P.C. (1994) Short-term and long-term effects of bacterivorous nematodes and nematophagous fungi on carbon and nitrogen mineralization in microcosms. *Biology and Fertility of Soils* 17, 249–256.

Bremner, J.M. (1965) Nitrogen availability index. In: Black, C.A., Evans, D.D., White, J.L., Ensminger, L.E. and Clark, F.E. (eds) *Methods of Soil Analysis Part 2*. Agronomy 9, American Society of Agronomy, Madison, Wisconsin, pp. 1324–1341.

Bringmark, E. and Bringmark, L. (1993) Standard respiration, a method to test the influence of pollution and environmental factors on a large number of samples. MATS Guideline – Test 03. In: Torstensson, L. (ed.) *Guidelines. Soil Biological Variables in Environmental Hazard Assessment*. Report No. 4262, Swedish EPA, Stockholm, pp. 34–39.

Bringmark, L. and Bringmark, E. (2001a) Soil respiration in relation to small-scale patterns of lead and mercury in mor layers of Southern Swedish forest sites. *Water, Air and Soil Pollution: Focus* 1, 395–408.

Bringmark, L. and Bringmark, E. (2001b) Lowest effect levels of lead and mercury on decomposition of mor layer samples in a long-term experiment. *Water, Air and Soil Pollution: Focus* 1, 425–437.

Brock, T.D. and Madigan, M.T. (1991) *Biology of Microorganisms*. Prentice-Hall, Englewood Cliffs, New Jersey.

Brookes, P.C. (1995) The use of microbial parameters in monitoring soil pollution by heavy metals. *Biology and Fertility of Soils* 19, 269–279.

Brookes, P.C. and McGrath, S.P. (1984) Effects of metal toxicity on size of the soil microbial biomass. *Journal of Soil Science* 35, 341–346.

Bundy, L.G. and Meisinger, J.J. (1994) Nitrogen availability indices. In: Weaver, R.W., Angle, J.S. and Bottomley, P.D. (eds) *Methods of Soil Analysis. Part 2. Microbiological and Biochemical Properties.* Soil Science Society of America, Madison, Wisconsin.

Burns, R.G. (1978) *Soil Enzymes*. Academic Press, London.

Burns, R.G. (1982) Enzyme activity in soil: location and a possible role in microbial ecology. *Soil Biology and Biochemistry* 14, 423–427.

Burton, D.L. and Beauchamp, E.G. (1984) Field techniques using the acetylene blockage of nitrous oxide reduction to measure denitrification. *Canadian Journal of Soil Science* 64, 555–562.

Buyanovsky, G.A., Kucera, C.L. and Wagner, G.H. (1987) Comparative analyses of carbon dynamics in native and cultivated ecosystems. *Ecology* 68, 2023–2031.

Chander, K. and Brookes, P.C. (1991) Effects of heavy metals from past application of sewage sludge on microbial biomass and organic matter accumulation in a sandy loam and silt loam U.K. soil. *Soil Biology and Biochemistry* 23, 927–932.

Christensen, H. (1991) Conversion factors relating thymidine uptake to growth rate with rhizosphere bacteria. In: Kleister, D.L. and Gregan, P.B. (eds) *The Rhizosphere and Plant Growth*. Kluwer Academic, Dordrecht, The Netherlands, pp. 99–102.

Christensen, H. and Christensen, S. (1995) [^{3}H]Thymidine incorporation technique to determine soil bacterial growth rate. In: Alef, K. and Nannipieri, P. (eds) *Methods in Applied Soil Microbiology and Biochemistry*. Academic Press, London, pp. 258–261.

Dalenberg, J.W and Jager, G. (1981) Priming effect of small glucose additions to ^{14}C-labelled soil. *Soil Biology and Biochemistry* 13, 219–223.

Dalenberg, J.W. and Jager, G. (1989) Priming effect of some organic additions to ^{14}C-labelled soil. *Soil Biology and Biochemistry* 21, 443–448.

Davidson, E.A., Stark, J.M. and Firestone, M.K. (1990) Microbial production and consumption of nitrate in an annual grassland. *Ecology* 71, 1968–1975.

Davidson, E.A., Hart, S.C., Shanks, C.A. and Firestone, M.K. (1991) Measuring gross nitrogen mineralisation, immobilisation and nitrification by ^{15}N isotopic pool dilution in intact soil cores. *Journal of Soil Science* 42, 335–349.

Dick, R.P. (1997) Soil enzyme activities as integrative indicators of soil health. In: Pankhurst, C., Doube, B. and Gupta, V. (eds) *Biological Indicators of Soil Health.* CAB International, Wallingford, UK, pp. 121–156.

Dilly, O. (2001) Microbial respiratory quotient during metabolism and after glucose amendment in soils and litter. *Soil Biology and Biochemistry* 33, 117–127.

Dilly, O. and Munch, J.C. (1996) Microbial biomass content, basal respiration and enzyme activities during the course of decomposition of leaf litter in a black alder (*Alnus glutinosa* (L.) Gaertn.) forest. *Soil Biology and Biochemistry* 28, 1073–1081.

Dilly, O., Bach, H.J., Buscot, F., Eschenbach, C., Kutsch, W.L., Middelhoff, U., Pritsch, K. and Munch, J.C. (2000) Characteristics and energetic strategies of the rhizosphere in ecosystems of the Bornhöved Lake district. *Applied Soil Ecology* 15, 201–210.

Dixon, J.L. and Turley, C.M. (2000) The effect of water depth on bacterial numbers, thymidine incorporation rates and C:N ratios in northeast Atlantic surficial sediments. *Hydrobiologia* 440, 217–225.

Doran, J.W. and Parkin, T.B. (1994) Defining and assessing soil quality. In: *Defining Soil Quality for a Sustainable Environment.* Special publication Number 35, Soil Science Society of America, Madison, Wisconsin, pp. 3–21.

Dumontet, S., Perucci, P. and Bufo, S.A. (1997) Les amendments organiques et traitements avec les herbicides: influence sur la biomasse microbienne du sol. In: *Congres du Groupment Francais des Pesticides, 21–22 Mai.* BRGM-Orleans, France.

Duxbury, J.M. (1986) Advantages of the acetylene method of measuring denitrification. In: Hauck, R.D. and Weaver, R.W. (eds) *Field Measurement of Dinitrogen Fixation and Denitrification.* Soil Science Society of America, Madison, Wisconsin, pp. 73–92.

Ellenbroek, F. and Cappenberg, T.E. (1991) DNA synthesis and tritiated thymidine incorporation by heterotrophic freshwater bacteria in continuous culture. *Applied and Environmental Microbiology* 57, 1675–1682.

Falchini, L., Muggianu, M., Landi, L. and Nannipieri, P. (2001) Influence of soluble organic C compounds present in root exudates on soil N dynamics. In: Benedetti, A. *et al.* (eds) *COST Action 831 – Joint Meeting of Working Groups Biotechnology of Soil Monitoring and Remediation.* Round Table: Defining Soil Quality. Rome, 10–11 December 1998. EUR 19225 - COST Action 831 - Report of 1997–1998. European Communities, Luxembourg, pp. 288–289.

Findlay, S., Meyer, J.L. and Edwards, R.T. (1984) Measuring bacterial production via rate of incorporation of [^{3}H]thymidine into DNA. *Journal of Microbiological Methods* 2, 57–72.

Focht, D.D. and Verstraete, W. (1977) Biochemical ecology of nitrification and denitrification. *Advances in Microbial Ecology* 1, 135–214.

Franzluebbers, A.J., Zuberer, D.A. and Hons, F.M. (1995) Comparison of microbiological methods for evaluating quality and fertility of soil. *Biology and Fertility of Soils* 19, 135–140.

Freeman, C. (1997) Using HPLC to eliminate quench interference in fluorogenic substrate assays of microbial enzyme activity. *Soil Biology and Biochemistry* 29, 203–205.

Freeman, C. and Nevison, G.B. (1999) Simultaneous analysis of multiple enzymes in environmental samples using methylumbelliferyl substrates and HPLC. *Journal of Environmental Quality* 28, 1378–1380.

Freeman, C., Liska, G., Ostle, N., Jones, S.E. and Lock, M.A. (1995) The use of fluorogenic substrates for measuring enzyme activity in peatlands. *Plant and Soil* 175, 147–152.

Fuhrman, J.A. and Azam, F. (1982) Thymidine incorporation as a measure of heterotrophic bacterioplankton production in marine surface waters: evaluation and field results. *Marine Biology* 66, 109–120.

Giller, K.E., Witter, E. and McGrath, S.P. (1998) Toxicity of heavy metals to microorganisms and microbial processes in agricultural soils: a review. *Soil Biology and Biochemistry* 30, 1389–1414.

Gong, P., Siciliano, S.D., Greer, C.W., Paquet, L., Hawari, J. and Sunahara, G.I. (1999) Effects and bioavailability of 2,4,6-trinitrotoluene in spiked and field-contaminated soils to indigenous microorganisms. *Environmental Toxicology and Chemistry* 18, 2681–2688.

Goyal, S., Chander, K., Mundra, M.C. and Kapoor, K.K. (1999) Influence of inorganic fertilizers and organic amendments on soil organic matter and soil microbial properties under tropical conditions. *Biology and Fertility of Soils* 29, 196–200.

Granhall, U. (1999) Microbial biomass determinations. Comparisons between some direct and indirect methods. *COST 831 – Working Group 4 Meeting*, Granada, Spain, 15–16 July, 1999.

Granli, T. and Bøckman, O.C. (1994) Nitrous oxide from agriculture. *Norwegian Journal of Agricultural Sciences Supplement* 12, 1–128.

Grisi, B., Grace, C., Brookes, P.C., Benedetti, A. and Dell'Abate, M.T. (1998) Temperature effects on organic matter and microbial dynamics in temperate and tropical soils. *Soil Biology and Biochemistry* 30, 1309–1315.

Guilbault, G.C. and Kramer, D.N. (1964) Fluorometric detection of lipase, acylase, alpha- and gamma-chymotrypsin and inhibitors of these enzymes. *Analytical Chemistry* 36, 409–412.

Hadas, A., Kautsky, L. and Portnoy, R. (1998) Mineralization of composted manure and microbial dynamics in soil as affected by long-term nitrogen management. *Soil Biology and Biochemistry* 28, 733–738.

Hansson, A.J., Andrén, O., Boström, S., Boström, U., Clarholm, M., Lagerlöf, J., Lindberg, T., Paustian, K., Pettersson, R. and Sohlenius, B. (1990) Structure of the agroecosystem. In: Andrén, O., Lindberg, T., Paustian, K. and Rosswall, T. (eds) *Ecology of Arable Land. Organisms, Carbon and Nitrogen Cycling. (Ecological Bulletin)* 40, 41–83.

Harris, D. and Paul, E.A. (1994) Measurement of bacterial growth rates in soil. *Applied Soil Ecology* 1, 277–290.

Hart, S.C. and Binkley, D. (1985) Correlations among indices for forest nutrient availability in fertilized and unfertilized loblolly pine plantations. *Plant and Soil* 85, 11–21.

Hart, S.C., Stark, J.M., Davidson, E.A. and Firestone, M.K. (1994) Nitrogen mineralization, immobilisation and nitrification. In: Weaver, R.W., Angle, J.S. and Bottomley, P.D. (eds) *Methods of Soil Analysis. Part 2, Microbiological and Biochemical Properties.* Soil Science Society of America, Madison, Wisconsin, pp. 985–1018.

Hatch, D.J., Jarvis, S.C. and Reynolds, S.E. (1991) An assessment of the contribution of net mineralization to N cycling in grass swards using a field incubation method. *Plant and Soil* 138, 23–32.

Hauck, R.D., Melsted, S.W. and Yankwich, P.E. (1958) Use of N-isotope distribution in nitrogen gas in the study of denitrification. *Soil Science* 86, 287–291.

Hayes, J.E., Richardson, A.E. and Simpson, R.J. (2000) Components of organic phosphorus in soil extracts that are hydrolysed by phytase and acid phosphatase. *Biology and Fertility of Soils* 32, 279–286.

Haynes, R.J. and Tregurtha, R. (1999) Effects of increasing periods under intensive arable vegetable production on biological, chemical and physical indices of soil quality. *Biology and Fertility of Soils* 28, 259–266.

Heinemeyer, O. and Kaiser, E.A. (1996) Automated gas injector system for gas chromatography: atmospheric nitrous oxide analysis. *Soil Science Society America Journal* 60, 808–811.

Hynes, R.K. and Knowles, R. (1983) Inhibition of chemoautotrophic nitrification by sodium chlorate and sodium chlorite: a reexamination. *Appied and Environmental Microbiology* 45, 1178–1182.

Jarvis, S.C., Hatch, D.J. and Lovell, R.D. (2001) An improved soil core incubation method for the field measurement of denitrification and net mineralization using acetylene inhibition. *Nutrient Cycling in Agroecosystems* 59, 219–225.

Jeffrey, W.H. and Paul, J.H. (1988) Underestimation of DNA synthesis by [^{3}H]thymidine incorporation in marine bacteria. *Applied and Environmental Microbiology* 54, 3165–3168.

Johansson, M., Pell, M. and Stenström, J. (1998) Kinetics of substrate-induced respiration (SIR) and denitrification: applications to a soil amended with silver. *Ambio* 27, 40–44.

Jones, R.B., Gilmour, C.C., Stoner, D.L., Weir, M.M. and Tuttle, J.H. (1984) Comparison of methods to measure acute metal and organometal toxicity to natural aquatic microbial communities. *Applied and Environmental Microbiology* 47, 1005–1011.

Juma, N.G. and Tabatabai, M.A. (1978) Distribution of phosphomonoesterases in soils. *Soil Science* 126, 101–108.

Kandeler, E., Marschner, P., Tscherko, D., Gahoonia, T. and Nielsen, N. (2002) Microbial community composition and functional diversity in the rhizosphere of maize. *Plant and Soil* 238, 301–312.

Keeney, D.R. (1982) Nitrogen availability indexes. In: Page, A.L. *et al.* (eds) *Methods of Soil Analysis. Part 2*, 2nd edn. Agronomy Monograph 9, Soil Science Society of America, Madison, Wisconsin, pp. 711–733.

Keeney, D.R. and Nelson, D.W. (1982) Nitrogen – inorganic forms. In: Black, C.A., Evans, D.D., White, J.L., Ensminger, L.E. and Clark, F.E. (eds) *Methods of Soil Analysis, Part 2*. American Society of Agronomy, Madison, Wisconsin, pp. 682–687.

Kilham, K. (1985) A physiological determination of the impact of environmental stress on the activity of microbial biomass. *Environmental Pollution* 38, 283–294.

Kilham, K. (1987) A new perfusion system for the measurement and characterization of potential rates of soil nitrification. *Plant and Soil* 97, 227–232.

Kirchman, D., K'Nees, E. and Hodson, R. (1985) Leucine incorporation and its potential as a measure of protein synthesis by bacteria in natural aquatic ecosystems. *Applied and Environmental Microbiology* 49, 559–607.

Kirkham, D. and Bartholomew, W.V. (1954) Equations for following nutrient transformations in soil utilizing data. *Soil Science Society American Proceedings* 18, 33–34.

Klemedtsson, L., Hansson, G. and Mosier, A.R. (1990) The use of acetylene for the quantification of N_2 and N_2O production from biological processes in soil. In: Revsbech, N.P. and Sorensen, J. (eds) *Denitrification in Soil and Sediment*, Vol. 56. Plenum Press, New York, pp. 167–180.

Koops, H.P. and Pommerening-Röser, A. (2001) Distribution and ecophysiology of the nitrifying bacteria emphasizing cultured species. *FEMS Microbial Ecology* 37, 1–9.

Kowalchuk, G.A. and Stephen, J.R. (2001) Ammonia-oxidizing bacteria: a model for molecular microbial ecology. *Annual Review in Microbiology* 55, 485–529.

Kowalchuk, G.A., Stephen, J.R., de Boer, W., Prosser, J.I., Embley, T.M. and Woldendorp, J.W. (1997) Analysis of ammonia-oxidizing bacteria of the beta subdivision of the class proteobacteria in coastal sand dunes by denaturing gradient gel electrophoresis and sequencing of PCR-amplified 16s ribosomal DNA fragments. *Applied and Environmental Microbiology* 63, 1489–1497.

Ladd, J.N., Forster, R.C., Nannipieri, P. and Oades, J.M. (1996) Soil structure and biological activity. *Soil Biochemistry* 9, 23–78.

Lober, R.W. and Reeder, J.D. (1993) Modified waterlogged incubation method for assessing nitrogen mineralization in soils and soil aggregates. *Soil Science Society American Journal* 57, 400–403.

Mader, T. (1994) Auswirkungen einer praxisüblichen Anwendung von Gardoprim (Terbuthylazin) auf mikrobielle und biochemische Stoffumsetzungen sowie sein abbauverhalten im Feld- und Laborversuch. *Hohenheimer Bodenkundliche Hefte*, 19.

Madsen, E.L. (1996) A critical analysis of methods for determining the composition and biogeochemical activities of soil microbial communities in situ. In: Stotzky, G. and Bollag, J.M. (eds) *Soil Biochemistry*, Vol. 9. Marcel Dekker, New York, pp. 287–370.

Malone, J.P., Stevens, R.J. and Laughlin, R.J. (1998) Combining the ^{15}N and acetylene inhibition techniques to examine the effect of acetylene on denitrification. *Soil Biology and Biochemistry* 30, 31–37.

Mamilov, A.Sh., Byzov, B.A., Zvyagintsev, D.G. and Dilly, O.M. (2001) Predation on fungal and bacterial biomass in a soddy-podzolic soil amended with starch, wheat straw and lucerne meal. *Applied Soil Ecology* 16, 131–139.

Mary, B., Recous, S. and Robin, D. (1998) A model for calculating nitrogen fluxes in soil using ^{15}N tracing. *Soil Biology and Biochemistry* 30, 1963–1979.

McCarty, G.W. and Bremner, J.M. (1986) Inhibition of nitrification in soil by acetylenic compounds. *Soil Science Society American Journal* 50, 1198–1201.

Meisinger, J.J. (1984) Evaluating plant available N in soil crop systems. In: Hauck, R.D. *et al.* (eds) *Nitrogen in Crop Production*. American Society of Agronomy, Madison, Wisconsin, pp 391–416.

Michel, P.H. and Bloem, J. (1993) Conversion factors for estimation of cell production rates of soil bacteria from tritiated thymidine and tritiated leucine incorporation. *Soil Biology and Biochemistry* 25, 943–950.

Mogge, B., Kaiser, E.A. and Munch, J.C. (1998) Nitrous oxide emissions and denitrification N-losses from forest soils in the Bornhöved Lake region (Northern Germany). *Soil Biology and Biochemistry* 30, 703–710.

Mogge, B., Kaiser, E.-A. and Munch, J.C. (1999) Nitrous oxide emissions and denitrification N-losses from agricultural soils in the Bornhöved Lake region: influence of organic fertilizers and land-use. *Soil Biology and Biochemistry* 31, 1245–1252.

Moreno, J., Garcia, C., Landi, L., Falchini, L., Pietramellara, G. and Nannipieri, P. (2001) The ecological dose value (ED50) for assessing Cd toxicity on ATP content and dehydrogenase and urease activities of soil. *Soil Biology and Biochemistry* 33, 483–489.

Moriarty, D.J.W. (1986) Measurement of bacterial growth rates in aquatic systems from rates of nucleic acid synthesis. *Advances in Microbial Ecology* 9, 245–292.

Mosier, A.R. and Schimel, D.S. (1993) Nitrification and denitrification. In: Knowles, R. and Blackburn, T.H. (eds) *Nitrogen Isotope Techniques.* Academic Press, San Diego, California, pp. 181–208.

Mosier, A.R., Guenzi, W.D. and Schweizer, E.E. (1986) Field denitrification estimation by nitrogen-15 and acetylene inhibition techniques. *Soil Science Society American Journal* 50, 831–833.

Myhre, G., Highwood, E.K., Shine, K.P. and Stordal, F. (1998) New estimates of radiative forcing due to well mixed greenhouse gases. *Geophysical Research Letters* 25, 2715–2718.

Myrold, D.D. (1990) Measuring denitrification in soils using ^{15}N techniques. In: Revsbech, N.P. and Sorensen, J. (eds) *Denitrification in Soil and Sediment*. Plenum Press, New York, pp. 181–198.

Myrold, D.D. and Tiedje, J.M. (1986) Simultaneous estimation of several nitrogen cycle rates using ^{15}N: theory and application. *Soil Biology and Biochemistry* 18, 559–568.

Nannipieri, P., Kandeler, E. and Ruggiero, P. (2002) Enzyme activities and microbiological and biochemical processes in soil. In: Burns, R.G. and Dick, R.P. (eds) *Enzymes in the Environment*. Marcel Dekker, New York, pp. 1–33.

Nicolaisen, M.H. and Ramsing, N.B. (2002) Denaturing gradient gel electrophoresis (DGGE) approaches to study the diversity of ammonia-oxidizing bacteria. *Journal of Microbiological Methods* 50, 89–203.

Nishio, T., Kanamori, T. and Fjuimoto, T.

(1985) Nitrogen transformation in an aerobic soil as determined by a $^{15}NH_4^+$ dilution technique. *Soil Biology and Biochemistry* 17, 149–154.

Obbard, J.P. (2001) Measurement of dehydrogenase activity using 2-*p*-iodophenyl-3-*p*-nitrophenyl-5-phenyltetrazolium chloride (INT) in the presence of copper. *Biology and Fertility of Soils* 33, 328–330.

Öhlinger, R., Beck, T., Heilmann, B. and Beese, F. (1996) Soil respiration. In: Schinner, F., Öhlinger, R., Kandeler, E. and Margesin, R. (eds) *Methods in Soil Biology*. Springer-Verlag, Berlin, pp. 93–110.

Orchard, V.A. and Cook, F.J. (1983) Relationships between soil respiration and soil moisture. *Soil Biology and Biochemistry* 15, 447–453.

Palmborg, C. and Nordgren, A. (1993) Soil respiration curves, a method to test the abundance, activity and vitality of the microflora in soils. In: Torstensson, L. (ed.) *Guidelines. Soil Biological Variables in Environmental Hazard Assessment*. Report 4262, Swedish EPA, Stockholm, pp. 149–156.

Pankhurst, C.E., Hawke, B.G., McDonald, H.J., Kirkby, C.A., Buckerfield, J.C., Michelsen, P., O'Brian, K.A., Gubta, V.V.S.R. and Double, B.M. (1995) Evaluation of soil biological properties as potential bioindicators of soil health. *Australian Journal of Experimental Agriculture* 35, 1015–1028.

Pant, H.K. and Warman, P.R. (2000) Enzymatic hydrolysis of soil organic phosphorus by immobilized phosphatases. *Biology and Fertility of Soils* 30, 306–311.

Parkin, T.B., Kaspar, H.F., Sexstone, A.J. and Tiedje, J.M. (1984) A gas-flow soil core method to measure field denitrification rates. *Soil Biology and Biochemistry* 16, 323–330.

Parkin, T.B., Sexstone, A.J. and Tiedje, J.M. (1985) Comparison of field denitrification rates by acetylene-based soil core and nitrogen-15 methods. *Soil Science Society American Journal* 49, 94–99.

Pasqual, J.A., Hernandez, T. and Garcia, C. (2001) In: Benedetti, A. *et al.* (eds) *COST Action 831 – Joint Meeting of Working Groups Biotechnology of Soil Monitoring, Conservation and Remediation*. Round Table: Defining Soil Quality. Rome, 10–11 December 1998. EUR 19225 – COST Action 831 – Report of 1997–1998. European Communities, Luxembourg, p. 310.

Paul, E.A. and Clark, F.E. (1989) *Soil Microbiology and Biochemistry*. Academic Press, Inc., San Diego, California, 275 pp.

Paustian, K., Bergström, L., Jansson, P.E. and Johnsson, H. (1990) Ecosystem dynamics. In: Andrén, O., Lindberg, T., Paustian, K. and Rosswall, T. (eds) *Ecology of Arable Land. Organisms, Carbon and Nitrogen Cycling*. (*Ecological Bulletin*) 40, 153–180.

Persson, T., Bååth, E., Clarholm, M., Lundkvist, B.E. and Sohlenius, H. (1980) Trophic structure, biomass dynamics and carbon metabolism of soil organisms in a Scots pine forest. In: Ågren, G.I., Andersson, F., Falk, S.O., Lohm, U. and Perttu, K. (eds) *Structure and Function of Northern Coniferous Forests. (Ecological Bulletin)* 32, 419–459.

Perucci, P., Vishetti, C. and Battistoni, F. (1999) Rimsulfuron in a silty clay loam soil: effects upon microbiological and biochemical properties under varying microcosm conditions. *Soil Biology and Biochemistry* 31, 195–204.

Perucci, P., Dumontet, S., Bufo, S.A., Mazzatura, A. and Casucci, C. (2000) Effects of organic amendment and herbicide treatment on soil microbial biomass. *Biology and Fertility of Soils* 32, 17–23.

Powlson, D.S. and Barraclough, D. (1993) Mineralization and assimilation in soil–plant system. In: Knowles, R. and Black, T.H. (eds) *Nitrogen Isotope Techniques*. Academic Press, San Diego, California, pp. 209–242.

Prescott, D.M. and James, T.W. (1955) Culturing of *Amoeba proteus* on *Tetrahymena*. *Experimental Cell Research* 8, 256–258.

Prosser, J.I. (1989) Autotrophic nitrification in bacteria. *Advances in Microbial Physiology* 30, 125–181.

Recous, S., Luxhoi, J., Fillery, I.R.P., Jensen,

L.S. and Mary, B. (2001) Soil temperature effects on N mineralisation, immobilisation and nitrification using ^{15}N dilution techniques. *Abstracts of the 11th Nitrogen Workshop, 9–12 September, Reims, France.*

Reddy, G.B., Faza, A. and Bennett, R. (1987) Activity of enzymes in rhizosphere and non-rhizosphere soils amended with sludge. *Soil Biology and Biochemistry* 19, 203–205.

Robarts, R.D. and Zohary, T. (1993) Fact or fiction – bacterial growth rates and production as determined by [methyl-^{3}H]-thymidine? *Advances in Microbial Ecology* 13, 371–425.

Rotman, B. and Papermaster, B.W. (1966) Membrane properties of living mammalian cells as studied by enzymatic hydrolysis of fluorogenic esters. *Proceedings of the National Academy of Science USA* 55, 134–141.

Rotthauwe, J.H., Witzel, K.P. and Liesack, W. (1997) The ammonia monooxygenase structural gene *amoa* as a functional marker: molecular fine-scale analysis of natural ammonia-oxidizing populations. *Applied and Environmental Microbiology* 63, 4704–4712.

Rudaz, A.O., Walti, E., Kyburz, G., Lehmann, P. and Fuhrer, J. (1999) Temporal variation in N_2O and N_2 fluxes from a permanent pasture in Switzerland in relation to management, soil water content and soil temperature. *Agriculture, Ecosystems and Environment* 73, 83–91.

Ryden, J.C., Lund, L.J., Letey, J. and Focht, D.D. (1979) Direct measurement of denitrification loss from soils. II. Development and application of field methods. *Soil Science Society American Journal* 43, 110–118.

Ryden, J.C., Skinner, J.H. and Nixon, D.J. (1987) Soil core incubation system for the field measurement of denitrification using acetylene-inhibition. *Soil Biology and Biochemistry* 19, 753–757.

Sahrawat, K.L. (1983) Mineralization of soil organic nitrogen under waterlogged condition in relation to other properties of tropical rice soils. *Australian Journal of Soil Research* 21, 133–138.

Sahrawat, K.L. and Ponnamperuma, F.N. (1978) Measurement of exchangeable NH_4^+ in tropical land soils. *Soil Science Society of America Journal* 42, 282–283.

Schimmel, E.L., Firestone, M.K. and Killham, K.S. (1984) Identification of heterotrophic nitrification in a sierran forest soil. *Applied and Environmental Microbiology* 48, 802–806.

Schinner, F., Oehlinger, R. and Kandeler, E. (1992) *Bodenbiologische Arbeitsmethoden.* Springer, Berlin.

Schmidt, E.L. and Belser, L.W. (1994) Autotrophic nitrifying bacteria. In: Weaver *et al.* (eds) *Methods in Soil Analysis Part 2, Microbiological and Biochemical Properties*, SSSA Book Series No. 5. Soil Science Society of America, Madison, Wisconsin, pp. 159–177.

Schneider, K., Turrion, M.-B. and Gallardo, J.-F. (2000) Modified method for measuring acid phosphatase activities in forest soils with high organic matter content. *Communications in Soil Science and Plant Analysis* 31, 3077–3088.

Servais, P., Martinez, J., Billen, G. and Vives-Rego, J. (1987) Determining [^{3}H]thymidine incorporation into bacterioplankton DNA: improvement of the method by DNase treatment. *Applied and Environmental Microbiology* 53, 1977–1979.

Shaw, L.J. and Burns, R.G. (1996) Construction and equilibration of a repacked soil column microcosm. *Soil Biology and Biochemistry* 28, 1117–1120.

Shaw, L.J. and Burns, R.G. (2003) Biodegradation of organic pollutants in the rhizosphere. *Advances in Applied Microbiology* 53, 1–60.

Sparling, G.P. (1997) Soil microbial biomass, activity and nutrient cycling. In: Pankhurst, C.E., Doube, B.M. and Gupta, V.V.S.R. (eds) *Biological Indicators of Soil Health.* CAB International, Wallingford, UK, pp. 97–120.

Speir, T.W. and Ross, D.J. (1978) Soil phosphatase and sulphatase. In: Burns, R.G. (ed.) *Soil Enzymes.* Academic Press, London, pp. 197–250.

Speir, T.W. and Ross, D.J. (2002) Hydrolytic enzyme activities to assess soil degradation and recovery. In: Burns, R.G. and Dick, R.P. (eds) *Enzymes in the Environment*. Marcel Dekker, New York, pp. 407–431.

Stanford, G. (1982) Assessment of soil nitrogen availability. In: Stevenson, F.J. (ed.) *Nitrogen in Agricultural Soils*. Agronomy Monograph 22. ASA and SSSA, Madison, Wisconsin.

Stanford, G. and Smith, S.J. (1972) Nitrogen mineralization potentials of soils. *Soil Science Society American Proceedings* 38, 99–102.

Stark, J.M. and Firestone, M.K. (1996) Kinetic characteristics of ammonium-oxidizer communities in a California oak woodland-annual grassland. *Soil Biology and Biochemistry* 28, 1307–1317.

Steen, E. (1990) Agricultural outlook. In: Andrén, O., Lindberg, T., Paustian, K. and Rosswall, T. (eds) *Ecology of Arable Land. Organisms, Carbon and Nitrogen Cycling. (Ecological Bulletin)* 40, 181–192.

Stenberg, B., Johansson, M., Pell, M., Sjödal-Svensson, K., Stenström, J. and Torstensson, L. (1998a) Microbial biomass and activities in soil as affected by frozen and cold storage. *Soil Biology and Biochemistry* 30, 393–402.

Stenberg, B., Pell, M. and Torstensson, L. (1998b) Integrated evaluation of variations in biological, chemical and physical soil properties. *Ambio* 27, 9–15.

Stenström, J., Stenberg, B. and Johansson, M. (1998) Kinetics of substrate-induced respiration (SIR): theory. *Ambio* 27, 35–39.

Stenström, J., Svensson, K. and Johansson, M. (2001) Reversible transition between active and dormant microbial states in soil. *FEMS Microbiology Ecology* 36, 93–104.

Stephen, J.R., Chang, Y.J., Macnaughton, S.J., Kowalchuk, G.A., Leung, K.T., Flemming, C.A. and White, D.C. (1999) Effect of toxic metals on indigenous soil p-subgroup proteobacterium ammonia oxidizer community structure and protection against toxicity by inoculated metal-resistant bacteria. *Applied and Environmental Microbiology* 65, 95–101.

Stevens, R.J. and Laughlin, R.J. (1998) Measurement of nitrous oxide and dinitrogen emissions from agricultural soils. *Nutrient Cycling in Agroecosystems* 52, 131–139.

Stevenson, F.J. (1985) *The Internal Cycle of Nitrogen in Soil. Cycle of Soil C, N, P, S, Micronutrients.* J. Wiley & Sons, New York, pp. 155–215.

Stotzky, G. (1965) Microbial respiration. In: Black C.A. *et al.* (eds) *Methods in Soil Analysis, Part 2*. Agronomy Monograph 9. ASA, Madison, Wisconsin, pp. 1550–1572.

Stoyan, H., de Polli, H., Bohm, S., Robertson, G.P. and Paul, E.A. (2000) Spatial heterogeneity of soil respiration and related properties at the plant scale. *Plant and Soil* 222, 203–214.

Svensson, K. (1999) Effect of different cropping systems on soil microbial activities. *COST 831 – Working Group 4 Meeting*, Granada, Spain, 15–16 July.

Svensson, K. and Pell, M. (2001) Soil microbial tests for discriminating between different cropping systems and fertilizer regimes. *Biology and Fertility of Soils* 33, 91–99.

Tabatabai, M.A. and Bremner, J.M. (1969) Use of *p*-nitrophenyl phosphate for assay of soil phosphatase activity. *Soil Biology and Biochemistry* 1, 301–307.

Tabatabai, M.A. and Dick, W.A. (2002) Enzymes in soil: research and developments in measuring activities. In: Burns, R.G. and Dick, R.P. (eds) *Enzymes in the Environment*. Marcel Dekker, New York, pp. 567–596.

Tarafdar, J.C. and Jungk, A. (1987) Phosphatase activity in the rhizosphere and its relation to the depletion of soil organic phosphorus. *Biology and Fertility of Soils* 3, 199–204.

Tate, R.L. (2002) Microbiology and enzymology of carbon and nitrogen cycling. In: Burns, R.G. and Dick, R.P. (eds) *Enzymes in the Environment*. Marcel Dekker, New York, pp. 227–248.

Taylor, J., Wilson, B., Mills, M. and Burns, R. (2002) Comparison of microbial numbers and enzymatic activities in surface soils and subsoils using various techniques. *Soil Biology and Biochemistry* 34, 387–401.

Terry, R.E. and Duxbury, J.M. (1985) Acetylene decomposition in soils. *Soil Science Society American Journal* 49, 90–94.

Thorn, P.M. and Ventullo, R.M. (1988) Measurement of bacterial growth rates in subsurface sediments using the incorporation of tritiated thymidine into DNA. *Microbial Ecology* 16, 3–16.

Tiedje, J.M., Simkins, S. and Groffman, P.M. (1989) Perspectives on measurement of denitrification in the field including recommended protocols for acetylene based methods. *Plant and Soil* 115, 261–284.

Torstensson, L. and Stenström, J. (1986) 'Basic' respiration rate as a tool for prediction of pesticide persistence in soil. *Toxicity Assessment: An International Quarterly* 1, 57–72.

Torstensson, M., Pell, M. and Stenberg, B. (1998) Need of strategy for evaluation of soil quality data: arable soil. *Ambio* 27, 4–8.

Trasar-Cepeda, M. and Gilsotres, F. (1987) Phosphatase-activity in acid high organic-matter soils in Galacia (NW Spain). *Soil Biology and Biochemistry* 19, 281–287.

Trasar-Cepeda, C., Leiros, M., Seoane, S. and Gil-Sotres, F. (2000) Limitations of soil enzymes as indicators of soil pollution. *Soil Biology and Biochemistry* 32, 1867–1875.

Trevors, J.T. (1984a) Effect of substrate concentration, inorganic nitrogen, O_2 concentration, temperature and pH on dehydrogenase activity in soil. *Plant and Soil* 77, 285–293.

Trevors, J.T. (1984b) Dehydrogenase activity in soil: a comparison between the INT and TTC assay. *Soil Biology and Biochemistry* 16, 673–674.

Trevors, J.T., Mayfield, C.I. and Inniss, W.E. (1982) Measurement of electron transport system (ETS) activity in soil. *Microbial Ecology* 8, 163–168.

Vekemans, X., Godden, B. and Penninckx, M.J. (1989) Factor-analysis of the relationships between several physiochemical and microbiological characteristics of some Belgian agricultural soils. *Soil Biology and Biochemistry* 21, 53–58.

Vischetti, C., Perucci, P. and Scarponi, L. (1997) Rimsulfuron in soil: effect of persistence on growth and activity of microbial biomass at varying environmental conditions. *Biogeochemistry* 39, 165–176.

vonMersi, W. and Schinner, F. (1991) An improved and accurate method for determining the dehydrogenase activity of soils with iodonitrotetrazolium chloride. *Biology and Fertility of Soils* 11, 216–220.

Waldrop, M.P., Balser, T.C. and Firestone, M.K. (2000) Linking microbial community composition to function in a tropical soil. *Soil Biology and Biochemistry* 32, 1837–1846.

Waring, S.A. and Bremner, J.M. (1964) Ammonium production in soil under waterlogged conditions as an index of nitrogen availability. *Nature* 201, 951–952.

Watts, S.H. and Seitzinger, S.P. (2000) Denitrification rates in organic and mineral soils from riparian sites: a comparison of N_2 flux and acetylene inhibition methods. *Soil Biology and Biochemistry* 32, 1383–1392.

Welp, G. (1999) Inhibitory effects of the total and water-soluble concentrations of nine different metals on the dehydrogenase activity of a loess soil. *Biology and Fertility of Soils* 30, 132–139.

Yakovchenko, V., Sikora, L.J. and Kaufmann, D.D. (1996) A biologically based indicator of soil quality. *Biology and Fertility of Soils* 21, 245–251.

Yakovchenko, V., Sikora, L.J. and Millner, P.D. (1998) Carbon and nitrogen mineralization of added particulate and macroorganic matter. *Soil Biology and Biochemistry* 30, 2139–2146.

Yoshinari, T. and Knowles, R. (1977) Acetylene inhibition of nitrous oxide reduction by denitrifying bacteria. *Biochem Biophys Res Commun* 69, 705–710.

Zibilske, L.M. (1994) Carbon mineralization. In: Weaver, R.W. *et al.* (eds) *Methods of Soil Analysis, Part 2. Microbiological and Biochemical Properties.* SSSA, Madison, Wisconsin, pp. 835–863.

8 Soil Microbial Diversity and Community Composition

8.1 Estimating Soil Microbial Diversity and Community Composition

JAN DIRK VAN ELSAS[1] AND MICHIEL RUTGERS[2]

[1]*Department of Microbial Ecology, Groningen University, Kerklaan 30, NL-9750 RA Haren, The Netherlands;* [2]*National Institute for Public Health and the Environment, Antonie van Leeuwenhoeklaan 9, NL-3721 MA Bilthoven, The Netherlands*

Soils are complex and very heterogeneous environments that may contain as many as 10 billion or more bacterial cells per gram, in addition to large numbers of other microorganisms such as fungi and protozoa, as well as several macroorganisms, collectively called macrofauna. Although microorganisms determine the chemical balance in the soil, interact with plants in positive and negative ways, and even influence soil structure, we have little understanding of the structure and dynamics of these microbial communities, as well as of their enormous diversity. However, over the past decade our perspective with respect to the microbial diversity and community structure of soils has improved enormously. This is in large part due to the rapid development and application of culture-independent methods that allow the characterization of soil microbial communities. For a long period, spanning almost 100 years, plating (counting colony-forming units, CFU) on different media was the technique of choice to investigate the microbial diversity of soil. However, the relative proportion of bacteria that grow readily on agar plates with common bacteriological media to those counted by microscopical approaches varies from 0.1–1% in pristine forest soils to 10% in environments such as arable soil. This implies that assessments of the microbial diversity of these habitats, in terms of species richness and abundance, are grossly underestimated (Amann *et al.*, 1995). Since so few soil microorganisms can be cultivated by standard techniques, the new culture-independent approaches, in particular cloning, sequencing and fingerprinting of ribosomal RNA (rRNA) genes, have already revealed evidence for the existence of an astonishing wealth of novel organisms, of which many are quite different from those known among the cultured isolates (Liesack and Stackebrandt, 1992). Still, as of today, any estimations of the extant numbers of, for example, prokaryotic species are mere guesses, but estimates of 4000 or more species per g of soil have been suggested (Torsvik *et al.*, 1990). In addition to continuously increasing our knowledge of the microbial diversity in soils, culture-independent methods, several of which are described in this chapter, also allow us to better understand the structure and dynamics of soil microbial communities. These methods allow us to monitor changes in either the overall microbial or bacterial community structure, or in more detail, changes in the prevalence of phylogenetic subgroups.

The ultimate goal of most assessments is to understand the overall functioning of soil microbial communities in terms of, for example, the flux of energy, as well as resources, through the system, and how this is influenced by natural environmental changes as well as human impact. Many methods are currently available that aid in the description of the diversity and functioning of soil microbial communities (summarized in Table 8.1). The methods described in this chapter form a subset of this wide range of methods. Together, they allow us to describe soil microbial communities in an indirect way, each with their limitations. However, they may be applied as routine methods in a comparative fashion, in order to provide a picture of soil microbial communities from three angles.

On the other hand, in order to understand the final outcome of the many functions carried out by different microbial populations, or by cooperation between such populations, we need to better understand the structure of the microbial communities of soil. Answers to questions such as: what species are there, what is the relative abundance of the different populations, are the organisms active or dormant and do they interact (via chemical signals, intermediate metabolites or by genetic interactions) are fundamental for a better understanding of the functioning and robustness of the microbial community. To answer these questions, methods are needed that can distinguish between the different populations present, quantify their abundance (population sizes) and, ideally, locate them *in situ*. Thus, there has been a perceived need for sensitive methods that can identify which microbial populations are affected by a certain environmental change (e.g. in time or upon human impact). One example is a recent study that has shown that long-term herbicide applications clearly affect the structure and diversity of specific microbial groups, as shown by 16S rRNA molecular community fingerprinting (Seghers *et al.*, 2003). The fingerprinting method, based on universal bacterial primers, did not show significant differences between the herbicide-treated and the control soil in this study. However, upon zooming in on specific populations or phylogenetic subgroups, such as the methanotrophs, differences in community structure were observed. Whether these differences in community structure also

Table 8.1. Methods to assess the microbial diversity of soils.

Type	Method	Description	Reference
Phenotypic	CLPP	Community-level physiological profiling	Garland and Mills (1991)
Cell components	FAME	Fatty acid methyl ester	Buyer and Drinkwater (1997); Zelles (1999)
	PLFA	Phospholipid fatty acid analysis	Frostegård *et al.* (1993); Zelles (1999)
Molecular/genetic	Cloning/sequencing of amplified 16S ribosomal RNA genes (rDNA)	Sequence analysis of clone library, resulting in overview of abundant clone types	Akkermans *et al.* (1995); Kowalchuk *et al.* (2004)
	16S rDNA-based PCR and fingerprintings: – DGGE/TGGE – SSCP – ARDRA – T-RFLP	PCR may be followed by any one of the fingerprinting methods to determine the diversity of the community on the basis of the 16S rRNA-based phylogenetic marker	Akkermans *et al.* (1995); Kowalchuk *et al.* (2004); van Elsas *et al.* (2000)
	Dot blot hybridization of 16S rRNA genes	Hybridization using short 16S rRNA gene-based fragments as probes; probes generated from V6 region are highly specific per strain	Heuer *et al.* (1999)
	Base composition profiles	Expressed as mole percent guanine + cytosine (% G + C)	Torsvik *et al.* (1990); Nüsslein and Tiedje (1998)
	Direct PCR detection	PCR amplification of target gene followed by detection on gel, or after hybridization	van Elsas *et al.* (1997)
	FISH	Fluorescent *in situ* hybridization of 16S rRNA	Akkermans *et al.* (1995); Amann *et al.* (1995)
	DNA reassociation	'C_0t curves': reassociation time is a measure for the genetic diversity in a sample	Torsvik *et al.* (1990)

translate into a change in function, or in a community that is less fit to deal with other forms of stress, is not known. It is known that there is a lot of functional redundancy in soils, which may serve as a buffering capacity in cases of severe disturbance. There is, however, no guarantee that the soil microbial community will completely take over the role of populations that have disappeared. How can we estimate or predict whether a slightly changed microbial community will be as resistant to future stressors as before the alteration?

During the past two decades, phenotypic and nucleic acid-based methods have been developed to better characterize the structure and diversity of microbial communities. Table 8.1 lists these methods. Three methods are discussed and described in detail in this chapter. They are:

1. Soil microbial community DNA- or RNA-based profiling methods.
2. Phospholipid-based methods (PLFA, phospholipid fatty acid).
3. Community-level physiological profiling (CLPP) tests based on the use of Biolog™ metabolic response plates.

All three methods are not based on isolated and purified organisms, but rather give a picture of different aspects of the microbial community based on a method that uses total macromolecules or cells extracted from soil. The different methods address different questions and therefore analyse either different sub-groups of the total microbial community or different aspects of the community. Together they form a solid basis for up-to-date soil microbial community analyses.

8.2 Soil Microbial Community Fingerprinting Based on Total Community DNA or RNA

JAN DIRK VAN ELSAS,[1] EVA M. TOP[2] AND KORNELIA SMALLA[3]

[1]*Department of Microbial Ecology, Groningen University, Kerklaan 30, NL-9750 RA Haren, The Netherlands;*
[2]*Department of Biological Sciences, 347 Life Sciences Building, University of Idaho, Moscow, ID 83844–3051, USA;*
[3]*Institute for Microbiology, Virology and Biosafety, Biologische Bundesanstalt, Messeweg 11/12, Braunschweig, Germany*

Introduction

Soil microbial community DNA and/or RNA extraction methods developed in the past decade have paved the way for direct, cultivation-independent, studies of microbial diversity in soils. The direct extraction and analysis of total microbial community DNA or RNA from soil was shown to be fundamental as the basis to describe the *in situ* soil microbiota and its diversity. After the pioneering work of Torsvik and co-workers in the early 1980s (e.g. Torsvik, 1980), the extraction of microbial DNA from soil has found its way into basically every soil microbiology laboratory in the world. During the past decade in particular, a multitude of different extraction protocols has been published. The *Molecular Microbial Ecology Manual*, first and second editions (Akkermans *et al.*, 1995; Kowalchuk *et al.*, 2004), contains examples of all of these protocols. This proliferation of protocols is often explained by the large variations between soils in chemical characteristics, which necessitates the application of different protocols to almost every new soil. However, in the past few years, many laboratories have replaced their often-laborious 'pet' protocols with rapid commercial kit-based protocols, which appear to work comparably well in almost every instance. Two current products, the Bio101 and MoBio soil DNA extraction kits, seem to cover most of the kit-based extraction protocols that are currently in use.

All the different protocols, even the kit-based ones, can be grouped in two basically different types of approaches:

1. Disruption of bonds between microbial cells and soil particles, resulting in the release of largely bacterial or archaebacterial cells, followed by cell lysis and extraction.
2. Direct cell lysis within the soil matrix in a slurry, and DNA extraction from the soil slurry (Ogram *et al.*, 1987).

While the former method provides DNA that is considered to be representative for the prokaryotic (bacterial or archaeal) fractions of the microbial community in soil, the latter has been shown to provide higher DNA yields, but is less specific for prokaryotes, as it also contains eukaryotic (fungal) and extracellular DNA (Steffan *et al.*, 1988; van Elsas *et al.*, 2000). Both methods coextract low levels of extracellular DNA from soil, which should be taken into account in the interpretation of results, i.e. whether positive detection indicates the presence of microbial cells carrying target DNA or merely extracellular target DNA. In spite of these potential problems, the direct lysis and extraction method has become the favourite DNA extraction method in many laboratories working with soil. On the other hand, recent advances have shown definite advantages of the indirect methods, with respect to the specificity of detection (Duarte *et al.*, 1998).

Direct community DNA extraction from soil, as developed by Ogram *et al.* (1987), has been shown to provide a substantial amount of total soil bacterial DNA, but was also found to be prone to unavoidable yield losses. Moré *et al.* (1994) showed that many direct protocols are likely to extract DNA mainly from easy-to-lyse cells, whereas, in particular, the minute microbial forms (dwarfs) that are abundant in soil are often excluded. Since the original protocols encompassed tedious caesium chloride (CsCl) gradient and/or hydroxyapatite chromatography purification steps, many more recent protocols have attempted to simplify the original protocol of Ogram *et al.* (van Elsas *et al.*, 2000). Examples of these simplified protocols can be found in the literature (Pillai *et al.*, 1991; Porteous and Armstrong, 1991; Selenska and Klingmüller, 1991a,b; van Elsas *et al.*, 1991; Tsai and Olson, 1992; Smalla *et al.*, 1993a; Duarte *et al.*, 1998; van Elsas *et al.*, 2000). Some protocols extract RNA in addition to DNA. Others (e.g. Smalla *et al.*, 1993a; Griffith *et al.*, 2000) have been selected as preferred ones in the *Molecular Microbial Ecology Manual* (Akkermans *et al.*, 1995; Kowalchuk *et al.*, 2004).

The nucleic acid extraction methods are thus useful for several purposes (Trevors and van Elsas, 1989). First, they provide insight into the prevalence and/or activity of specific genes in microbial communities in soil ecosystems, resulting in a better understanding of the functioning and selection of such genes under specific soil conditions. Secondly, by using 16S/18S or 23S/25S ribosomal RNA gene sequences as 'signature molecules' (biomarkers), overall community DNA analysis can be the basis for description of microbial community structures. This can be achieved by applying fingerprinting techniques such as temperature or denaturing gradient gel electrophoresis (TGGE or DGGE), terminal restriction fragment length polymorphism (T-RFLP), or single-strand conformational polymorphism (SSCP). These and other methods have been described in the *Molecular Microbial Ecology Manual* (Akkermans *et al.*, 1995; Kowalchuk *et al.*, 2004). In all three methods, PCR products that represent a 'picture' of the soil microbial community are generated with sets of conserved primers, and are subsequently separated by either one of the aforementioned fingerprinting methods. The basis of the fragment separation differs between these methods, as follows. SSCP takes advantage of the conformational changes that occur stochastically in single-stranded DNA (or RNA) molecules, and result in different migration behaviour in polyacrylamide gels. T-RFLP is based on the detection, by a fluorescent label, of a terminal fragment produced by enzymatic restriction of the mixed amplicons, thereby identifying phylotypes with a unique terminal restriction fragment (T-RF). DGGE and TGGE are based on the use of denaturing or temperature gradients, respectively. Both gradient types separate double-stranded DNA molecules by their specific migration distance, which is determined by their melting behaviour resulting from the nucleotide sequence. All of these methods yield a fingerprint pattern that is a representation of the microbial community structure in the soil. Depending on the primers used in the initial PCR amplification, different microbial groups can be targeted. For instance, a range of primer sets has become available that target total bacteria, total fungi or specific bacterial groups, such as the α- or β-subgroups of the proteobacteria, the high-G+C% Gram-positives, the pseudomonads, the bacilli, and *Burkholderia* spp. (Van Overbeek *et al.*, 2005). Primers for specific fungal groups, such as different vesicular arbuscular mycorrhizae, are also under development or have been published (de Souza, personal communication).

Thus, the analysis of the various fingerprints obtained from these microbial community DNA extracts allows us to make inferences about the types of microorganisms present within the extractable cell fraction of the soil. This obviously includes the non- or poorly culturable cells, which represent the largest fraction of cells that can be found in most soils (silent majority). The new angle on soil microbial diversity offered by the direct DNA-based methods has already led to the discovery of a wealth of novel organisms. For instance, several new deep-branching groups of proteobacteria have recently been described, based on the analysis of 16S ribosomal RNA gene clones generated from DNA of total soil communities.

In this section, we discuss the use of two nucleic acid extraction methods, one DNA extraction method developed and used routinely in our laboratories (method I; modified from Smalla *et al.*, 1993a), and one miniaturized DNA/RNA extraction method (method II) recently implemented in our labs (Gomes *et al.*, 2004). We then describe the use of 16S ribosomal RNA gene (rDNA)-based PCR coupled to DGGE analysis to describe the microbial diversity in soil. It has been shown that restrictable and PCR-amplifiable DNA of relatively high molecular weight can be obtained with the majority of nucleic acid extraction protocols. The microbial diversity measurements are, to some extent, dependent on the DNA extraction method used, and hence it is important to use one standard protocol in routine analyses.

Principle of the Methods

In many laboratories, soil nucleic acid extraction is nowadays performed by using commercially available kits. The main driving force of this development has been the relative ease of use and speed offered by these kits.

Several groups have shown that such kits extract microbial community DNA which is roughly, but not completely, similar to that obtained by the more traditional approaches. However, for specific purposes, e.g. when novel results should be compared to previous results obtained by traditional extraction, and in the light of their robustness, the traditional protocols are still in use. Below, we present one robust and highly versatile traditional approach (method I) and one approach modified from a highly accepted commercial kit-based method (method II).

Soil nucleic acid extraction method I

This method will primarily extract DNA from soil, although RNA is sometimes coextracted. The method is based on an efficient lysis of bacterial cells in the soil matrix in a slurry, followed by quick removal of soil particles and humic compounds, using different rapid purification steps. Removal of humic material, proteins, RNA and polysaccharides, as well as other soil compounds (minerals), is required to obtain DNA of sufficient purity for hybridization, restriction or PCR amplification analysis, as well as for cloning purposes (Steffan *et al.*, 1988; Smalla *et al.*, 1993a,b; Tebbe and Vahjen, 1993). The protocol is based on the direct extraction protocol of Ogram *et al.* (1987) for the extraction of DNA from sediments, as adapted by Smalla *et al.* (1993a), with omission of the laborious and often inefficient purification via hydroxyapatite chromatography. We adopted bead beating of soil slurries in a Braun's cell homogenizer (B. Braun Diessel Biotech, Melsungen, Germany) as the method of choice for cell lysis, since this strategy was shown to yield higher quantities of DNA than freeze/thaw lysis (Smalla *et al.*, 1993a). However, in laboratories that do not possess a bead beater, freeze/thaw-assisted cell lysis (either using or not using lytic enzymes) may be used as an alternative. Careful control of the bead-beating time and conditions was essential to obtain DNA of large fragment size. This is important, since severely sheared DNA is unsuitable for PCR-based detection of specific genes or analysis of community structure, e.g. using bacterial 16S ribosomal RNA gene sequences as targets. Following cell lysis, extraction with cold phenol in the presence of soil particles separated the DNA from contaminating compounds, while offering protection from nucleases. Subsequent precipitation steps with CsCl and potassium acetate (KAc) were included to further remove impurities (proteins, RNA, humic material) from the DNA. For some soils, e.g. several silt loams, DNA preparations thus obtained are of sufficient purity to serve as targets for restriction, amplification or hybridization analysis. For other soils, e.g. a high organic matter loamy sand, a final clean-up step was required, performed by adsorption/elution over commercially available glassmilk (Geneclean II kit, Bio 101, La Jolla, California, USA) or resin spin columns (Wizard DNA Clean-Up System, Promega, Madison, Wisconsin, USA). In our experience, this flexible protocol allows for the extraction and purification of high-quality DNA from virtually any soil type (van Elsas *et al.*, 1997).

Soil nucleic acid extraction method II

This nucleic acid extraction protocol, which will produce a mixture of DNA and RNA, is based on the use of a commercially available kit (Bio 101, La Jolla, California, USA), which allows an upscaling of the extraction and working in small reaction vessels. The protocol (Gomes *et al.*, 2004) is a modification of the method described by Hurt *et al.* (2001). Because harsh cell lysis is crucial to obtain nucleic acids representing the microbial community in bulk and rhizosphere soils, a bead-beating step was added to the protocol of Hurt *et al.* (2001) to ensure the efficient disruption of the cells. Smalla *et al.* (1993a) had already shown that bead beating yields higher amounts of DNA than freeze/thaw lysis. To process many samples in parallel, a miniaturization of the protocol was required. The whole procedure was scaled down to 0.5 g (wet weight) of soil. Materials, volumes and equipment were adapted to this miniaturized scale, making the protocol simple, fast and suitable for processing large numbers of samples within a short period of time. However, degradation of RNA was still observed with the modifications mentioned above. Therefore, an RNA-protecting substance, ethanol or isopropanol, was added before breaking up the cells (Gomes *et al.*, 2004), as it is known that this reduces the degree of RNA degradation. We recommend the addition of ethanol as the most efficient step to achieve this. The extraction buffer of Hurt *et al.* (2001) was, thus, slightly modified and kept in the incubation step after bead beating.

PCR amplification

Following extraction and purification of the soil nucleic acids, either the DNA is subjected directly to PCR using standard protocols, or RNA is first reverse-transcribed, after which the copy DNA (cDNA) produced is PCR-amplified. PCR has become the method of choice for the enrichment of specific target sequences for subsequent detection, or even cloning (Akkermans *et al.*, 1995; Kowalchuk *et al.*, 2004). The method is based on the cyclic enzymatic 'inward' extension of primers at two opposite ends of a DNA template, resulting in the generation of numerous copies of this template. The amplification cycle, which consists of template denaturing, primer-to-target annealing and primer extension steps, is achieved by concerted changes in reaction temperature, most easily performed in a programmable thermal cycler. Due to the high denaturing temperature (often 94°C), DNA polymerases used in the PCR have to be thermostable, for example the frequently used *Thermus aquaticus Taq* polymerase.

To achieve the desired specificity, primer choice is fundamental, as discussed above. In addition, the primers should be checked as to whether they actually perform well in a soil DNA background, as differences with respect to inhibition of the action of the polymerase have been found between different primer sets. Following PCR, the amplicons obtained should be carefully checked for quality and quantity by electrophoresis in agarose gel, using standard procedures (Sambrook *et al.*, 1989).

Denaturing gradient gel electrophoresis (DGGE)

In order to obtain a fingerprint of the soil microbial community targeted, the PCR products obtained with soil DNA as the target are subjected to electrophoresis over a polyacrylamide gel containing a gradient of denaturing substances via standard procedures. This method is described fully in the *Molecular Microbial Ecology Manual* (Akkermans *et al.*, 1995; Kowalchuk *et al.*, 2004) and is summarized under 'Procedures' (below). The method is able to separate DNA fragments of the same length but with different nucleotide sequences, such as those generated by PCR with 16S or 18S ribosomal RNA gene-based primers. The separation is based on the differentially decreased mobility of the partially melted DNA molecules in a linearly increasing gradient of denaturants (urea and formamide). The melting occurs in discrete melting domains of the molecule. Once the melting condition of a particular region is reached, the helical structure of the double strand turns into a partially melted structure with greatly reduced migration in the gel. Differences in the sequences of the molecular types will cause their migration behaviour to differ. This results in a banding pattern, in which each band, in principle, represents a different molecular type.

In practical terms, a GC-clamp (a 40 bp high-guanine-plus-cytosine-containing stretch) has to be attached to the 5′-end of one primer. The clamp prevents complete melting of the molecules in the denaturing gradient, which results in partially melted molecules of which the migration is almost completely halted.

The DGGE gel has to be run under constant voltage for 4–16 h, after which the gel is stained by ethidium bromide, SYBR green or SYBR gold, or by silver nitrate. The stained gel can be visualized under UV light and gel pictures can be further analysed using relevant computer programs, such as GelCompar (Rademaker, 1995). To calibrate the method and the analysis, a marker containing amplicons of known position in the gel needs to be run in parallel (defining the relative positions of bands). If needed, specific bands from the patterns can be excised from the gel, re-amplified and subjected to sequencing.

Materials and Apparatus

Soil DNA extraction

- Bead beater (Braun's cell homogenizer) or similar
- Glass beads (0.10–0.11 mm)
- Common laboratory ware (glass or plastic tubes, microcentrifuge tubes, vials, pipettes)
- Gel electrophoresis apparatus and electricity source

PCR

- Reaction tubes or arrays
- Thermal cycling apparatus, e.g. Perkin-Elmer (Nieuwerkerk a/d IJssel, NL) or similar
- Gel electrophoresis apparatus and source

Denaturing gradient gel electrophoresis (DGGE)

- DGGE apparatus, e.g. Ingeny (Goes, The Netherlands), PhorU and power
- Gradient maker
- Software program for analysis of banding patterns, e.g. GELCOMPAR (Applied Maths, Sint-Martens-Latem, Belgium)

Chemicals and Solutions

Soil DNA extraction

- Tris-buffered phenol, pH 8.0
- 120 mM sodium phosphate buffer, pH 8.0
- Lysozyme
- 20% sodium dodecyl sulphate (SDS)
- Chloroform:iso-amylalcohol (24:1)
- 5 M NaCl
- Ice-cold 96% ethanol
- TE buffer pH 8.0 (10 mM Tris–HCl pH 8.0, 1 mM EDTA pH 8.0; Sambrook *et al.*, 1989)
- 8 M KAc
- CsCl
- 0.1% diethyl-pyrocarbonate (DEPC) solution

PCR

- Sterile deionized water
- Buffer for *Taq* DNA polymerase Stoffel fragment (Perkin Elmer) 10×
- Each dNTP (1 mM) mix, 5×
- $MgCl_2$ 25 mM
- Primer 1 10 μM
- Primer 2 10 μM
- Formamide
- T4 gene 32 protein (5 mg/ml)
- *Taq* DNA polymerase Stoffel fragment (10 U/μl)

DGGE

- Formamide (Merck, Darmstadt, Germany), deionized using standard procedure (e.g. by 5% AG 501-X8 resin treatment). *Avoid contact with skin and eyes*
- AG 501-XS Resin (BioRad Laboratories, Veenendaal, The Netherlands)
- Tris base (Boehringer, Mannheim, Germany)
- Acetic acid, anhydrous (Merck)
- 30% Acrylamide (4K Mix 37.5:1). *May cause cancer. Toxic if in contact with skin and if swallowed*
- Urea
- Ammonium persulphate (APS) (BioRad) 20% in Milli-Q water
- TEMED (*N,N,N,N'*-tetramethyl ethylenediamine) (Sigma, Zwijndrecht, The Netherlands)
- SYBR Gold 1, 10,000× concentrated in dimethyl sulphoxide (DMSO) (Molecular Probes, Leiden, The Netherlands), per gel:
  - 250 μl 20× Tris–acetate–EDTA (TAE)
  - 9.75 ml Milli-Q water
  - 2.5 μl SYBR Gold 1
- 20× TAE:
  - 97 g Tris base in 500 ml Milli-Q water; set at pH 7.8 with acetic acid
  - 32.8 g sodium acetate
  - 40 ml 0.5 M EDTA (pH 8)
  - adjust volume to 1 l with Milli-Q water
- Solution A:
  - 100 ml 30% acrylamide 4K mix (37.5:1)
  - 5 ml 50× TAE
  - adjust volume to 500 ml with sterile Milli-Q water
  - store in dark at room temperature
- Solution B:
  - 100 ml 30% acrylamide 4K mix (37.5:1)
  - 5 ml 50× TAE
  - add 50 ml of sterile Milli-Q water
  - 168 g urea (Wm = 60.06 g/mole)
  - 160 ml formamide, deionized
  - adjust volume to 500 ml with sterile Milli-Q water, heat 1 h at 37°C
  - store in dark at room temperature
- Stacking gel solution:
  - 26.67 ml 30% acrylamide (37.5:1)
  - 1 ml 50× TAE
  - adjust volume to 100 ml with sterile Milli-Q water
  - store in dark at room temperature
- Loading buffer (6×):
  - 0.05% (w/v) bromophenol blue
  - 40% (w/v) sucrose
  - 0.1 M EDTA (pH 8.0)
  - 0.5% (w/v) SDS

Procedures

Soil nucleic acid extraction method I

The procedure described will primarily extract microbial DNA from soil. It is based on the extraction of 10 g of soil. It can be scaled down easily to accommodate 1–5 g soil samples. The full procedure is often needed to obtain restrictable and amplifiable DNA from a loamy sand soil with high organic matter content, whereas purification until (and including) purification step I can be sufficient for DNA from soils with low organic matter content (e.g. silt loam types).

1. Resuspend 10 g of soil in 15 ml 120 mM sodium phosphate buffer pH 8.0 (Sambrook *et al.*, 1989) in a 50 ml polypropylene tube.

2. *Lysozyme treatment* (optional):
(i) Add 75 mg lysozyme to the soil suspension, homogenize and incubate for 15 min at 37°C.
(ii) Chill on ice.

3. *Bead-beating lysis*:
(i) Transfer the soil suspension obtained under **1** (above) to a bead-beating vial containing 15 g glass beads (0.09–0.13 mm diameter).
(ii) Homogenize three times for 90 s in the bead beater (MSK cell homogenizer, B. Braun Diessel Biotech) at 4000 oscillations/min with intervals of 15–30 s.
(iii) Transfer the lysate to a 45 ml centrifuge or 50 ml polypropylene tube.
(iv) Add 900 μl of 20% sodium dodecyl sulphate (SDS) and mix well.
(v) Leave either on ice for 1 h to enhance lysis, or at room temperature for 15 min.

4. *Freeze/thaw lysis: As an alternative to bead beating, freeze/thaw lysis can be applied. However, we strongly recommend bead beating, as it results in more complete lysis.*
(i) Add 900 μl of 20% SDS to the soil slurry obtained under **2** (above) and mix.
(ii) Freeze at –20°C (or –80°C) for 1 h, then keep at 37°C for 30–45 min.
(iii) Repeat freeze/thaw cycle twice.

5. *Extraction and precipitation*:
(i) Add an equal volume of Tris-buffered phenol pH 8.0 (Sambrook *et al.*, 1989) to the lysed cell slurry obtained under **3** (above).
(ii) Mix well (manually) and centrifuge for 5 min at 10,000 × *g* (or 15 min at 3000 × *g* for polypropylene tubes) at room temperature.
(iii) Recover the aqueous (upper) phase in a new centrifuge (or polypropylene) tube.
(iv) Back-extract the phenol/soil mixture with 5 ml 120 mM sodium phosphate buffer (pH 8.0).
(v) Pool the aqueous phases.
(vi) Extract the pooled aqueous phases with an equal volume of chloroform/iso-amylalcohol (24:1).

(vii) Recover the upper aqueous phase. If a heavy interphase is present, extract the aqueous phase again with an equal volume of chloroform/iso-amylalcohol.
(viii) Add 0.1 volume of 5 M NaCl and two volumes of ice-cold 96% ethanol. Keep at –80°C for 20 min or at –20°C for at least 1 h (often overnight).
(ix) Centrifuge for 5–10 min at 10,000 × *g* (or 15–20 min at 3000 × *g* for polypropylene tubes).
(x) Discard the supernatant and wash the pellet with ice-cold 70% ethanol. Air-dry pellet.
(xi) Resuspend pellet in 1–1.5 ml (sterile) TE buffer pH 8.0 (10 mM Tris–HCl pH 8.0, 1 mM EDTA pH 8.0). Due to the volume of the pellet, the final volume may be larger.
(xii) The solution at this stage is called the crude extract.

6. *Purification step I: CsCl and KAc precipitations.* Perform all steps at room temperature, unless stated otherwise.
(i) Add 0.5 g CsCl to 500 μl crude extract.
(ii) Incubate for 1–3 h.
(iii) Centrifuge for 20 min at maximum speed in an Eppendorf centrifuge.
(iv) Recover the supernatant (≈500 μl) in a 10 ml tube.
(v) Add 2 ml deionized water and 1.5 ml iso-propanol. Mix and incubate for a minimum of 5 min at room temperature.
(vi) Centrifuge for 15 min at 10,000 × *g*. When polypropylene tubes are employed, use maximally at 3000 × *g* for 20 min. Check degree of pelleting.
(vii) Discard supernatant and resuspend pellet in 500 μl TE buffer (pH 8.0) and transfer the suspension to a new Eppendorf tube.
(viii) Add 100 μl 8 M KAc, mix and incubate for 15 min at room temperature.
(ix) Centrifuge for 15 min at maximum speed in an Eppendorf centrifuge.
(x) Recover supernatant and add 0.6 volume of iso-propanol, mix and incubate for 5 min at room temperature.
(xi) Centrifuge for 15 min at full speed in an Eppendorf centrifuge.
(xii) Wash pellet with ice-cold 70% ethanol, dry and resuspend it in 500 μl TE buffer (pH 8.0).

7. *Purification step II: Wizard DNA Clean-Up System* (Promega, Madison, Wisconsin, USA). The system is based on DNA adsorption to clean-up resin in 6 M guanidine thiocyanate, washing with 80% iso-propanol and elution with TE buffer or deionized water. The reagents can be kept at room temperature protected from exposure to direct sunlight.
(i) Add 1 ml of DNA clean-up resin to 250 μl partially purified DNA extract (after purification step I) in an Eppendorf tube and mix by gently inverting several times.
(ii) Attach the syringe barrel of a 2.5 ml disposable luer-lock syringe to the extension of the Wizard minicolumn.
(iii) Pipette the suspension from the Eppendorf tube into the syringe barrel. Using the plunger, push the slurry slowly into the minicolumn.

(iv) To wash the column, pipette 2 ml of 80% iso-propanol into the syringe barrel. Gently push the solution through the minicolumn.
(v) Remove syringe and place the minicolumn containing the loaded resin on top of an Eppendorf tube. Centrifuge for 20 s at full speed in an Eppendorf centrifuge to dry the resin. Leave the minicolumn at room temperature for 5–15 min to evaporate traces of iso-propanol still present.
(vi) Transfer the minicolumn to a new Eppendorf tube. Apply 125 μl of prewarmed (60–70°C) TE buffer (pH 8.0) or deionized water and leave for 5–10 min. The DNA will remain intact on the minicolumn for up to 30 min. Centrifuge for 20 s at full speed in an Eppendorf centrifuge to elute the bound DNA.
(vii) Repeat step (vi) using the same Eppendorf tube. The total volume of eluate will be about 250 μl.
(viii) Discard minicolumn. The purified DNA may be stored at 4°C or –20°C.
(ix) If needed (as judged by colour and/or suitability for restriction digestion or amplification), repeat steps (i)–(viii).

Soil nucleic acid extraction method II

Before starting the extraction, it is important to prepare RNase-free solutions and materials. Non-disposable materials must be immersed in 0.1% diethyl-pyrocarbonate (DEPC) solution and autoclaved. Alternatively, glassware and metal spatulas can be baked at 400°C for 4 h to inactivate RNases, or cleaned with RNase Away (Molecular Bio-Products, San Diego, California, USA). All solutions must be prepared with deionized water previously treated with 0.1% (vol/vol) DEPC. Add DEPC to the water and incubate for at least 2 h at 37°C to inactivate RNase. The water must be autoclaved after treatment to destroy DEPC.

1. Add soil samples (up to 0.5 g wet weight) into 2 ml microcentrifuge tubes containing 0.4 g glass beads (0.10–0.11 mm). Bacterial pellets (up to 0.5 g wet weight) extracted from environmental samples or microbial cultures can also be processed by following the next steps. Keep samples on ice.
2. Add up to 0.8 ml of ethanol to the samples until they are totally immersed. Homogenize the samples twice by using the FastPrep FP120 bead-beating system (Bio 101, Vista, California, USA) at 5.5 m/s for 30 s. Alternatively, bead beating can be performed by using a cell homogenizer (B. Braun Diessel Biotech) twice for 30 s (4000 oscillations/min). Samples should be kept on ice (30 s) between the bead-beating steps.
3. Centrifuge the tubes for 5 min at 16,000 × *g*. Discard the supernatant and add 1.0 ml of extraction buffer pH 7.0 (1% hexadecyl trimethylammonium bromide (CTAB), 2% SDS, 1.5 M NaCl, 100 mM sodium phosphate buffer pH 7.0, 10 mM Tris–HCl pH 7.0 and 1 mM EDTA pH 8.0). Mix the samples thoroughly and incubate for 30 min at 65°C, mixing carefully every 10 min.

4. Centrifuge the tubes at 16,000 × *g* for 5 min and transfer the supernatant into 2 ml tubes containing 1 ml aliquots of chloroform:isoamyl alcohol (24:1) previously chilled on ice. Mix the samples thoroughly and centrifuge at 16,000 × *g* for 5 min.
5. Transfer the upper phase into new microcentrifuge tubes. Precipitate the nucleic acid by addition of 0.6 volume of isopropanol, incubation for at least 30 min at room temperature, and centrifugation at 16,000 × *g* for 20 min. Pelleted nucleic acids can be either washed with 70% (vol/vol) ethanol for subsequent RNA/DNA separation or directly stocked into 0.5 ml 70% (vol/vol) ethanol at –70°C until use.
6. Before starting the RNA/DNA recovery procedure, centrifuge suspensions at 16,000 × *g* for 5 min, carefully remove the ethanol and air-dry the pellet containing the nucleic acids for 5 min (do not dry the pellet completely). Dissolve the nucleic acids in 200 µl of DEPC-treated deionized water.
7. Collect 100 µl of the extracted nucleic acids for RNA purification using the RNeasy Mini Kit (QIAGEN GmbH, Germany) according to the manufacturer's instructions. If necessary, this step can be repeated for better RNA purification. Use the other 100 µl aliquot for DNA purification using the GENECLEAN spin kit (Q Biogene, USA) according to the manufacturer's instructions. Alternatively, to obtain higher yields, DNA and RNA can be separated at once from the same sample by using the QIAGEN RNA/DNA mini kit (QIAGEN GmbH, Germany). However, according to our experience, further purification steps might be necessary before using the nucleic acids recovered for PCR applications and gene expression analysis.

PCR

Since DNA amplification via PCR is extremely sensitive, care should be taken not to contaminate the sample material or new PCR reaction mixes with target DNA or PCR products resulting from previous reactions, since this might lead to false positives. Aerosols, which may form when samples containing target DNA or PCR product are handled, are notorious sources of contamination. Therefore, sample preparation and processing, setting up of the PCR reaction mixes, and, in particular, handling of the PCR products, all have to be performed with extreme care. To avoid the occurrence of false-positive results, we find it adequate to have separate sample preparation, PCR and product analysis rooms, with separate equipment, including pipettes. In addition, all glass- and plasticware used in the PCR room should be exclusively handled there. Further, PCR reagents should only be prepared in the PCR room, and divided in aliquots, which are then stored at –20°C.

A notorious problem when using PCR amplification systems based on conserved regions of the 16S ribosomal RNA gene is the presence of these bacterial sequences as contaminants in many commercially available enzymes. To avoid amplification of this instead of target DNA, resulting in false results when amplifying soil DNA with conserved 16S ribosomal RNA gene-based primers, enzyme solutions as well as PCR reaction mixes should be treated so as to remove any 16S ribosomal gene sequences. Several strategies have been developed for this, e.g. treatment with psoralen, UV irradiation or treatment with DNase (Steffan *et al.*, 1988). We commonly treat the PCR reaction mixes and the *Taq* polymerase with DNase I, as outlined below.

The PCR protocol described in the following will address the amplification of soil DNA using eubacterial primers to generate fingerprints of bacterial communities of soil. For amplification of other microbial groups, such as the fungi or specific bacterial groups, the reader is referred to the relevant literature.

Treatment of (Taq) *DNA polymerase with DNase I to remove contaminating bacterial DNA*

1. Add 10 µl DNase I (10 U/µl) to PCR reaction mix (without *Taq* polymerase and primers) for ten reactions.
2. Add 0.1 µl DNase I to 5 µl *Taq* DNA polymerase Stoffel fragment (10U/µl).
3. Incubate tubes for 30 min at 37°C, then inactivate DNase I by heating at 98°C for 10 min. The reaction mix as well as enzyme can be used directly for PCR.

The enzyme used is AmpliTaq DNA Polymerase, Stoffel fragment (Perkin Elmer/Cetus). In our laboratory, this enzyme was shown to be least inhibited by soil impurities. However, several other enzymes may work as well on soil-derived DNA.

Procedure

1. Prepare master mix of reagents for any number of 50 µl reactions. Each mixture contains (final amounts in 50 µl final reaction volume):

Sterile deionized water	23.45 µl
Stoffel buffer 10×	5 µl (final conc. 1×)
Each dNTP (1 mM) mix – 5×	10 µl (final conc. 200 µM)
$MgCl_2$ 25 mM	7.5 µl (final conc. 3.75 mM)
Primer 1 10 µM	1 µl (final conc. 0.2 µM)
Primer 2 10 µM	1 µl (final conc. 0.2 µM)
Formamide*	0.5 µl (final conc. 1%)
T4 gene 32 protein (5 mg/ml)*	0.05 µl (final conc. 0.25 µg/50 µl)
Taq DNA polymerase Stoffel fragment (10 U/µl)	0.5 µl (final conc. 0.1 U/µl)

Total volume 49 μl. Cover with two drops of heavy mineral oil. An aliquot of 1 μl soil DNA extract is added through the oil to the reaction mixture containing all other components in the 'hot start' procedure (see **3** below) prior to starting thermal cycling.

*Formamide (Merck) is added to enhance specific primer annealing. T4 gene 32 protein (Boehringer) is added since it enhances the stability of single-stranded DNA, facilitating primer annealing (Tebbe and Vahjen, 1993). We found a sixfold lower concentration than that used by Tebbe and Vahjen (1993) to be optimal.

2. Prepare positive and negative controls, as indicated above.
3. 'Hot start' procedure: pre-heat the PCR mixture at 95°C for 2 min prior to adding enzyme or target DNA, in order to denature any bonds adventitiously or erroneously formed between primers and target.
4. Run PCR in a programmable thermal cycler, for 25–40 cycles. Often, the machine can be set so as to run the amplification cycles overnight.
5. After thermal cycling, perform a final primer extension at 72°C (e.g. 10 min), and keep reaction mixtures at 4°C until analysis.

Denaturing gradient gel electrophoresis (DGGE) of PCR products

Denaturing gradient gel electrophoresis (DGGE) separates DNA molecules of similar size by virtue of differences in their internal sequence (i.e. their melting behaviour) in a gradient of increasing denaturant (formamide and urea) strength. The gels are run at 60°C in a DGGE apparatus for 16 h.

Denaturing polyacrylamide gels

Polyacrylamide gels are formed by the polymerization of monomeric acrylamide into polymeric acrylamide chains and the cross-linking of these chains by *N,N'*-methylene bisacrylamide. The polymerization reaction is initiated by the addition of ammonium persulphate (APS), and the reaction is accelerated by TEMED (*N,N,N',N'*-tetramethylethylenediamine), which catalyses the formation of free radicals from APS. The porosity of the gel is determined by the length of the chains and the degree of cross-linking. The length of the chains is determined by the concentration of acrylamide in the polymerization reaction.

The protocol below is described for use with the Ingeny PhorU system, which is recommended for ease of use.

1. Choose the steepness of the desired gradient, e.g. 50–75%, using the suggested mixtures for the 'upper' solution (e.g. 50%), and for the lower solution (e.g. 75%, see Table 8.2).
2. Prepare 25 ml of each of these solutions in a 50 ml bluecap tube. Also, place 5 ml of solution A in a 12 ml bluecap tube (for stacking gel). Leave at room temperature for at least 30 min.
3. Thaw a tube with 10% APS from freezer (–20°C).

4. Prepare set-up for casting the gel (see 'Step by step' manual Ingeny PhorU-2, pp. 8–9). Clean glass plates and spacers, respectively, with KOH/methanol, soap and water, deminerilized water and ethanol. Dry the plates.
5. Add 150 µl 10% APS to the 'upper' and the 'lower' solution.
6. Add 12 µl TEMED to the 'upper' and the 'lower' solution, mix well, and proceed *immediately* with the following steps (**7** and **8**).
7. Place 25 ml of 'lower' solution (highest denaturant concentration, e.g. 75%) in the chamber of the gradient mixer connected to the pump. Open tap to the other chamber to let the air in the tap escape, and close. Transfer liquid from the other chamber back to the right chamber using a Pasteur pipette. Add magnetic stirring rod.
8. Place 25 ml of 'upper' solution (lowest denaturant concentration, e.g. 50%) in the other, empty chamber.
9. Switch on magnetic stirrer (position 4) to stir the 'lower' solution. Switch on pump and immediately carefully open the tap connecting the two chambers. Pump speed should be 4–6 ml/min. Gel should be cast in 15 min. Leave for 15–30 min to solidify.
10. Add 50 µl 10% APS and 7 µl TEMED to 5 ml solution A (stacking gel – 8% acrylamide). Using a 5 ml syringe with a long needle, slowly pour the top part of the gel by hand. Avoid air bubbles. Add the comb. Wait for at least 1 h for the gel to polymerize completely.
11. Switch on buffer tank at 60°C (takes 45 min to heat up). The buffer tank contains 15 l of 0.5× TAE.
12. After gel polymerization, all screws should be adjusted to just touch the plexiglass pressure unit. Place the cassette in the buffer tank. Push the U-shaped spacer all the way down now. Little air bubbles under the gel can be removed by holding the cassette at an angle. Remove the comb. Tighten the upper two screws again. Connect buffer flow to connector on cassette. Fill upper buffer tank and close the tap.
13. Connect the electrical plugs. Add the samples. Set power supply to the desired voltage (e.g. 75 V) and time (e.g. 16 h). Wait 10 min, while electrophoresis is in progress, before opening the upper buffer flow again. The reservoir should not overflow.
14. After running, remove gel. Stain with SYBR Gold (1 h) and rinse. Observe on UV transilluminator.

Calculation

The molecular community profiles can be loaded into a software program such as GelCompar – see Rademaker (1995) for details. The program will digitize the profiles and can assign values to each band, in accordance with its intensity. The matrix of band intensity values per lane can then be used in any statistical approach that allows rigid comparison of the profiles. Depending on the purpose of the comparisons, principal components

Table 8.2. Mixtures of solutions A (0% denaturants) and B (80% denaturants) to achieve desired denaturant concentrations.

Denaturant concentration (%)	Solution A (ml)	Solution B (ml)
25	17.2	7.8
30	15.6	9.4
35	14.1	10.9
40	12.5	12.5
45	11	14
50	9.4	15.6
65	4.7	20.3
70	3.1	21.9
75	1.6	23.4

analysis, canonical correspondence/variant analysis or discriminant analysis can be employed on the basis of the collective data. See relevant literature as well as Section 8.3 of this volume.

Discussion

The DNA extraction protocols selected and described here are very suitable for the detection of specific microorganisms and their genes in soil (Smalla *et al.*, 1993a; Duarte *et al.*, 1998; van Elsas *et al.*, 2000). However, as discussed, there may still be some doubt as to the localization of such DNA sequences, i.e. inside cells or extracellularly. As this is a potential problem inherent to all soil DNA extraction protocols, including the cell extraction-based ones, results obtained should be carefully interpreted as to their meaning for the actual cell populations present. Furthermore, even though the bead beater is known to lyse a major part of bacterial species efficiently, including many Gram-positives, there is no absolute certainty that cell lysis is representative for the soil bacterial community. The findings of Moré *et al.* (1994) suggest that lysis may well be confined to the larger cell-size fraction of the microbial community, leaving a major fraction of minute cells unlysed. This feature also has to be taken into account when the protocol is to be used for microbial community structure studies in soil. In fact, it is clear that with this, or any other, DNA extraction protocol, only a limited view of the soil microbiota can be obtained, due to the fact that the extraction/lysis efficiency is commonly below 100% of the total detectable cell populations. The primers used for PCR amplification of soil DNA are obviously determinative for the level at which different microbial groups can be assessed. Thus, diversity indices can be produced for broad microbial groups, such as all bacteria or all fungi. At a finer level of resolution, a picture of the diversity of specific groups (such as the α- or β-proteobacteria, the high-G+C% actinomycetes or the pseudomonads) can be obtained. The advantage of using the group-specific approach is that the enormous complexity of the microflora in soil can be reduced to a level that is better interpretable. In fact, one can focus on those microbial groups that are less numerous in the soil, and would have been overlooked in pictures obtained for the broad groupings, as they would be minority organisms there.

The fingerprints obtained via DGGE separation are often robust and reliable for a given soil or soil treatment, reporting on the microbial diversity status of the sample. The fingerprint obtained quite often provides a representative picture of the numerically dominant types in the microbial group targeted. However, as with any method applied to soil, the DGGE data are not without caveats. These caveats have been described extensively in the literature, and will be discussed only briefly here. First, whereas PCR amplification will often be without bias, that is, each particular sequence will be amplified at about the same rate, sometimes aberrations from this common rate have been observed. This results in a phenomenon called 'preferential amplification', in which a particular sequence is amplified at a (much) higher rate than others, leading to a distorted picture of the relative abundance of the different microbial types present. Secondly, a fraction of the bands that make up the DGGE profile are actually chimeric sequences, which are the result of an aberrant PCR process; the percentage at which chimeras can be formed varies, but low frequencies (a few percent) to frequencies as high as 30% have been reported. To get a handle on this PCR artefact, one can use sequence analysis followed by the feature 'Check_chimera' available in several sequence analysis programs.

Other features of PCR-DGGE analysis of soil DNA are the known facts that some bacterial types can produce multiple bands on DGGE (a result of the presence of several slightly different rRNA operons in the same cell), and that fragments of different sequence can migrate to the same gel positions. Therefore, a note of caution is in place with respect to the interpretation of the profiles generated by DGGE, and the use of these profiles to produce diversity indices.

With all these cautionary remarks in mind, one can safely state that the current protocol opens a wide window for studying the diversity of total microbial populations in soil at various different levels of resolution, an ability that one could only dream of just a decade ago. It is foreseen that developments will continue and that the current methodology will be refined, or even superceded, by other direct methods, such as the use of DNA microarray technology. However, we would like to propose the currently described method as the method of choice for routine assessments of the microbial diversity of soil, possibly at two levels, i.e. that of total bacteria (Kowalchuk *et al.*, 2004) and of total fungi (Vainio and Hantula, 2000).

8.3 Phosphoplipid Fatty Acid (PLFA) Analyses

Ansa Palojärvi

MTT – Agrifood Research Finland, FIN-31600 Jokioinen, Finland

Introduction

The analysis of ester-linked (EL) phospholipid fatty acids (PLFAs) is an acknowledged biochemical approach to microbial community characterization. PLFAs are constituents of all cell membranes, and have no storage function. Under the conditions expected in naturally occurring communities, phospholipids represent a relatively consistent fraction of cell mass, even though some changes are detected in pure cultures due to changes in growth media composition or temperature (Harwood and Russel, 1984). Lipid analysis offers an alternative that does not rely upon the cultivation of microorganisms. Current extraction and derivatization methods permit effective recovery of PLFAs from living organisms (White, 1988). PLFAs also degrade quickly upon an organism's death (White, 1988), if the degrading enzymatic activity is not inhibited (see Zelles *et al.*, 1997), which allows the detection of rapid changes in microbial populations.

Extraction and subsequent analysis by gas chromatography and mass spectrometry provides precise resolution, sensitive detection and accurate quantification of a broad array of PLFAs. Each analysis yields a profile composed of numerous PLFAs defined on the basis of compound structure and the quantity of each compound present in the sample. The advantage of the method is that the amount of total PLFAs can be used as an indicator for viable microbial biomass, and that further characterization can be done based on specific signature biomarker fatty acids. Taxonomically, the method does not include *Archaea*, since they have ether-linked rather than ester-linked phospholipid fatty acids in their cell membranes. There are no indicated restrictions to the application of the method for any kinds of soil and environmental samples and it is relatively time- and cost-competitive.

The PLFA method has been applied for various environmental questions. Changes in the PLFA patterns have been detected according to the different levels of metal contamination (Pennanen, 2001). Agricultural management practices (Petersen *et al.*, 1997; Schloter *et al.*, 1998) and pesticide use (Widmer *et al.*, 2001) have been shown to cause shifts in the PLFA patterns. Changes in the PLFA patterns have been detected after chloroform fumigation of soil (Zelles *et al.*, 1997). Validation of PLFA in the determination of microbial community structure has been reviewed, e.g. by Zelles (1999) and Pennanen (2001).

Certain PLFAs can serve as unique signatures for specific functional groups of microorganisms. Such biomarkers cannot detect individual species of microorganisms due to overlapping PLFA patterns. Nevertheless, comparison of total community PLFA profiles accurately mirrors shifts in community composition and provides a way to link community composition to specific metabolic and environmental conditions. Signature fatty acids are listed by several authors (Morgan and Winstanley, 1997; Zelles, 1999; Kozdrój and van Elsas, 2001a). For fungal biomarkers see Frostegård and Bååth (1996), Miller *et al.* (1998) and Olsson (1999).

Principle of the Methods

The phospholipid fatty acid (PLFA) analysis is based on the single-phase extraction of lipids described by Bligh and Dyer (1959). The lipids are fractionated into different lipid classes: neutral lipids, glycolipids and phospholipids (Fig. 8.1). Neutral and glycolipids are normally handled as waste fractions and disregarded in microbial community analysis, although they can be used for other purposes, e.g. to describe the nutritional status of microbes (White *et al.*, 1998).

The phospholipid fraction is then methylated to give fatty acid methyl esters (FAME) and analysed by gas chromatograph (GC) with flame-ionization detector (FID) or GC coupled with a mass spectrometer (GC–MS). Due to the very variable contents of different PLFA in cell membranes, the method without further fractionation (Fig. 8.1A) is able to detect the most abundant ester-linked PLFA only. In most cases, 20–40 PLFA are identified. They make up most of the biomass, but not of the number of PLFA in the cells (see Zelles, 1999). This method has been successful in separating microbial communities in various experiments and it is applicable for monitoring.

Extended PLFA analysis discovers a very wide variety of cellular fatty acids, including both ester-linked and non-ester-linked PLFAs (Fig. 8.1B), and offers good potential for the use of signature fatty acids.

Whole-cell fatty acid patterns are based on FAME analysis after direct saponification and methylation of lipids without fractionation to different fatty acid groups (Fig. 8.1C). The method was originally designed for microbial identification of pure cultures (the commercially available Microbial Identification System (MIDI or MIS); Haack *et al.*, 1994; Kozdrój and van Elsas, 2001b). Even though both the whole-cell fatty acid patterns and PLFA patterns are based on FAME analysis, it should be noted that the methods are not comparable. The whole-cell fatty acid patterns (often called the FAME method) comprise lipids derived from non-living organic matter (Petersen *et al.*, 2002), and include storage lipids, which are more sensitive to growth conditions. The PLFA and FAME methods have been reviewed by Zelles (1999).

In this section, a procedure for PLFA analysis is described which is based on White *et al.* (1979), Frostegård *et al.* (1993) and Palojärvi *et al.* (1997), with slight modifications.

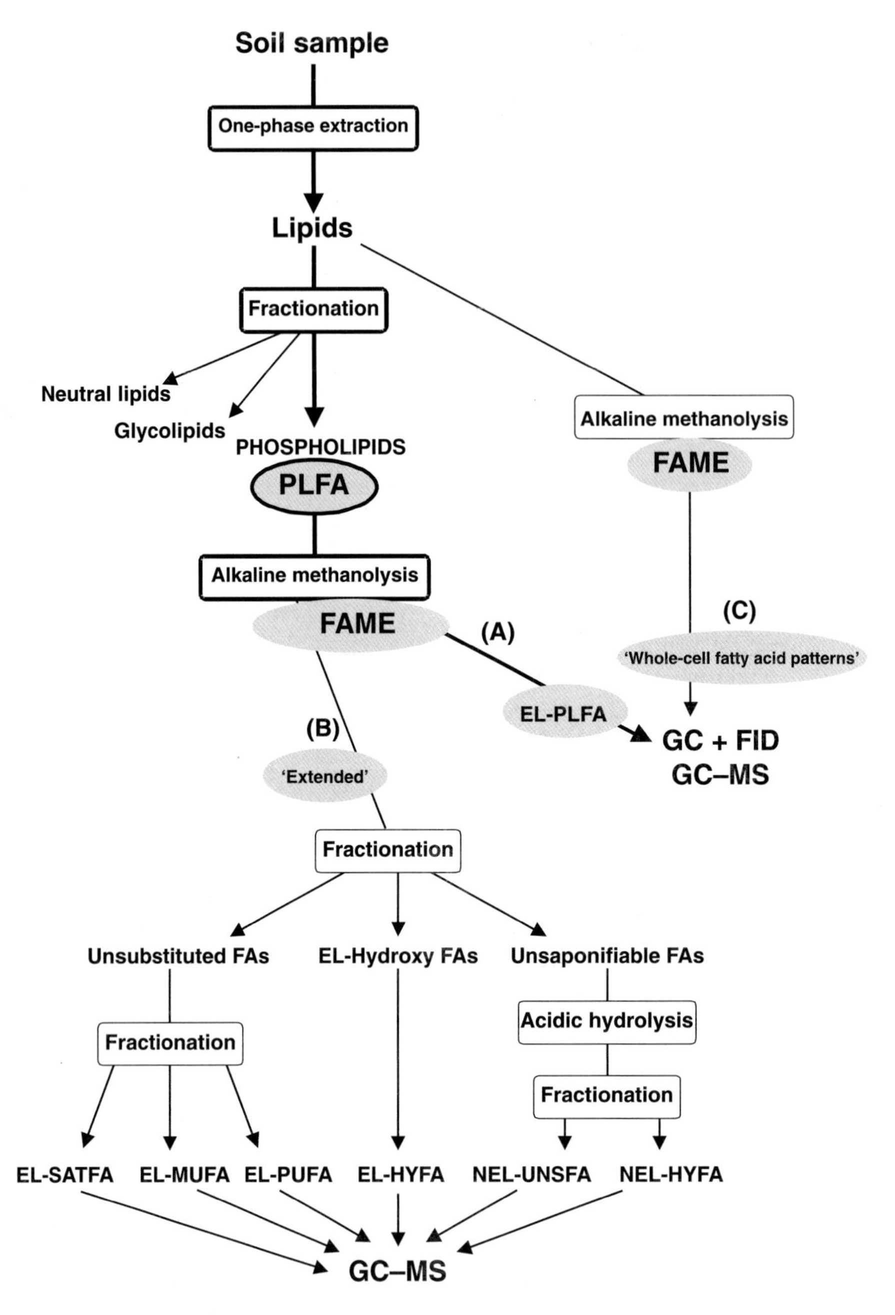

Fig. 8.1. Flow diagram of the extraction and fractionation steps of (A) ester-linked phospholipid fatty acid (EL-PLFA); (B) extended PLFA; and (C) whole-cell fatty acid analyses. Abbreviations: FAME, fatty acid methyl esters; GC, gas chromatograph; FID, flame-ionization detector; MS, mass spectrometer; FA, fatty acid; EL, ester linked; SATFA, saturated FA; MUFA, monounsaturated FA; PUFA, polyunsaturated FA; HYFA, hydroxy FA; NEL, non-ester linked; UNSFA, unsubstituted FA. Figure modified after Zelles and Bai (1994) and Palojärvi and Albers (1998).

Materials and Apparatus

In addition to standard laboratory equipment, the following supplies and apparatus are needed:

Supplies

- Glass test tubes and small glass bottles with Teflon-lined screw caps (ca. 50 ml and 10 ml test tubes for centrifuge, 4–10 ml bottles for various solvent fractions)
- Disposable glass pipettes (Pasteur, Micro pipettes) and compatible pipettors
- Nitrogen gas
- Silicic acid (e.g. Unisil 100–200 mesh) or commercial solid-phase extraction columns (e.g. Varian Bond Elut SILICA Si 500 mg or 2 g)

Apparatus

- Fume hood
- Vortex mixer
- (Orbital) shaker
- Centrifuge
- Nitrogen evaporator
- Gas chromatograph (GC) with flame-ionization detector (FID) or, preferably, GC coupled with a mass spectrometer (GC–MS). The GC should be equipped with a splitless injector and a long non-polar capillary column (e.g. HP-5; 0.2 mm internal diameter, 0.33 μm film thickness, 50 m column length)

Chemicals and solutions

- Citrate buffer (0.15 M $Na_3C_6H_5O_7{\cdot}2H_2O$, pH 4.0), or phosphate buffer (50 mM K_2HPO_4, pH 7.4)
- Solvents (chloroform stabilized with ethanol, methanol, acetone, toluene, hexane, iso-octane); analytical grade or higher
- Bligh-and-Dyer solution (chloroform:methanol:buffer, 1:2:0.8; v/v/v)
- Methanol:toluene (1:1; v/v)
- Methanolic KOH (0.2 M; make fresh daily)
- Hexane:chloroform (4:1; v/v)
- Acetic acid (1.0 M)
- Internal standard 19:0 (methyl nonadeconoate; Sigma)

- Fatty acid methyl ester standards (e.g. Sigma, Supelco, Nu-Chek-Prep)

Note: All chemicals are analytical grade or higher; chloroform should be stabilized with ethanol. Phosphate buffer is widely used for PLFA extractions, but Nielsen and Petersen (2000) showed that citric buffer gives the highest PLFA yields.

Procedure

Lipid extraction

Lipid extractions are carried out on a few grams of fresh, frozen or freeze-dried soil. Fresh soil can be kept in a refrigerator for a few weeks (Petersen and Klug, 1994). The proper amount of soil should always be established beforehand. The amount of PLFA in the sample should be high enough to be above the detection limit. On the other hand, the presence of very high amounts of PLFA may cause problems during the analyses. In experiments with different soils or treatments, it is recommended to standardize the amounts of material to give similar levels of microbial biomass. The single-phase extraction mixture contains chloroform, methanol and buffer in the ratio 1:2:0.8. The water content of the sample is calculated and a sufficient amount of buffer is added. The following procedure is suitable for 1–4 g of soil.

- The soil sample is weighed into a 50 ml test tube.
- Add buffer so that the total water content of the soil is 1.5 ml.
- Add 1.9 ml chloroform, 3.7 ml methanol and 2.0 ml Bligh-and-Dyer solution.
- Mix well using a Vortex mixer.
- Leave the sample in a (orbital) shaker at a low speed (ca. 200 rpm) for at least for 4 h, or overnight, and centrifuge (*c*. 1000 $\times$ g for 10 min).
- Transfer the supernatant to another 50 ml test tube, wash the soil pellet with 2.5 ml Bligh-and-Dyer solution, repeat the mixing and centrifugation, and combine the supernatants.
- To separate the solvent phase, add 3.1 ml chloroform and 3.1 ml buffer, and mix well with the Vortex (*c*. 1 min).
- Let the mixture stand overnight to separate the lower organic phase (chloroform) from the upper water–methanol phase.
- Transfer the organic phase into a small glass bottle.
- The samples are placed in a water or sand bath (40°C) and dried under a stream of N_2 gas until near dryness.
- The lipid samples are stored in a freezer.

Lipid fractionation

- The lipids are separated into neutral, glyco- and phospholipids on columns containing silicic acid by eluting with chloroform, acetone and methanol, respectively.
- Prior to separation, commercial columns (500 mg) must be activated with 6 ml chloroform.
- The dried lipid samples are transferred on to the columns in 3 × 100 μl chloroform.
- Elute neutral lipids with 6 ml chloroform and glycolipids with 12 ml acetone. For PLFA analysis, these are waste fractions.
- Phospholipids are eluted with 6 ml methanol into a 10 ml test tube.
- The methanol fraction is reduced until dryness under N_2 gas in a water or sand bath (maximum temperature 30°C).

Internal standards

- A known amount of methyl nonadecanoate (19:0; Sigma) is added to the phospholipid fraction as an internal standard.
- Methyl tridecanoate (13:0; Sigma) can be used as an additional internal standard to indicate possible losses of short-chain fatty acids.

Mild alkaline methanolysis

- Samples are dissolved in 1 ml methanol:toluene solution.
- Methanolic KOH (1 ml) is added, and the mixture is incubated in a water bath (37°C, 15 min). Add immediately 2 ml hexane:chloroform solution and 0.3 ml acetic acid to neutralize the solution (pH can be checked from the lower phase; pH 5–7).
- Mix well with Vortex (*c.* 1 min) and centrifuge (*c.* 1000 × *g* for 5 min).
- Transfer the upper, organic phase into a small glass bottle.
- Wash the mixture once with 2 ml hexane:chloroform and combine the upper, organic phase with the former one.
- Reduce the sample with N_2 gas without heating. Place the sample, re-dissolved in 100 μl iso-octane (or hexane), in a GC vial with a glass insert.

GC–MS or GC + FID analysis

The resulting fatty acid methyl esters (FAME) of ester-linked phospholipid fatty acids (EL-PLFA) are separated, quantified and identified by gas chromatography (GC) coupled with a mass spectrometer unit for peak identification (GC–MS). Alternatively, the analysis can be carried out with a GC equipped with a flame-ionization detector (FID); in this case peak identification is based on retention times only. The GC + FID analysis must be confirmed from time to time against GC–MS. Helium is typically used as a carrier gas. The flow rate and temperature programming of the column should be adjusted for the individual instrument. An example of settings: flow rate 0.9 ml/min, initial temperature 70°C for 2 min, increase 30°C/min until 160°C, increase 3°C/min until 280°C, final temperature 280°C for 10 min. The identification and the response factors of different PLFA compounds are based on fatty acid methyl ester standards. Double-bond positions can be determined with their dimethyl disulphide adducts (Nichols *et al.*, 1986).

Fatty acid nomenclature

A short-hand nomenclature used to characterize fatty acids is as follows. In the expression X:YωZ, X indicates the number of carbon atoms in the fatty acid, while Y indicates the degree of unsaturation (= the number of carbon–carbon double bonds). The position of the first double bond from the methyl (or aliphatic; 'ω') end of the molecule is represented by Z. Alternatively, double-bond positions from the carboxyl ('Δ') end are sometimes given. The suffixes 'c' and 't' indicate *cis* and *trans* geometry, respectively. Because *cis* geometry is most common, c is often omitted. The prefixes 'i' and 'a' refer to iso (the second carbon from methyl end) and anteiso (the third carbon from methyl end) branching; 'br' indicates unknown methyl branching position. Other methyl branching is indicated by the position of the additional methyl carbon from the carboxyl end followed by 'Me' (i.e. 10Me18:0). Some authors include the carbon of methyl group in X and use the prefix 'p'. This means, for example, that 10Me16:0 is identical to p10–17:0. The number before the prefix 'OH' indicates the position of a hydroxy group from the carboxyl end (i.e. 3-OH14:0). α and β are sometimes used to indicate a hydroxy substitution at position 2 or 3 from the carboxyl end, respectively. Cyclopropane fatty acids are designated by the prefix 'cy'.

Calculation

The calculation of the concentrations of PLFA is shown in Eq. (8.1):

$$C_x(\text{nmol/g}) = \frac{A_x \times c_i[\mu\text{g}] \times f \times 1000}{A_i \times W[\text{g}] \times M[\mu\text{g}/\mu\text{mol}]} \tag{8.1}$$

where: C_x = concentration of the fatty acid studied; A_x = peak area of the fatty acid studied; A_i = peak area of the internal standard; c_i = absolute amount of internal standard in the vial (μg); f = response factors of different PLFA compounds (peak area to concentration ratio compared to internal standard; if not known, then = 1); W = amount of soil (g); and M = molecule weight of the fatty acid (µg/µmol).

Note: The results can also be expressed as micrograms PLFA per gram dry soil. Often the relative contribution of different PLFA is of major interest ('PLFA fingerprint') and mole fractions (in percentage) are calculated.

Statistical analyses

Multivariate analyses can be applied to PLFA profiles. Principal components analysis (PCA) is most widely used (e.g. Palojärvi *et al.*, 1997; Petersen *et al.*, 1997). Canonical correspondence analysis enables determination of the influence of different environmental factors on the PLFA patterns (Bossio *et al.*, 1998). Non-metric multidimensional scaling (NMDS; Siira-Pietikäinen *et al.*, 2001) and neural computing methods (Noble *et al.*, 2000) have also been applied recently.

Discussion

PLFA analysis is applicable for monitoring and detecting changes in the soil microbial communities. Additionally, viable microbial biomass estimates and information on several biomarker PLFAs can be obtained. Several comparisons have shown that PLFA can detect rapid changes and produce results on microbial characterization comparable to other community-level methods (e.g. Widmer *et al.*, 2001).

Further methodological perspectives: Macnaughton *et al.* (1997) suggested a pressurized hot solvent extraction to enable rapid and improved extraction of lipids from large numbers of environmental samples. Intact phospholipid profiling (IPP), using liquid chromatography/electrospray ionization/mass spectrometry, is an advanced alternative to EL-PLFA analysis by GC–MS (Fang *et al.*, 2000). Further methods have been developed for archaeal ether-linked lipids by Bai and Zelles (1997) and Fritze *et al.* (1999). Different isotope and radiolabelling techniques of PLFA have been applied to study specific biogeochemical processes and microbial activity (Boschker *et al.*, 1998; Roslev *et al.*, 1998).

8.4 Substrate Utilization in Biolog™ Plates for Analysis of CLPP

MICHIEL RUTGERS,[1] ANTON M. BREURE[1] AND HERIBERT INSAM[2]

[1]*National Institute for Public Health and the Environment, Antonie van Leeuwenhoeklaan 9, NL-3721 MA Bilthoven, The Netherlands;* [2]*Universität Innsbruck, Institut für Mikrobiologie, Technikerstraße 25, A-6020 Innsbruck, Austria*

Introduction

For characterization of the microbiology of soil samples, functional aspects related to substrate utilization are as useful as taxonomic or structural investigations based on DNA or RNA analysis (Grayston *et al.*, 1998). An understanding of the functional or metabolic diversity of microbial communities, particularly defined by the substrates used for energy metabolism, is integral to our understanding of biogeochemistry (Hooper *et al.*, 1995). Thus, diversity at the functional level rather than at the taxonomic level may be crucial for the long-term stability of an ecosystem (Pankhurst *et al.*, 1996).

In addition, microbial community analysis on the basis of the best available techniques only addresses a limited number of dominant features. For instance, using PCR-DGGE analysis based on the 16S ribosomal RNA gene sequence, up to about 100 dominant DNA sequences can be recognized. The same holds for other techniques, such as phospholipid fatty acid (PLFA) profiling, morphological characterizations and metabolic diversity analyses (Rutgers and Breure, 1999). Estimates of the numbers of species in microbial communities in soil and sediment range from 10^4–10^5 different species/g dry weight (Torsvik *et al.*, 1990; Dykhuizen, 1998). Consequently, irrespective of the technique, the danger exists that important parts of the community will be overlooked. The use of complementary techniques, such as those offered by the combination of DNA profiling and metabolic diversity analyses, reduces this danger.

Since first proposed by Garland and Mills (1991), *in vitro* community-level physiological profiles (CLPPs) have been used frequently to characterize microbial communities of different habitats, ranging from sediments to seawater, and from oligotrophic groundwater to soils and composts (e.g. Garland and Mills, 1991, 1996; Grayston and Campbell, 1995; Insam *et al.*, 1996; Grayston *et al.*, 1998, 2001).

This section describes the determination of CLPP using 96-well plates for metabolic substrate utilization. Microplate plates allow for easy measurement of many microbial responses, making community analysis practical and open for laboratory automation.

Principle of the method

CLPP involves direct inoculation of environmental samples into Biolog™ microplates, incubation and spectrophotometric detection of heterotrophic microbial activity. Its simplicity and rapidity is attractive to the microbial ecologist, but it requires careful data acquisition, analysis and interpretation. The Biolog™ system was initially developed for characterization of pure isolates of Gram-positive (using GP2 plates, formerly GP plates) and Gram-negative (using GN2 plates, formerly GN plates) heterotrophic isolates (Bochner, 1989). Development of a method for CLPP started with the application of GN2 plates and, to a lesser extent, GP2 plates, both containing 95 different carbon sources. Recently, so-called ECO plates were developed for CLPP of terrestrial communities (Insam, 1997). These plates contain 31 different carbon sources and allow for a triplicate experiment in one plate. GN2 and GP2 plates were recently introduced after GN and GP plates and now include a gellum-forming agent. ECO plates contain this agent too. There are some indications of a different colour-forming regime between GN and GN2 plates (O'Connell *et al.*, 2001).

Several approaches have been used to account for biases related to inoculum density, inoculum activity, incubation time and micro-environment (Garland, 1997; Preston-Mafham *et al.*, 2002). For instance, standardization of initial inoculum density is commonly used (Garland and Mills, 1991), although it is laborious and there is still dispute about an appropriate cell enumeration method. Normalization of optical density (OD) readings by dividing by average well colour development (AWCD) is bound to a number of conditions (Haack *et al.*, 1995; Heuer and Smalla, 1997; Konopka *et al.*, 1998). Single time-point readings and integration of the OD over time are still most widely used, but bear the effect of inoculum density (Garland and Mills, 1991). Following a proposition of continuous plate reading for analysing the kinetics rather than the degree of colour development at a given time, a sigmoidal growth model has been developed (Haack *et al.*, 1995; Garland, 1997; Lindstrom *et al.*, 1998). Again, some kinetic parameters are dependent on inoculum density and need to be normalized prior to statistical analysis. Recently, a normalization procedure that employs integrated OD values derived from a number of dilutions of the same sample, rather than a single dilution level, has been developed as an inoculum-density- and time-independent method of analysis (Gamo and Shoji, 1999; Garland and Lehman, 1999; Breure and Rutgers, 2000; Franklin *et al.*, 2001). The inoculum-density-independent approach is being applied in the Netherlands Soil Monitoring Network. This network consists of about 300 sampling locations (Bloem and Breure, 2003).

In this section, some analytical aspects of sampling and of the analysis of communities by CLPP are described. Considerations for sampling and storing soil, sediment or surface water samples for CLPP analysis are discussed. The overview is neither complete nor exhaustive. For this, the reader is referred to standard textbooks of soil microbial methods. Two different methods for establishing CLPPs are described. The first method is most often used and requires just one plate for a CLPP analysis. This method requires careful inoculum standardization and subsequent data analysis in order to develop a CLPP which is independent of small changes in inoculum size and activity. The second method is based on inoculation of serial sample dilutions, and requires a set of plates for just one CLPP analysis. This method is essentially inoculum-density independent, and can be used for comparing samples of different origin.

Sampling, storage and extraction

As for most biological investigations, fresh samples are often superior to stored samples. However, one has to realize that the microbial communities reflect the climatic conditions just before sampling, especially in shallow soil horizons. This aspect is particularly true for physiology-based methods such as CLPP. Consequently, with remote sampling locations, or for monitoring purposes, it is recommended to include an equilibration period under standardized conditions in the laboratory. For instance, in the Netherlands Soil Monitoring Network, sieved soil samples (2–4 mm mesh) are equilibrated for 4 weeks at 10°C, under 50% water-holding capacity (WHC). This procedure is thought to minimize the effect of gross climatic differences (such as rainfall and temperature) in the weeks before sampling (Bloem and Breure, 2003). If storage is necessary, samples may be stored up to 10 days at 4°C; for longer periods, freezing is recommended.

Extraction procedures range from simple horizontal shaking, head-over-head shaking, or blending of soil or sediment slurries in water, buffer or salt solutions, to elaborate sequential extractions (Hopkins *et al.*, 1991). Usually, a few grams of soil, sediment or compost are sufficient. However, it is strongly recommended that soil sampling, storage and extraction procedures are standardized in order to make comparisons between CLPPs more reliable.

Materials and apparatus

Microplates and their inoculation

Two options are possible, the use of either Biolog™ ECO plates or GN2 plates, containing 31 or 95 different C sources, respectively, plus a water well (Biolog Inc., Hayward, California, USA). ECO plates contain three replicates of the carbon substrate and control. The plates are inoculated with 130 μl suspension per well, diluted (in $\frac{1}{4}$ strength Ringer solution) to obtain a cell density of approximately 1×10^8 cells/ml (determined, for example, by acridine orange direct count (AODC); Bloem *et al.*, 1995). The plates are then incubated at 20°C in the dark and subsequent colour development is measured every 12 h for 5 days (592 nm).

For the inoculum-density-independent approach, serial threefold dilutions of the bacterial suspension are produced (3^0–3^{-11}; with dense bacterial suspensions it is better to use a more diluted series, e.g. 3^{-2}–3^{-13}). The dilution series should essentially give a complete range of colour formation levels from > 95% to < 5% average colour in the plate. Each dilution is inoculated in a section of an ECO plate (Fig. 8.2). The plates are then incubated at 20°C in the dark under 90 ± 5% air humidity to avoid undesired evaporation. Colour development is measured at 590 nm and 750 nm every 8 h or 12 h for 7 days, using an autosampler and a microplate reader or spectrophotometer.

Microplate stackers and autosamplers might be used, since experiments for the analysis of CLPPs usually contain many plates. Commercially available autosamplers and stackers can handle stacks of 20 to more than 100 plates in one run.

Procedures

According to specific needs or availability of equipment, extraction procedures may be modified, inoculation densities may be altered, or single-point data reading may be replaced by continuous readings (alternatives see above). In the following, subheadings preceded by an 'A' relate to the simple method with one Biolog™ plate per sample. Subheadings preceded by a 'B' relate to the inoculum-density-independent method using more than one plate per sample.

Example of an extraction procedure

1. Blend 5 g fresh soil with 20 ml or 50 ml of 0.1% (w/v) sodium cholate solution, 8.5 g cation exchange resin (Dowex 50WX8, 20–50 mesh, Sigma) and 30 glass beads.
2. (Optional) Shake the suspension on a head-over-head shaker (2 h, 4°C).
3. Centrifuge at low speed (500 × *g* to 800 × *g*) for 2 min.
4. Decant the supernatant into a sterilized flask.
5. (Optional) Resuspend the pellet in 10 ml Tris buffer (pH 7.4) and shake for 1 h.
6. Centrifuge as above and add the supernatant to the earlier extract.
7. (Optional) If the extract is turbid or dark (due to clay or humic particles) centrifuge the resulting supernatant another time.

A. Dilution, inoculation and incubation (single-plate procedure)

1. Dilute the samples with $\frac{1}{4}$ strength Ringer solution ten (sediment)- to 1000 (composts)-fold and check the cell density by acridine orange direct counting (AODC). As an alternative, substrate-induced respiration

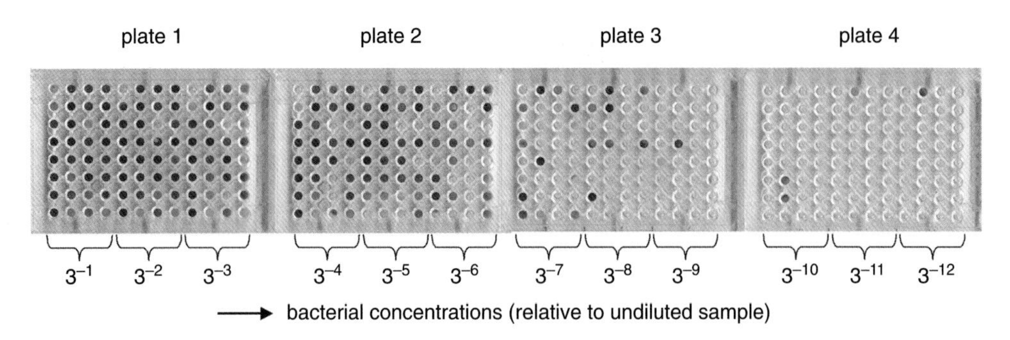

Fig. 8.2. Series of Biolog™ plates (ECO) demonstrating the loss of colour development after stepwise dilution of the inoculum (dilution factor per step is 3 in this case). The loss of colour per well is indicative of the community-level physiological profile (CLPP).

(Anderson and Domsch, 1978) is recommended to determine microbial biomass of the soil samples. Appropriate dilutions of the extracts may then be used to obtain similar inocula.

2. Dilute samples appropriately to obtain a cell density of approximately 1 $\times$ 10^8 cells/ml. In the case of background coloration of the extract, further dilutions are recommended.

3. Inoculate the plate with 100 µl bacterial suspension per well (Biolog™ recommends 150 µl per well, but this is somewhat difficult to handle).

4. Cover the plates with a lid and incubate at 20°C in the dark (other temperatures are possible). Prevent undesired evaporation by putting the plates in polyethylene bags or in a humidified incubation chamber.

5. Measure colour development (592 nm) every 12 h for 5 days using an automated plate reader. Readings may be terminated if the average well colour density reaches an optical density of 2. If curve parameters or the area under the curve are determined, make sure you always have the same reading intervals and the same number of readings (e.g. 10). In case of a long lag-time, or low incubation temperature, incubation may be prolonged (reading interval up to 24 h).

B. Dilution, inoculation and incubation (inoculum-density-independent procedure)

1. Make a dilution series of the bacterial suspension in physiological salts solution or $\frac{1}{4}$ strength Ringer, using threefold dilution steps.

2. Inoculate a minimum of 100 µl of 3^0–3^{-11} diluted bacterial suspension (mineral soil horizons, sandy soils) in 4 ECO plates (1 dilution per section = 32 wells). Depending on the initial concentration of bacteria in the suspension, another set of dilutions might be used (e.g. 3^{-2}–3^{-13} in the case of organic soils and clay).

3. Cover the plates with a lid and incubate at 20°C in the dark (other temperatures are possible). Prevent undesired evaporation by putting the plates in polyethylene bags or in a humidified incubation chamber.

4. Measure colour development (592 nm) every 8 h for 7 days using an automated plate reader. Readings may be terminated if the average well colour density reaches an optical density of 2. If curve parameters or the area under the curve are determined, make sure you always have the same reading intervals and the same number of readings (e.g. 20). In the case of a long lag-time, or low incubation temperature, incubation may be prolonged (reading interval up to 24 h).

Calculation

A. Calculation and data management (single-plate procedure)

Correct raw OD data by blanking response wells against the well showing the minimum absorbance value (R-minimum; Insam *et al.*, 1996). This blanking avoids negative values when compared to subtracting the control well from the response well.

Two alternatives are suggested:

1. Calculate the AWCD from each plate at each reading time. For each plate, those time points of reading are selected that have an AWCD closest to 0.6. Alternatively, other AWCDs (e.g. 0.30, 1.00) may be chosen.
2. Normalize data by dividing each well OD by AWCD (Garland and Mills, 1996). This is particularly important when inoculum densities are not standardized prior to inoculation. Data analysis may be further elaborated by calculating the area under the curve for each well OD for the entire period of incubation (Guckert *et al.*, 1996).

Another procedure that has been used successfully is the estimation of kinetic parameters (K, r, s) by fitting the curve of OD versus time to a density-dependent logistic growth equation (Lindstrom *et al.*, 1998):

$$Y = OD_{592} = \frac{K}{1 + e^{-r(t-s)}} \qquad (8.2)$$

where: K is the asymptote (or carrying capacity); r determines the exponential rate of OD change; t is the time following inoculation of the microplates; and s is the time when the midpoint of the exponential portion of the curve (i.e. when $Y = K/2$) is reached.

For statistical testing of results, and in particular if the use of MANOVA is planned, inoculate a sufficient number of replicates (one ECO plate contains three replicates) according to $n_i \times q = n \geq 31 + q + 2$, where: q is the number of groups to be compared and n_i is the replicate number required per group (sample sizes are equal in each group). For example, if two groups are compared, an $n_i > 17$ is required (i.e. 6 ECO plates; Insam and Hitzl, 1999).

B. Calculation and data management (inoculum-density-independent procedure)

It is generally recognized that the inoculum density highly affects the outcome of the CLPP, and different methods exist to correct for this (see above). We propose a method to produce a CLPP which is independent of the concentration of microorganisms in the environmental sample or in the inoculum. This is done by inoculating a stack of Biolog™ plates with a series of three- or tenfold dilutions of the inoculum. The pattern of attenuation of colour development along the dilution gradient is regarded as characteristic for the CLPP. The final CLPP is then derived from the relative abundances of all Biolog™ substrate conversions of a sample, which is the difference between the amount of inoculum required for 50% response of a specific substrate conversion in the Biolog™ plate and the amount of inoculum required for 50% response in the Biolog™ plate on average. In this way, the problem of inoculum standardization is solved.

To determine this difference, the following concept has been established (Fig. 8.3A). The amount of inoculum (e.g. in log CFU/ml or in a dilution factor) in the sample that causes 50% of the maximal theoretical average response in a Biolog™ plate is determined from a dilution series. Thereafter, the amount of inoculum that causes 50% of the maximal theoretical response for a specific substrate conversion is compared to that value (both log transformed values), resulting in a value for the relative abundance of that specific substrate conversion. According to this procedure, the final CLPP consists of 95 (in case of GN2 plates) or 31 (in case of ECO plates) relative abundance values, and the average abundance is zero.

The rationale for performing dilution experiments is that the response in the Biolog™ system cannot be related directly to the amount of organisms in the wells, due to non-linearities (Garland, 1997). Consequently, application of a range of cell concentrations in the experimental procedure is advised to escape from estimating the Biolog™ response at extrapolated cell concentrations (Garland and Lehman, 1999; Breure and Rutgers, 2000).

The response (A, for instance normalized well colour development, normalized area under the curve) per substrate (s) is then plotted against the bacterial concentrations in the wells (for instance CFU/ml), which leads to a curve that can be fitted with a log normal distribution (upper line in Fig. 8.3A):

$$A_s = \frac{t}{1+10^{h(\log \mathrm{CFU50_s} - \log \mathrm{CFU})}} \tag{8.3}$$

where: t is the asymptotical maximum of the curve at infinite cell concentration; log CFU is the logarithm of the number of colony-forming units; log $\mathrm{CFU50_s}$ is the inflexion point of the curve; and h is the Hill slope (dimensionless). The log $\mathrm{CFU50_s}$ gives a measure for the amount of inoculum in the sample necessary for 50% response of the specific activity.

It is recommended to standardize the responses by dividing the observed response by the maximum theoretical response, because t (the maximum) can then be set to 1. The maximum theoretical response can be

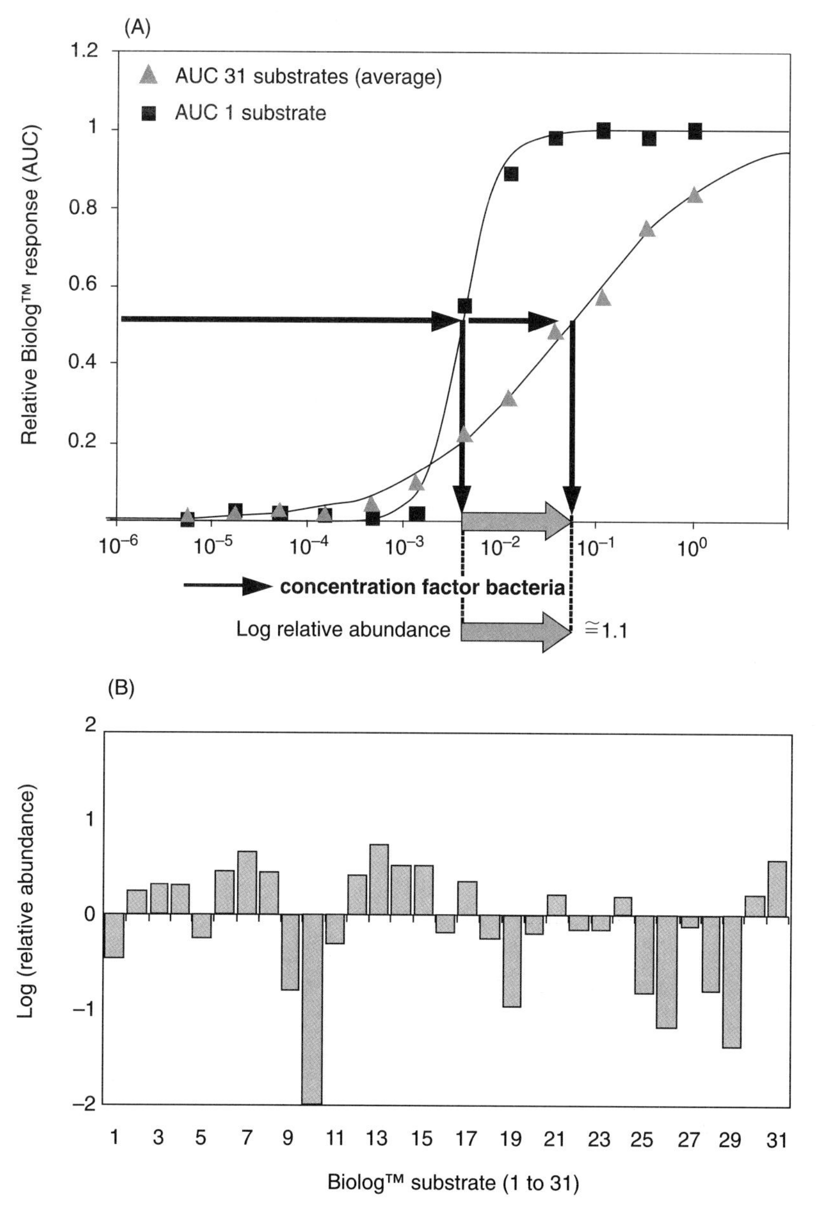

Fig. 8.3. Outline of the procedure to construct a community-level physiological profile (CLPP) which is essentially independent of the inoculum density. (A) A range of inoculum densities is inoculated in a series of Biolog™ plates. Upon dilution, the average response in the Biolog™ plate, or the response of an individual well (e.g. well colour development, or the integrated colour over time), decreases. The dilution of a specific response is fitted with a sigmoidal curve (Eq. 8.3). The log relative abundance (log RA) is given by the difference between the curve for the average response in the plate and the response for a specific substrate; in this example approximately +1.1. (B) In the case of ECO plates, the CLPP consists of 31 relative abundance values; the average abundance is zero. In this example, there was no colour formation in well number 10 at the highest inoculum concentration. Accordingly, this value was artificially set to –2 (see text).

derived from a collation of all previously performed experiments with the same experimental set-up. If this is impossible (e.g. because it is the first experiment) be alert for unrealistic fits with respect to the value of t.

The rationale for using Eq. (8.3) resides in the assumption that maximum response can be reached by infinitely concentrating the sample, while zero response can be reached by an infinite dilution of the sample. The sequential dilution of a discrete number of catalytic units (e.g. microorganisms) is distributed according to Poisson statistics. Although the sigmoidal log-normal distribution curve (Eq. 8.3) does not exactly describe this relationship, it is often used as a robust and adequate approximation (e.g. Haanstra *et al.*, 1985).

The average response in the whole Biolog™ plate ($A_{average}$) was calculated from the responses (A_s). $A_{average}$ was then plotted as a function of the cell concentration and fitted with Eq. (8.3), yielding values for the inflexion point (log $CFU50_{average}$) and the Hill slope (lower line in Fig. 8.3A). The Hill slope gives a measure for the evenness of the Biolog™ substrate conversions in a sample. Discussion of this parameter, although valuable for biodiversity studies, is beyond the scope of this chapter (but see Garland and Lehman, 1999; Breure and Rutgers, 2000; Franklin *et al.*, 2001).

Per well in the Biolog™ plate the log $CFU50_s$ is used to determine the relative abundance (RA) of an activity in the community:

$$\log RA_s = \log CFU50_s - \log CFU50_{average} \quad (8.4)$$

where RA is the number of organisms (in CFU/ml) able to perform 50% activity of the specific substrate conversion relative to the number of organisms converting 50% of all substrates in the Biolog™ plate. Consequently, the average RA is zero, a positive value indicates that the potency to convert this substrate is above average in the particular sample, a negative value vice versa. In cases where there is insufficient colour formation even at the highest inoculum concentrations, this value is artificially set to −2 (substrate number 10 in Fig. 8.3.B). In cases of too much colour formation at the lowest inoculum concentrations, this value is artificially set to +2. In this way, unrealistically high and low values for the relative abundance (on a logarithmic scale) will be avoided.

Ultimately, this procedure yields a CLPP of relative abundancies for all 31 (ECO plates) or 95 substrate (GN2 plates) conversions (Fig. 8.3B), which is essentially independent of the number of cells in the inoculum. The only prerequisite is that the highest concentration contains a sufficient amount of cells to colour the plate for at least about 60% of the theoretical maximum. The complete calculation procedure can be automated, for instance by using visual basic programming in Excel. Such visual basic software has been developed in the Netherlands Soil Monitoring Network (Bloem and Breure, 2003).

Data analysis (single-plate and inoculum-density-independent procedure)

The CLPPs of samples can be collated in multivariate analysis using discriminant analysis (DA), principal component analysis (PCA) and/or redundancy analysis (RDA) for exploratory data analysis, and MANOVA for statistical testing (applied on DA, PCA, RDA factors, or single substrates, or on substrate groups (such as carbohydrates, amino acids, carboxylic acids, etc.)).

Discussion

Microbial biomass forms a major part of the total biomass in soil. The number of species in soil is so high that it is not possible to characterize this biomass by determination of the abundance and composition of the species present (e.g. Torsvik *et al.*, 1990; Dykhuizen, 1998). Here, the use of Biolog™ plates to characterize the microbial biomass in soil is described as a specific technique to characterize a sample of the microbial community, i.e. that part which is able to proliferate in the wells of the Biolog™ plate.

By determination of CLPPs in Biolog™ plates, it is possible to distinguish different microbial communities by use of standardized media. It shows the metabolic activity of substrate-responsive microbial cells, after extraction from soil in the medium. The profile, therefore, is dependent on the extraction efficiency, and on how well the organisms are able to metabolize in the medium used. By standardization of the procedure of extraction and incubation, it is possible to distinguish between communities in a reproducible way.

This is especially attractive when stability or succession of communities in time, or due to environmental changes, is investigated. It was seen earlier that CLPP is dependent on the density of the inoculum. Therefore, it was proposed to standardize the amount of cells in the inoculum (Garland and Mills, 1991; Garland, 1997; Preston-Mafham, 2002). An easy method for achieving this is described in this section.

Inoculum standardization does not always provide the solution to differences in cell density, because it is inherently impossible to exactly predict the response in Biolog™ plates from any enumeration technique, and also due to non-linearity between the cell number and the Biolog™ response. The alternative is to use an inoculum-density-independent method, based on a series of inoculum concentrations. Consequently, this method consumes more than one Biolog™ plate per sample. It has been applied in a national survey to determine the quality of soils in The Netherlands and has shown good reproducibility and a strong discriminating power between microbial communities in different soil types and soil management (Breure and Rutgers, 2000; Schouten *et al.*, 2000; Bloem and Breure, 2003).

CLPPs give information on stability of, and changes in, the metabolic capacities of microbial biomass, which may relate to microbial community structure. However, if changes occur, it is not possible to give a causal relationship between (changes in) environmental conditions and changes in community structure. Statistical techniques have to be invoked to couple ecological effects (and stress) and CLPP responses.

In conclusion, the diversity of soil microbial communities is generally so high that it cannot be captured by one single method. The progress made in the use of methods based on ribosomal RNA and their genes has helped to better understand microbial communities, but such data should be carefully evaluated. For example, most methods only look at the 'tip of the iceberg' of the microbial community, or, in other words, only detect or analyse the most dominant populations, in terms of numbers of organisms, biomolecules or physiological characteristics. This is fine as a first attempt to understand the system, as long as the conclusions drawn from these data consider these limitations. Several pitfalls of these methods, especially when based on PCR amplification, have already been recognized and described (Von Wintzingerode *et al.*, 1997). There is clearly a need for more sensitive methods that give an estimate of the entire diversity, including the populations that are present in lower numbers but could play a crucial role in the habitat. In addition, all of these methods do not necessarily tell us which organisms are most strongly involved in the mainstream energy flux of the ecosystem. To achieve this, molecular ecology approaches should be complemented with metabolic mass balance studies and new techniques to correlate soil microbial diversity and structure with soil function. Given the extremely fast development of these newer molecular methods during only the past two decades, current and future research will continue to generate new methods that will allow us to further improve our understanding of the structure and function of soil microbial communities.

References for Chapter 8

Akkermans, A.D.L., van Elsas, J.D. and De Bruijn, F.J. (1995) *Molecular Microbial Ecology Manual.* Kluwer Academic, Dordrecht, The Netherlands.

Amann, R.I., Ludwig, W. and Schleifer, K.H. (1995) Phylogenetic identification and in situ detection of individual microbial cells without cultivation. *Microbiological Reviews* 59, 143–169.

Anderson, J.P.E. and Domsch, K.H. (1978) A physiological method for the quantitative measurement of microbial biomass in soil. *Soil Biology and Biochemistry* 10, 215–221.

Bai, Q.Y. and Zelles, L. (1997) A method for determination of archaeal ether-linked glycerolipids by high performance liquid chromatography with fluorescence detection as their 9-anthroyl derivatives. *Chemosphere* 35, 263–274.

Bligh, E.G. and Dyer, W.J. (1959) A rapid method of total lipid extraction and purification. *Canadian Journal of Biochemistry and Physiology* 37, 911–917.

Bloem, J. and Breure, A.M. (2003) Microbial indicators. In: Markert, B.A., Breure, A.M. and Zechmeister, H.G. (eds) *Bioindicators and Biomonitors.* Elsevier, Amsterdam, pp. 259–282.

Bloem, J., Bolhuis, P.R., Veninga, M.R. and Wieringa, J. (1995) Microscopic methods for counting bacteria and fungi in soil. In: Alef, K. and Nannipieri, P. (eds) *Methods in Applied Soil Microbiology and Biochemistry.* Academic Press, Toronto, pp. 162–191.

Bochner, B. (1989) *Instructions for the use of Biolog GP and GN microplates.* Biolog Inc., Hayward, California.

Boschker, H.T.S., Nold, S.C., Wellsbury, P., Bos, D., de Graaf, W., Pel, R., Parkes, R.J.

and Cappenberg, T.E. (1998) Direct linking of microbial populations to specific biogeochemical processes by ^{13}C-labelling of biomarkers. *Nature* 392, 801–805.

Bossio, D.A., Scow, K.M., Gunapala, N. and Graham, K.J. (1998) Determinants of soil microbial communities: effects of agricultural management, season, and soil type on phospholipid fatty acid profiles. *Microbial Ecology* 36, 1–12.

Breure, A.M. and Rutgers, M. (2000) The application of Biolog plates to characterise microbial communities. In: Benedetti, A., Tittarelli, F., de Bertoldi, S. and Pinzari, F. (eds) *Proceedings of the COST Action 831, Joint Working Group Meeting Biotechnology of Soil, Monitoring, Conservation and Bioremediation.* 10–11 December 1998, Rome, Italy, (EUR 19548), pp. 179–185.

Buyer, J.S. and Drinkwater, L.E. (1997) Comparison of substrate utilization assay and fatty acid analysis of soil microbial communities. *Journal of Microbiological Methods* 30, 3–11.

Duarte, G.F., Rosado, A.S. and van Elsas, J.D. (1998) Extraction of ribosomal RNA and genomic DNA from soil for studying the diversity of the indigenous bacterial community. *Journal of Microbiological Methods* 32, 21–29.

Dykhuizen, D.E. (1998) Santa Rosalia revisited: why are there so many species of bacteria. *Antonie van Leeuwenhoek* 73, 25–33.

Fang, J., Barcelona, M.J. and Alvarez, P.J.J. (2000) A direct comparison between fatty acid analysis and intact phospholipid profiling for microbial identification. *Organic Geochemistry* 31, 881–887.

Franklin, R.B., Garland, J.L., Bolster, C.H. and Mills, A.L. (2001) Impact of dilution on microbial community structure and functional potential: comparisons of numerical simulations and batch culture experiments. *Applied and Environmental Microbiology* 67, 702–712.

Fritze, H., Tikka, P., Pennanen, T., Saano, A., Jurgens, G., Nilsson, M., Bergman, I. and Kitunen, V. (1999) Detection of Archaeal diether lipid by gas chromatography from humus and peat. *Scandinavian Journal of Forest Research* 14, 545–551.

Frostegård, Å. and Bååth, E. (1996) The use of phospholipid fatty acid analysis to estimate bacterial and fungal biomass in soil. *Biology and Fertility of Soils* 22, 59–65.

Frostegård, Å., Bååth, E. and Tunlid, A. (1993) Shifts in the structure of soil microbial communities in limed forests as revealed by phospholipid fatty acid analysis. *Soil Biology and Biochemistry* 25, 723–730.

Gamo, M. and Shoji, T. (1999) A method of profiling microbial communities based on a most-probable-number assay that uses Biolog plates and multiple sole carbon sources. *Applied and Environmental Microbiology* 65, 4419–4424.

Garland, J.L. (1997) Analysis and interpretation of community-level physiological profiles in microbial ecology. *FEMS Microbiology Ecology* 24, 289–300.

Garland, J.L. and Lehman, R.M. (1999) Dilution/extinction of community phenotypic characters to estimate relative structural diversity in mixed communities. *FEMS Microbiology Ecology* 30, 333–343.

Garland, J.L. and Mills, A.L. (1991) Classification and characterization of heterotrophic microbial communities on the basis of patterns of community-level-sole-carbon-source-utilization. *Applied and Environmental Microbiology* 57, 2351–2359.

Garland, J.L. and Mills, A.L. (1996) Patterns of potential C source utilization by rhizosphere communities. *Soil Biology and Biochemistry* 28, 223–230.

Gomes, N.C.M., Costa, R. and Smalla, K. (2004) Rapid simultaneous extraction of DNA and RNA from bulk and rhizosphere soil. In: Kowalchuk, G.A., De Bruijn, F.J., Head, I.M., Akkermans, A.D. and van Elsas, J.D. (eds) *Molecular Microbial Ecology Manual,* 2nd edn. Kluwer Academic, Dordrecht, The Netherlands, pp. 159–170.

Grayston, S.J. and Campbell, C.D. (1995) Functional biodiversity of microbial communities in the rhizopheres of hybrid larch (*Larix eurolepis*) and Sitka spruce (*Picea sitchensis*). *Tree Physiology* 16, 1031–1038.

Grayston, S.J., Wang, S., Campbell, C.D. and Edwards, A.C. (1998) Selective influence of plant species on microbial diversity in the rhizosphere. *Soil Biology and Biochemistry* 30, 369–378.

Grayston, S.J., Griffith, G.S., Mawdsley, J.L., Campbell, C.D. and Bardgett, R.D. (2001) Accounting for variability in soil microbial communities of temperate upland grassland ecosystems. *Soil Biology and Biochemistry* 33, 533–551.

Griffith, R.I., Whiteley, A.S., O'Donnell, A.G. and Bailey, M.J. (2000) Rapid method for coextraction of DNA and RNA from natural environments for analysis of ribosomal DNA- and rRNA-based microbial community composition. *Applied and Environmental Microbiology* 66, 5488–5491.

Guckert, J.B., Carr, G.J., Johnson, T.D., Hamm, B.G., Davidson, D.H. and Kumagai, Y. (1996) Community analysis by Biolog: curve integration for statistical analysis of activated sludge microbial habitats. *Journal of Microbiological Methods* 27, 183–197.

Haack, S.K., Garchow, H., Odelson, D.A., Forney, L.J. and Klug, M.J. (1994) Accuracy, reproducibility, and interpretation of fatty acid methyl ester profiles of model bacterial communities. *Applied and Environmental Microbiology* 60, 2483–2493.

Haack, S.K., Garchow, H., Klug, M.J. and Forney, L.J. (1995) Analysis of factors affecting the accuracy, reproducibility, and interpretation of microbial community carbon source utilizations. *Applied and Environmental Microbiology* 61, 1458–1468.

Haanstra, L., Doelman, P. and Oude Voshaar, J.H. (1985) The use of sigmoidal dose response curves in soil ecotoxicological research. *Plant and Soil* 84, 293–297.

Harwood, J.L. and Russel, N.J. (1984) *Lipids in Plants and Microbes*. George Allen & Unwin Ltd, London.

Heuer, H. and Smalla, K. (1997) Evaluation of community-level catabolic profiling using Biolog GN microplates to study microbial community changes in potato phyllosphere. *Journal of Microbiological Methods* 30, 49–61.

Heuer, H., Hartung, K., Wieland, G., Kramer, I. and Smalla, K. (1999) Polynucleotide probes that target a hypervariable region of 16S rRNA genes to identify bacterial isolates corresponding to bands of community fingerprints. *Applied and Environmental Microbiology* 65, 1045–1049.

Hooper, D., Hawksworth, D. and Dhillion, S. (1995) Microbial diversity and ecosystem processes. In: Heywood, V.H. and Watson, R.T. (eds) *Global Biodiversity Assessment*. Cambridge University Press, Cambridge, pp. 433–443.

Hopkins, D.W., MacNaughton, S.J. and O'Donnell, A.G. (1991) A dispersion and differential centrifugation technique for representatively sampling microorganisms from soil. *Soil Biology and Biochemistry* 23, 217–225.

Hurt, R.A., Qiu, X., Wu, L., Roh, Y., Palumbo, A.V., Tiedje, J.M. and Zhou, J. (2001) Simultaneous recovery of RNA and DNA from soils and sediments. *Applied and Environmental Microbiology* 67, 4495–4503.

Insam, H. (1997) A new set of substrates proposed for community characterization in environmental samples. In: Insam, H. and Rangger, A. (eds) *Microbial Communities*. Springer-Verlag, Heidelberg, pp. 259–260.

Insam, H. and Hitzl, W. (1999) Data evaluation of community level physiological profiles: a reply to the letter of PJA Howard. *Soil Biology and Biochemistry* 31, 1198–1200.

Insam, H., Amor, K., Renner, M. and Crepaz, C. (1996) Changes in functional abilities of the microbial community during composting of manure. *Microbial Ecology* 31, 77–87.

Konopka, A., Oliver, L. and Turco, J.R. (1998) The use of carbon substrate utilization patterns in environmental and ecological microbiology. *Microbiology Ecology* 35, 103–115.

Kowalchuk, G.A., De Bruijn, F.J., Head, I.M., Akkermans, A.D.L. and van Elsas, J.D. (2004) *Molecular Microbial Ecology Manual*, 2nd edn. Kluwer Academic, Dordrecht, The Netherlands.

Kozdrój, J. and van Elsas, J.D. (2001a) Structural diversity of microorganisms in chemically perturbed soil assessed by molecular and cytochemical approaches.

Journal of Microbiological Methods 43, 197–212.

Kozdrój, J. and van Elsas, J.D. (2001b) Structural diversity of microbial communities in arable soils of a heavily industrialised area determined by PCR-DGGE fingerprinting and FAME profiling. *Applied Soil Ecology* 17, 31–42.

Lindstrom, J.E., Barry, R.P. and Braddock, J.F. (1998) Microbial community analysis: a kinetic approach to constructing potential C source utilization patterns. *Soil Biology and Biochemistry* 30, 231–239.

Macnaughton, S.J., Jenkins, T.J., Wimpee, M.H., Cormiér, M.R. and White, D.C. (1997) Rapid extraction of lipid biomarkers from pure culture and environmental samples using pressurized accelerated hot solvent extraction. *Journal of Microbiological Methods* 31, 19–27.

Miller, M., Palojärvi, A., Rangger, A., Reeslev, M. and Kjoller, A. (1998) The use of fluorogenic substrates to measure fungal presence and activity in soil. *Applied and Environmental Microbiology* 64, 613–617.

Moré, M.I., Herrick, J.B., Silva, M.C., Ghiorse, W.C. and Madsen, E.L. (1994) Quantitative cell lysis of indigenous microorganisms and rapid extraction of microbial DNA from sediment. *Applied and Environmental Microbiology* 60, 1572–1580.

Morgan, J.A.W. and Winstanley, C. (1997) Microbial biomarkers. In: van Elsas, J.D., Trevors, J.T. and Wellington, E.M.H. (eds) *Modern Soil Microbiology*. Marcel Dekker, New York, pp. 331–352.

Nichols, P.D., Guckert, J.B. and White, D.C. (1986) Determination of monounsaturated fatty acid double-bond position and geometry for microbial monocultures and complex consortia by capillary GC–MS of their dimethyl disulphide adducts. *Journal of Microbiological Methods* 5, 49–55.

Nielsen, P. and Petersen, S.O. (2000) Ester-linked polar lipid fatty acid profiles of soil microbial communities: a comparison of extraction methods and evaluation of interference from humic acids. *Soil Biology and Biochemistry* 32, 1241–1249.

Noble, P.A., Almeida, J.S. and Lovell, C.R. (2000) Application of neural computing methods for interpreting phospholipid fatty acid profiles of natural microbial communities. *Applied and Environmental Microbiology* 66, 694–699.

Nüsslein, K. and Tiedje, J.M. (1998) Characterization of the dominant and rare members of a young Hawaiian soil bacterial community with small-subunit ribosomal DNA amplified from DNA fractionated on the basis of its guanine and cytosine composition. *Applied and Environmental Microbiology* 64, 1283–1289.

O'Connell, S.P., Lehman, R.M. and Garland, J.L. (2001) Biolog media formulation (GN2 vs. GN) affects microbial community-level physiological profiles. *Proceedings of the 9th International Symposium on Microbial Ecology*. Amsterdam, p. 164.

Ogram, A., Sayler, G.S. and Barkay, T.J. (1987) DNA extraction and purification from sediments. *Journal of Microbiological Methods* 7, 57–66.

Olsson, P.A. (1999) Signature fatty acids provide tools for determination of the distribution and interactions of mycorrhizal fungi in soil. *FEMS Microbiology Ecology* 29, 303–310.

Palojärvi, A. and Albers, B. (1998) Extraktion und Bestimmung membrangebundener Phospholipidfettsäuren. In: Remde, A. and Tippmann, P. (eds). *Mikrobiologische Charakterisierung aquatischer Sedimente–Methodensammlung. Vereinigung für Allgemeine und Angewandte Mikrobiologie (VAAM)*. R. Oldenbourg Verlag, München, pp. 187–195.

Palojärvi, A., Sharma, S., Rangger, A., von Lützow, M. and Insam, H. (1997) Comparison of Biolog and phospholipid fatty acid patterns to detect changes in microbial community. In: Insam, H. and Rangger, A. (eds) *Microbial Communities. Functional versus Structural Approaches*. Springer Verlag, Berlin, pp. 37–48.

Pankhurst, C.E., Ophel-Keller, K., Doube, B.M. and Gupta, V.V.S.R. (1996) Biodiversity of soil microbial communities in agricultural systems. *Biodiversity and Conservation* 5, 197–209.

Pennanen, T. (2001) Microbial communities in boreal coniferous forest humus exposed to heavy metals and changes in soil pH – a summary of the use of phospholipid fatty acids, Biolog® and ^{3}H-thymidine incorporation methods in field studies. *Geoderma* 100, 91–126.

Petersen, S.O. and Klug, M.J. (1994) Effects of sieving, storage, and incubation temperature on the phospholipid fatty acid profile of a soil microbial community. *Applied and Environmental Microbiology* 60, 2421–2430.

Petersen, S.O., Debosz, K., Schjonning, P., Christensen, B.T. and Elmholt, S. (1997) Phospholipid fatty acid profiles and C availability in wet-stable macro-aggregates from conventionally and organically farmed soils. *Geoderma* 78, 181–196.

Petersen, S.O., Frohne, P.S. and Kennedy, A.C. (2002) Dynamics of a soil microbial community under spring wheat. *Soil Science Society of America Journal* 66, 826–833.

Pillai, S.D., Josephson, K.L., Bailey, R.L., Gerba, C.P. and Pepper, I.L. (1991) Rapid method for processing soil samples for polymerase chain reaction amplification of specific gene sequences. *Applied and Environmental Microbiology* 57, 2283–2286.

Porteous, L.A. and Armstrong, J.L. (1991) Recovery of bulk DNA from soil by a rapid, small-scale extraction method. *Current Microbiology* 22, 345–348.

Preston-Mafham, J., Boddy, L. and Randerson, P.F. (2002) Analysis of microbial community functional diversity using sole-carbon source utilisation profiles, a critique. FEMS *Microbiology Ecology* 42, 1–14.

Rademaker, J. (1995) Analysis of molecular fingerprints. In: Akkermans, A.D.L., van Elsas, J.D. and De Bruijn, F.J. (eds) *Molecular Microbial Ecology Manual.* Kluwer Academic, Dordrecht, The Netherlands.

Roslev, P., Iversen, N. and Henriksen, K. (1998) Direct fingerprinting of metabolically active bacteria in environmental samples by substrate specific radiolabelling and lipid analysis. *Journal of Microbiological Methods* 31, 99–111.

Rutgers, M. and Breure, A.M. (1999) Risk assessment, microbial communities, and pollution-induced community tolerance. *Human and Ecological Risk Assessment* 5, 661–670.

Sambrook, J., Fritsch, E.F. and Maniatis, T. (1989) *Molecular Cloning. A Laboratory Manual,* 2nd edn. Cold Spring Harbor Laboratory Press, Cold Spring Harbor, New York.

Schloter, M., Zelles, L., Hartmann, A. and Munch, J.C. (1998) New quality of assessment of microbial diversity in arable soils using molecular and biochemical methods. *Journal of Plant Nutrition and Soil Science* 161, 425–431.

Schouten, T., Bloem, J., Didden, W.A.M., Rutgers, M., Siepel, H., Posthuma, L. and Breure, A.M. (2000) Development of a biological indicator for soil quality. *Setac Globe* 1, 30–33.

Seghers, D., Verthé, K., Reheul, D., Bulcke, R., Siciliano, S.D., Verstraete, W. and Top, E.M. (2003) Effect of long-term herbicide applications on the bacterial community structure and function in an agricultural soil. *FEMS Microbiology Ecology* 46, 139–146

Selenska, S. and Klingmüller, W. (1991a) Direct detection of *nif*-gene sequences of *Enterobacter agglomerans* in soil. *FEMS Microbiology Letters* 80, 243–246.

Selenska, S. and Klingmüller, W. (1991b) DNA recovery and direct detection of *Tn*5 sequences from soil. *Letters in Applied Microbiology* 13, 21–24.

Siira-Pietikäinen, A., Haimi, J., Kanninen, A., Pietikäinen, J. and Fritze, H. (2001) Responses of decomposer community to root-isolation and addition of slash. *Soil Biology and Biochemistry* 33, 1993–2004.

Smalla, K., Cresswell, N., Mendonca-Hagler, L.C., Wolters, A.C. and van Elsas, J.D. (1993a) Rapid DNA extraction protocol from soil for polymerase chain reaction-mediated amplification. *Journal of Applied Bacteriology* 74, 78–85.

Smalla, K., Van Overbeek, L.S., Pukall, R. and van Elsas, J.D. (1993b) Prevalence of *npt*II and *Tn*5 in kanamycin-resistant bacteria from different environments. *FEMS Microbiology Ecology* 13, 47–58.

Steffan, R.J., Goksoyr, J., Bej, A.K. and Atlas, R.M. (1988) Recovery of DNA from soils and sediments. *Applied and Environmental Microbiology* 54, 2908–2915.

Tebbe, C.C. and Vahjen, W. (1993) Interference of humic acids and DNA extracted directly from soil in detection and transformation of recombinant DNA from bacteria and a yeast. *Applied and Environmental Microbiology* 59, 2657–2665.

Torsvik, V.L. (1980) Isolation of bacterial DNA from soil. *Soil Biology and Biochemistry* 12, 18–21.

Torsvik, V., Goksøyr, J. and Daae, F.L. (1990) High diversity of DNA of soil bacteria. *Applied and Environmental Microbiology* 56, 782–787.

Trevors, J.T. and van Elsas, J.D. (1989) A review of selected methods in environmental microbial genetics. *Canadian Journal of Microbiology* 35, 895–902.

Tsai, Y.-L. and Olson, B.H. (1992) Detection of low numbers of bacterial cells in soils and sediments by polymerase chain reaction. *Applied and Environmental Microbiology* 58, 754–757.

Vainio, E.J. and Hantula, J. (2000) Direct analysis of wood-inhabiting fungi using denaturing gradient gel electrophoresis of amplified ribosomal DNA. *Mycology Research* 104, 927–936.

van Elsas, J.D., Van Overbeek, L.S. and Fouchier, R. (1991) A specific marker, *pat*, for studying the fate of introduced bacteria and their DNA in soil using a combination of detection techniques. *Plant and Soil* 138, 49–60.

van Elsas, J.D., Mantynen, V. and Wolters, A.C. (1997) Soil DNA extraction and assessment of the fate of *Mycobacterium chlorophenolicum* strain PCP-1 in different soils via 16S ribosomal RNA gene sequence based most-probable-number PCR and immunofluorescence. *Biology and Fertility of Soils* 24, 188–195.

van Elsas, J.D., Smalla, K. and Tebbe, C.C. (2000) Extraction and analysis of microbial community nucleic acids from environmental matrices. In: Jansson, J., van Elsas, J.D. and Bailey, M. (eds) *Tracking Genetically Engineered Micro-organisms.* Landes Bioscience, Austin, Texas, pp. 29–52.

Van Overbeek, L.S., Van Vuurde, J.W.L. and van Elsas, J.D. (2005) Application of molecular fingerprinting techniques for studying shifts in bacterial endophytic communities. In: Schultz, B. and Boyle, C. (eds) *Bacterial Endophytes.* Springer Verlag, Heidelberg (in press).

Von Wintzingerode, F., Göbel, U.B. and Stackebrandt, E. (1997) Determination of microbial diversity in environmental samples: pitfalls of PCR-based rRNA analysis. *FEMS Microbiology Reviews* 21, 213–229.

White, D.C. (1988) Validation of quantitative analysis for microbial biomass, community structure, and metabolic activity. *Archiv für Hydrobiologie Beihefte* 31, 1–18.

White, D.C., Davis, W.M., Nickels, J.S., King, J.C. and Bobbie, R.J. (1979) Determination of the sedimentary microbial biomass by extractible lipid phosphate. *Oecologia* 40, 51–62.

White, D.C., Flemming, C.A., Leung, K.T. and MacNaughton, S.J. (1998) In situ microbial ecology for quantitative appraisal, monitoring, and risk assessment of pollution remediation in soils, the subsurface, the rhizosphere and in biofilms. *Journal of Microbiological Methods* 32, 93–105.

Widmer, F., Fließbach, A., Laczko, E., Schulze-Aurich, J. and Zeyer, J. (2001) Assessing soil biological characteristics: a comparison of bulk soil community DNA-, PLFA-, and Biolog™-analyses. *Soil Biology and Biochemistry* 33, 1029–1036.

Zelles, L. (1999) Fatty acid patterns of phospholipids and lipopolysaccharides in the characterisation of microbial communities in soil: a review. *Biology and Fertility of Soils* 29, 111–129.

Zelles, L. and Bai, Q.Y (1994) Fatty acid patterns of phospholipids and lipopolysaccharides in environmental samples. *Chemosphere* 28, 391–411.

Zelles, L., Palojärvi, A., Kandeler, E., von Lützow, M., Winter, K. and Bai, Q.Y. (1997) Changes in soil microbial properties and phospholipid fatty acid fractions after chloroform fumigation. *Soil Biology and Biochemistry* 29, 1325–1336.

9 Plant–Microbe Interactions and Soil Quality

9.1 Microbial Ecology of the Rhizosphere

PHILIPPE LEMANCEAU,[1] PIERRE OFFRE,[1] CHRISTOPHE MOUGEL,[1] ELISA GAMALERO,[2] YVES DESSAUX,[3] YVAN MOËNNE-LOCCOZ[4] AND GRAZIELLA BERTA[2]

[1]UMR 1229 INRA/Université de Bourgogne, 'Microbiologie et Géochimie des Sols', INRA-CMSE, 17 rue Sully, BP 86510 21065, Dijon cedex, France; [2]Università del Piemonte Orientale 'Amedeo Avogadro', Department of Science and Advanced Technology, Corso Borsalino 54, 15100 Alessandria, Italy; [3]Institut des Sciences du Végétal, CNRS, UPR2355, Bâtiment 23, Avenue de la Terrasse, 91198 Gif-sur-Yvette, France; [4]UMR CNRS 5557 Ecologie Microbienne, Université Claude Bernard (Lyon 1), 43 bd du 11 Novembre, 69622 Villeurbanne cedex, France

Soils are known to be oligotrophic environments, whereas the soil microbiota is mostly heterotrophic, and consequently microbial growth in soil is mainly limited by the scarce sources of readily available carbon (Wardle, 1992). Therefore, in soils, the microbiota is mostly in stasis (fungistasis/bacteriostasis) (Lockwood, 1977). In contrast, plants are autotrophic organisms responsible for most of the primary production resulting from photosynthesis. Significant amounts of photosynthetates are released from the plant roots to the soil, through a process called rhizodeposition. These products comprise exudates, lysates, mucilage, secretions and compounds released from dead/dying sloughed-off cells, as well as gases including respiratory CO_2 and ethylene. Depending on plant species, age and environmental conditions, rhizodeposits can account for up to 40% of net fixed carbon (Lynch and Whipps, 1990). On average, 17% of net fixed carbon appears to be released by the roots (Nguyen, 2003). This significant release of carbohydrates in the soil by the plant roots stimulates the density and activity of the microbiota located closely to the roots but also affects the physico-chemical properties of the neighbouring soil (Rovira, 1965; Curl and Truelove, 1986; Lynch and Whipps, 1990). Altogether, the modifications of the biological and physico-chemical properties of the soil induced by plant roots are commonly called the rhizosphere effect; the rhizosphere, as proposed one century ago by Hiltner (1904), being the volume of soil surrounding roots in which the microflora is influenced by these roots.

Besides trophic interactions, the relations between plants and microorganisms are also mediated by toxic compounds involved in plant defence reactions against soilborne pathogens (Bais *et al.*, 2004); similarly, various soilborne microbial populations have the ability to produce antibiotics contributing to their ecological fitness, which is especially important in a competitive environment such as the rhizosphere (Mazzola *et al.*, 1992).

Altogether these characteristics of the rhizosphere lead to the fact that not all microbial groups and populations are equally stimulated by the rhizodeposits. Specific microbial groups and populations have clearly been shown to be preferentially associated with plant roots (Mavingui *et al.*, 1992; Lemanceau *et al.*, 1995; Edel *et al.*, 1997). These populations, which are selected by the plant, are better adapted than the others to the rhizosphere environment. As an example, populations of fluorescent pseudomonads associated with the roots differ from soil counterparts in terms of carbon and energetic metabolism: they are more frequently able to use specific organic compounds such as trehalose (electron donors) and they show a higher ability to mobilize ferric iron and to dissimilate nitrogen oxides (electron acceptors) (Lemanceau *et al.*, 1988, 1995; Clays-Josserand *et al.*, 1995; Frey *et al.*, 1997). Populations able to develop strategies to counteract toxic compounds produced by the host plant or by other root-associated microbial populations will have a competitive advantage over the others (Duffy *et al.*, 2003).

In plant-selected microbial populations, the rapid adaptation of the metabolism to the changing conditions of the rhizosphere relies on their ability to perceive the variations of the environment. Among the perception systems described so far, two-component systems allow bacteria to sense and respond to a wide range of environmental changes. The minimal system consists of two proteins: a sensor and a transducer (Stock *et al.*, 1989). Signalling in the rhizosphere also involves microbe–microbe communication. This is the case of the so-called 'quorum sensing system', which allows bacterial populations to sense their densities and to regulate cell physiology, including enzyme and antibiotic synthesis at the population level, making their response to the environmental variation more efficient and concerted (Whitehead *et al.*, 2001). Signalling also occurs between plants and microbes. Various regulatory signals are perceived by receptors and transduced via downstream effectors, and they mediate information that can influence plant–microbe interactions. The corresponding signal molecules are difficult to detect and so far only a few of them have been described (Hirsch *et al.*, 2003). However, well-known examples are the isoflavonoids and flavonoids present in root exudates of various leguminous plants, which activate *Rhizobium* genes responsible for the nodulation process (Peters, 1986; Bais *et al.*, 2004).

Trophic and toxic-mediated interactions, together with communications, lead to the selection of the most adapted populations, to variations of their physiology, and then to shifts in the structure and activity of the microbial community. These variations are known to affect plant health and growth. Depending on its persistence, the rhizosphere effect may not only affect the growth and health of the current crop, but may also affect the soil quality.

The rhizosphere, being a major hot-spot for soil microorganisms, is expected to affect various parameters relevant for soil quality: soil (micro)structure, degradation and mineralization of organic xenobiotic compounds, mobilization and speciation of heavy metals, contribution to the phytoextraction of these toxic elements, enrichment of symbiotic and free-living microbial populations favourable to plant growth and health. The soil quality determined by the indigenous microbial populations will affect the growth and health of the plants. In this way, plant–microbe interactions may be considered as a feedback loop, in which: (i) the plant modifies the environment of soilborne microorganisms through the release of various compounds; (ii) these variations are perceived by the soil microbiota; (iii) this perception by specific populations leads to a variation of their physiology and the selection of the populations most adapted, and in turn to a shift in the microbial community and activity; and (iv) these variations affect the growth and health of the plant and then root exudation patterns, in such a way that the rhizosphere environment is being modified. This feedback loop is a dynamic process submitted to a continuous incrementation, which, in the long term, drives the coevolution of plant and associated microorganisms.

A major challenge in the study of rhizosphere ecology is to make progress in our knowledge of the impact of the plant on the soilborne microbiota. Indeed, the rhizosphere effect appears to be plant- and even cultivar-specific. These studies have to take into account not only specific microbial groups and culturable populations, but also microbial communities, in an untargeted way. The expected output of the corresponding research will be to monitor the indigenous microbiota via the cultivation of specific plant genotypes and the application of particular agricultural practices (crop rotation, intercropping) in order to favour beneficial indigenous populations and disfavour those that are detrimental for the plant and, more generally, for soil quality.

In the present chapter, information will be given on methods to: (i) assess soil quality in relation to the contribution of nodulating symbiotic bacteria promoting growth of legumes via nitrogen fixation; (ii) use arbuscular mycorrhiza for characterization of soil quality in relation to ecotoxicology; (iii) assess the phytosanitary soil quality resulting from both antagonistic and pathogenic microbial populations; and (iv) assess indigenous populations of free-living plant-beneficial microorganisms.

9.2 Nodulating Symbiotic Bacteria and Soil Quality

Alain Hartmann,[1] Sylvie Mazurier,[1] Dulce N. Rodríguez-Navarro,[2] Francisco Temprano Vera,[2] Jean-Claude Cleyet-Marel,[3] Yves Prin,[3] Antoine Galiana,[3] Manuel Fernández-López,[4] Nicolás Toro[4] and Yvan Moënne-Loccoz[5]

[1]UMR 1229 INRA/Université de Bourgogne, 'Microbiologie et Géochimie des Sols', INRA-CMSE, BP 86510, 21065 Dijon cedex, France; [2]Centro de Investigación y Formación Agraria (CIFA), Las Torres y Tomejil, Aparatdo Oficial, E-41200 Alcalá del Río (Sevilla), Spain; [3]Laboratoire des Symbioses Tropicales et Méditerranéennes, UMR INRA/IRD/CIRAD/ AGRO-M 1063, Campus International de Baillarguet, 34398 Montpellier cedex 5, France; [4]Grupo de Ecología Genética, Estación Experimental del Zaidín, Consejo Superior de Investigaciones Científicas, Profesor Albareda 1, 18008 Granada, Spain; [5]UMR CNRS 5557 Ecologie Microbienne, Université Claude Bernard (Lyon 1), 43 bd du 11 Novembre, 69622 Villeurbanne cedex, France

Introduction

Several bacterial taxa are involved in symbiotic interactions with plant roots, in which the microorganism nodulates the root and fixes nitrogen. These diazosymbiotic bacteria include five genera (*Rhizobium*, *Mesorhizobium*, *Sinorhizobium*, *Azorhizobium* and *Bradyrhizobium*; often collectively referred to as rhizobia) and close to 30 species of the Rhizobiaceae family (Young *et al.*, 2001), which are associated with plants (designated as legumes) from the three botanical families *Papilionoideae*, *Mimosoideae* and *Cesalpinioideae*, as well as an actinobacterial genus (*Frankia*), which nodulates non-legume plants such as *Alnus* or *Casuarina*. The diversity of rhizobial populations in soil has been studied extensively. For a majority of rhizobial species, the ability to engage in a symbiotic relationship with a legume implies plasmid-borne genes. Recently, the documented range of bacterial partners capable of nitrogen fixation within legume nodules has been revised with the inclusion of another α-proteobacterium (*Methylobacterium nodulans*; Sy *et al.*, 2001) and of β-proteobacteria (Moulin *et al.*, 2001). Here, we will focus on diazosymbionts from the α-proteobacteria interacting with legumes (Table 9.1), and adopt the usual term 'rhizobia' to designate these bacteria.

Table 9.1. List of known rhizobia–legume symbiotic interactions. Names of legumes nodulated by more than one rhizobial species are in bold.

Bacterial symbiont (genus, species and biovar)	Principal host legumes
Rhizobium	
R. leguminosarum	
bv. *viciae*	*Pisum* (pea), *Lens, Lathyrus, Vicia*
bv. *trifolii*	***Trifolium***
bv. *phaseoli*	***Phaseolus* (bean)**
R. tropici	***Phaseolus* (bean)**, *Leucaena*, etc.
R. etli	
bv. *phaseoli*	***Phaseolus* (bean)**
bv. *mimosae*	***Mimosa affinis***
R. gallicum	
bv. *gallicum*	***Phaseolus* (bean)**, *Onobrychis, Leucaena*, etc.
bv. *phaseoli*	***Phaseolus* (bean)**
R. giardinii	
bv. *giardinii*	***Phaseolus* (bean)**, *Leucaena*, etc.
bv. *phaseoli*	***Phaseolus* (bean)**
R. galegae	*Galega* spp.
R. hainanense	*Desmodium* spp., etc.
R. mongolense	***Medicago ruthenica***
R. huautlense	*Sesbania* spp.
R. undicola	*Neptunia natans*
R. yanglingense	*Amphicarpaea, Coronilla*, etc.
Sinorhizobium	
S. meliloti	***Medicago***, *Trigonella, Melilotus*
S. fredii	***Glycine max* (soybean)**, etc.
S. teranga	
bv. *sesbaniae*	***Sesbania***, etc.
bv. *acaciae*	***Acacia***
S. saheli	
bv. *sesbaniae*	***Sesbania***, etc.
bv. *acaciae*	***Acacia***
S. medicae	***Medicago***
S. morelense	***Leucaena leucocephala***
Mesorhizobium	
M. loti	***Lotus***
M. huakuii	***Astragalus sinicus***
M. ciceri	***Cicer arietinum***
M. mediterraneum	***Cicer arietinum***
M. tianshanense	***Glycine max* (soybean)**, *Glycyrrhiza*, etc.
M. plurifarium	***Acacia, Leucaena***
M. amorphae	*Amorpha fruticosa*
M. chacoense	***Prosopis alba***
Bradyrhizobium	
B. japonicum	***Glycine max* (soybean)**
B. elkanii	***Glycine max* (soybean)**
B. liaoningensis	***Glycine max* (soybean)**
Azorhizobium	
A. caulinodans	*Sesbania rostrata*

The legume–rhizobia symbiosis is extremely important for the nitrogen cycle in natural terrestrial ecosystems, as well as under agronomic conditions. Indeed, in traditional crop rotation systems, it has been used to enhance soil fertility and the productivity of non-legume crops since ancient times (Tilman, 1998), long before the identification by Hellriegel and Wilfarth (1888) of rhizobia as the source of fixed nitrogen in root nodules. Certain legumes are major crops throughout the world, and grain legumes are grown over almost 1.5 million km^2 of land each year, half of which is cultivated with soybean. Forage legumes cover about 30 million km^2, the three dominant genera being *Trifolium, Lotus* and *Medicago.* Legumes are also grown in mixed cropping systems (for example a legume and a cereal), or can be used as green manure. In addition, ligneous legumes are useful for restoring degraded soils or establishing a plant cover on nitrogen-poor mineral substrates. Compared with the use of nitrogen fertilizers, legume-based biological nitrogen fixation (BNF) is more environmentally friendly; thus, optimizing its potential may contribute to improved soil quality and soil preservation.

BNF can fix up to 250 kg N/ha, even more for some legumes, and it is a less expensive source of nitrogen compared with nitrogen fertilizers. In certain situations, however, the symbiosis does not function well, due to the absence, or low numbers, of appropriate rhizobia in soil, or the presence of nodulating rhizobia that are poor nitrogen fixers. Inoculation of the specific and efficient rhizobial symbiont can greatly improve BNF and consequently the yield of the corresponding legume. Soil and climatic conditions (e.g. extreme soil pH, salinity, tillage, temperature stress, drought, availability of mineral nutrients and chemical residues) can affect the numbers and/or the diversity of rhizobial populations, and the functioning of BNF. For instance, the detrimental effects of heavy metals on indigenous rhizobial populations are well established.

In this section, methods available to assess the legume–rhizobia symbiotic interaction are presented. Relevant objectives when addressing the various components of soil quality include the detection (i.e. soil nodulating potential; pp. 233–236) and enumeration (by a MPN approach; pp. 236–240) of nodulating rhizobia indigenous to soil, the assessment of the symbiotic efficacy of indigenous rhizobial populations (pp. 241–243), and the estimation of the amount of plant nitrogen derived from symbiotic fixation (pp. 243–247).

Detection of Indigenous Rhizobia in Soil and Assessment of Soil Nodulating Potential

Introduction

Some legume species are nodulated exclusively by one rhizobial species (e.g. lucerne) or even by a given biovar of one species (e.g. pea, clover). Other legumes are more permissive, as they can be nodulated by strains belonging to different species or even genera of rhizobia (e.g. soybean,

bean). Conversely, rhizobial strains may exhibit a narrow or wide host range (Amarger, 2001). The soil nodulating potential corresponds to the ability of soil to support legume nodulation, which is a prerequisite for the expression of beneficial effects from symbiotic nitrogen fixation. Thus, the soil nodulating potential can be estimated as the ability to nodulate a given legume, or by the occurrence of nodulating strains from a given rhizobial species.

Different methods can be chosen to detect (and/or enumerate) rhizobia from soil. Plant infection tests have been used extensively, since there is no specific medium for direct counting of rhizobia in soil. Experimental designs and protocols for plant infection tests vary with the legume studied, and have been described in the literature (Vincent, 1970; Somasegaran and Hoben, 1994). These methods are presented here, particularly in the case of large-seeded legumes such as soybean or bean, for the detection (and, in the next section, the enumeration) of rhizobial symbionts.

Principle of the methods

When the objective is merely to assess whether or not nodulating rhizobia are present in soil, the occurrence of nodules on roots of the appropriate legume is a sufficient criterion. Since certain rhizobial strains can nodulate, but fix little or no nitrogen (i.e. ineffective strains), it may be useful to use scoring systems that integrate nodule repartition, size and/or internal colour (Dommergues *et al.*, 1999), because these three parameters may be indicative of symbiotic efficacy. For instance, nitrogen fixation is more likely to occur in big, pink nodules than in small, white nodules. Indeed, such nodulation scores often correlate well with indices of plant growth such as foliage dry matter and nitrogen content (Brockwell *et al.*, 1982). When fields are grown with legumes, this approach may be implemented directly by sampling the root system of field-grown legumes and scoring the extent of nodulation. Nodulation scores can be particularly useful to compare nodulation of the same legume in neighbouring fields or when farming practices differ, as well as when comparing legume cultivars or characterizing the impact of plant protection products (Fettell *et al.*, 1997; Moënne-Loccoz *et al.*, 1998). If the legume is not grown in the soil of interest, the assessment can be done after sampling the soil and performing plant infection tests, based on the capacity of specific rhizobia to nodulate a given species of legume: the legume is grown in the soil sampled and the extent of plant nodulation is recorded.

Materials and apparatus

When assessing field-grown legumes:

- Digging equipment such as shovels, etc.
- Scalpels (to split the nodules open)

When assessing soil with plant infection tests:

- Calcinated clay (Chemsorb; CONEX) or perlite
- 16 mm diameter clay beads (Argi-16TBF, TBF, France)
- 1-l plastic containers
- Legume seeds (soybean or bean)
- Controlled-environment cabinet (or growth chamber)
- Sieve

Chemicals and solutions

When assessing soil with plant infection tests:

- Sterile deionized water
- Saturated calcium hypochlorite solution

Procedure and calculation

When assessing field-grown legumes: plants (at least 15 individual plants) need to be sampled when the legumes are expected to be fully nodulated, e.g. shortly before flowering. The root systems and surrounding soil are carefully dug up (to a depth of at least 20 cm, preferably more). The soil is removed by shaking, followed, if need be, by dipping in water, and nodulation is scored (Table 9.2). The average nodulation score is computed.

When assessing soil by use of plant infection tests: in order to avoid inconsistent results due to the spatial heterogeneity of rhizobial soil populations, assessment needs to be carried out on composite soil samples to be agronomically meaningful. Composite soil samples are obtained by pooling and mixing six to ten soil cores taken at random over the field plot. All equipment used for sampling must be free of rhizobia, and thus the use of disinfected instruments and disposable plastic bags is recommended. Soil samples are sieved (2–4 mm). They may be used immediately or stored at 4°C for up to 1 month, if necessary. The method described hereafter is optimized for soybean and bean, but may also be used for other legumes. Plants are grown in soil, or in a mixture of soil and sterile, substrate like calcinated clay or perlite (i.e. siliceous sand expanded at 1200°C). One-litre plastic containers are filled with the following layers (from bottom to top): 200 ml of 16 mm diameter Argi-16TBF clay beads, 400 ml of rhizobia-free substrate (calcinated clay or perlite), 100 g of soil, 200 ml of rhizobia-free substrate. Seeds (soybean or bean) are disinfected for 5 min in a saturated calcium hypochlorite solution and rinsed six times with sterile deionized water. Four seeds are sown per container. Containers are watered with sterile deionized water. Plants are grown in a controlled-environment cabinet (16 h light at 240 μE/s, 8 h dark, at 22°C). Nodulation of the plants, which may be achieved even when soil contains low population levels of

Table 9.2. Scoring system to assess the extent of nodulation (derived from Brockwell *et al.*, 1982).

Nodule score	Number of presumably effective[a] nodules	
	Crown[b]	Elsewhere
7	> 7	> 9
6	> 7	5–9
5	> 7	0–4
4	1–7	> 9
3	1–7	5–9
2	1–7	1–4
1	1–7	0
0	0	0

[a]Presumably effective nodules are identified on the basis of size and internal pigmentation.
[b]The crown is defined as the top 5 cm of the root system.

nodulating rhizobia, is recorded at 1 month, by counting nodules or using a scoring table.

Discussion

The occurrence of indigenous rhizobia in most soils has been extensively documented (Amarger, 1980). For instance, rhizobia nodulating peas and clovers are ubiquitous in most French and Spanish soils. Conversely, rhizobia nodulating lucerne are absent in soils with a pH below 6, whereas rhizobia nodulating lupin are not found in soil with a pH above 6 (Amarger, 1980). Different analyses of the *Sinorhizobium meliloti* population in agricultural soils of Spain indicate that highly competitive strains, genetically similar to the well-studied strain GR4, are often present, noticeably in soils with a high lucerne yield (Villadas *et al.*, 1995; Velázquez *et al.*, 1999). In similar soils from the same geographical region, but with poor lucerne yield, the GR4-like subpopulation is typically replaced by other, less-effective strains (our unpublished results), which suggests that the presence of competitive, effective strains can be used as a soil quality indicator. The growth of non-native legumes often requires the inoculation of appropriate rhizobial symbionts (e.g. for soybean outside Asia), since they are naturally absent from soils. These introduced rhizobia are likely to survive and adapt in soil (Revellin *et al.*, 1996).

Enumeration of Indigenous Rhizobia by MPN Plant Infection Counts

Introduction

The number of indigenous rhizobia can vary to a large extent when comparing different soils or farming practices. Unfortunately, selective growth media are not available to enumerate rhizobia directly from soil. Therefore, alternative procedures, based on most probable number (MPN) determination, are needed. The MPN is a statistical estimate derived from the number of nodulated plants obtained following inoculation with aliquots from a dilution series of the soil under study.

Principle of the methods

The plant infection count is based on the most probable number (MPN) method. Axenic plantlets are inoculated with diluted soil suspensions, tubes containing nodulated plants are recorded as positive, and rhizobial numbers are deduced from a statistical table (Fisher and Yates, 1963).

Materials and apparatus

- Erlenmeyer flasks
- Rotary shaker
- Seeds
- Sterile perlite
- Gibson glass tubes (220 × 22 mm)
- Filter paper strip
- Aluminium foil
- Growth chamber or a greenhouse

Chemicals and solutions

- Sterile deionized water
- Saturated calcium hypochlorite solution
- Nitrogen-free plant nutrient solution

Procedure

1. A soil suspension is prepared by transferring 10 g of soil to a sterile flask containing 90 ml of sterile water, and flasks are shaken for 30 min at 150 rpm.
2. A tenfold dilution series is prepared (ten dilution steps, including the initial soil suspension, are necessary to provide dilutions with and without rhizobia). One ml of each dilution is used to inoculate axenic seedlings grown in test tubes (four tubes per dilution), prepared as follows:

- In the case of soybean or bean, seeds are disinfected for 5 min in a saturated calcium hypochlorite solution and rinsed six times with sterile deionized water.
- Seeds are germinated in sterile perlite for 3 days at 28°C.
- Axenic seedlings are planted in Gibson tubes (Gibson, 1963) containing a strip of filter paper as a wick and root support. The tubes are filled with quarter-strength Jensen nitrogen-free solution (Jensen, 1942) and capped with aluminium foil, as described by Vincent (1970) (Fig. 9.1).

3. Plants are grown for 4 weeks in a growth chamber (16 h light at 240 μE/s, 8 h dark, at 22°C) or a greenhouse. Plants are scored as positive (at least one nodule present) or negative (no nodule). Narrowing the dilution series (four- or twofold dilutions) decreases the confidence limit of the MPN.

Calculation

The MPN of rhizobia can be deduced from the McCrady table (Vincent, 1970) (Table 9.3).

Discussion

Depending on the legume considered, numbers of nodulating rhizobia below 10^2–10^4 (10^4 for soybean) per g of dry soil can be a limiting factor for the symbiotic growth of legumes, and inoculation is recommended. However, it needs to be kept in mind that MPN counts are not sufficient to assess whether nodulating rhizobia are effective at fixing nitrogen (see pp. 241–243). One limitation to the use of plant infection counts is that it takes several weeks to obtain the results, and a growth chamber or greenhouse is required. In addition, it must be kept in mind that experimental conditions do not well reflect those under field conditions, and soil, climatic and biotic factors prevailing in the soil environment also influence nodulation (as well as the efficiency of symbiotic nitrogen fixation). Molecular tools may also be used to enumerate rhizobia, e.g. by plating soil suspensions on semi-selective media and identifying rhizobia by colony hybridization, but the procedure is tedious (Laguerre *et al.*, 1993; Bromfield *et al.*, 1995; Hartmann *et al.*, 1998a). Species-specific molecular markers (detected by real-time quantitative PCR) seem promising for direct estimation of *Bradyrhizobium japonicum*, but they require further calibration.

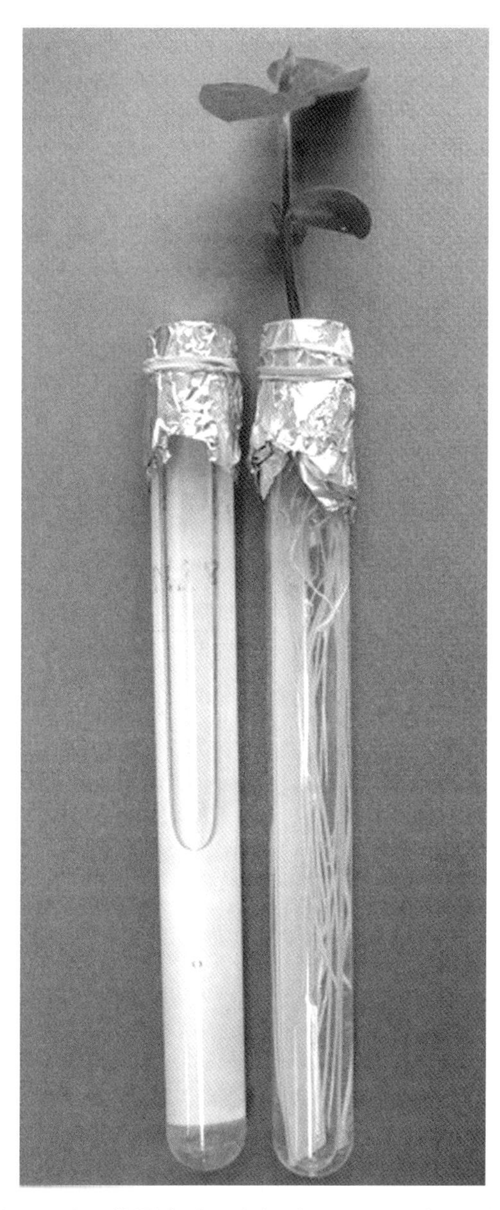

Fig. 9.1. Most probable number (MPN) plant infection counts of rhizobia nodulating soybean in Gibson tubes (220 × 22 mm). The tubes are filled with quarter-strength Jensen nitrogen-free solution and capped with aluminium foil. The filter-paper wick can be seen clearly in the left tube, which does not contain a plant. The right tube contains a soybean seedling.

Table 9.3. Number of rhizobia estimated by the plant infection count (from Vincent, 1970). Tenfold dilutions, four tubes per dilution step.

Positive tubes	Number of dilution steps (s)			
	$s = 10$			
40	$> 7 \times 10^8$			
39	$> 7 \times 10^8$			
38	6.9×10^8			
37	3.4×10^8			
36	1.8×10^8			
35	1.0×10^8			
34	5.9×10^7			
33	3.1×10^7	$s = 8$		
32	1.7×10^7	$> 7 \times 10^6$		
31	1.0×10^7	$> 7 \times 10^6$		
30	5.8×10^6	6.9×10^6		
29	3.1×10^6	3.4×10^6		
28	1.7×10^6	1.8×10^6		
27	1.0×10^6	1.0×10^6		
26	5.8×10^5	5.9×10^5		
25	3.1×10^5	3.1×10^5	$s = 6$	
24	1.7×10^5	1.7×10^5	$> 7 \times 10^4$	
23	1.0×10^5	1.0×10^5	$> 7 \times 10^4$	
22	5.8×10^4	5.8×10^4	6.9×10^4	
21	3.1×10^4	3.1×10^4	3.4×10^4	
20	1.7×10^4	1.7×10^4	1.8×10^4	
19	1.0×10^4	1.0×10^4	1.0×10^4	
18	5.8×10^3	5.8×10^3	5.9×10^3	
17	3.1×10^3	3.1×10^3	3.1×10^3	$s = 4$
16	1.7×10^3	1.7×10^3	1.7×10^3	$> 7 \times 10^2$
15	1.0×10^3	1.0×10^3	1.0×10^3	$> 7 \times 10^2$
14	5.8×10^2	5.8×10^2	5.8×10^2	6.9×10^2
13	3.1×10^2	3.1×10^2	3.1×10^2	3.4×10^2
12	1.7×10^2	1.7×10^2	1.7×10^2	1.8×10^2
11	1.0×10^2	1.0×10^2	1.0×10^2	1.0×10^2
10	5.8×10^1	5.8×10^1	5.8×10^1	5.9×10^1
9	3.1×10^1	3.1×10^1	3.1×10^1	3.1×10^1
8	1.7×10^1	1.7×10^1	1.7×10^1	1.7×10^1
7	1.0×10^1	1.0×10^1	1.0×10^1	1.0×10^1
6	5.8	5.8	5.8	5.8
5	3.1	3.1	3.1	3.1
4	1.7	1.7	1.7	1.7
3	1.0	1.0	1.0	1.0
2	0.6	0.6	0.6	0.6
1	< 0.6	< 0.6	< 0.6	< 0.6

The confidence interval at a 0.95 probability is calculated by multiplying and dividing the number by 3.8.

Assessment of the Symbiotic Efficacy of Indigenous Soil Rhizobial Populations

Introduction

Successful nitrogen fixation is expected to translate into better plant development, and this may be assessed by analysing plant development and biomass. In the method presented on pp. 233–236, the plant may obtain significant amounts of combined nitrogen from the soil and effective nitrogen fixation does not always lead to significantly higher plant biomass, which prevents satisfactory assessment of the symbiotic efficacy of the indigenous rhizobia unless dealing with chemically poor soils. To circumvent this limitation, assessment can be carried out using a rhizobia-free soil-less system, in which legumes are inoculated with diluted suspensions of the soil under study (whole soil inocula technique; Somasegaran and Hoben, 1994). Thus, both nodulation and the symbiotic effectiveness of the nodulating population can be characterized, thereby providing a more complete estimate of the symbiotic potential of rhizobial populations present in the soil.

Principle of the method

The method involves extracting indigenous microorganisms from soil and using the resulting soil suspension directly as an inoculum for legumes grown in a soil-less system from which the plant cannot take up significant amounts of nitrate and/or ammonium (Fig. 9.2). Plant biomass is assessed and compared with that in the absence of inoculation or in the non-inoculated positive control (inoculation with an efficient strain).

Material and apparatus

- 5-l plastic containers
- 16-mm diameter clay beads (Argi-16TBF)
- Perlite
- Surface-sterilized seeds
- Greenhouse
- Oven (105°C)
- Balance
- Basic microbiology equipment

Fig. 9.2. Efficiency tests for *Bradyrhizobium japonicum* strains in 5-l soil-less containers. The two containers at the front are non-inoculated controls (pale plants), whereas the containers at the back are inoculated with efficient strains.

Chemicals and solutions

- N-free mineral nutritive solution (per litre: K_2HPO_4, 0.14 g; $MgSO_4.7H_2O$, 0.25 g; $CaCl_2$, 0.28 g; K_2SO_4, 0.12 g; Sequestrene 138 Fe Novartis, 10 mg; H_3BO_3, 2 mg; $MnSO_4.H_2O$, 1.80 mg; $ZnSO_4.7H_2O$, 0.2 mg; $CuSO_4.5H_2O$, 0.08 mg; $Na_2MoO_4.2H_2O$, 0.25 mg)
- Fresh culture of an efficient rhizobial strain

Procedure

- The N-free soil-less system is prepared as follows (Fig. 9.2). 5-l plastic containers are filled with the following layers (from bottom to top): 1 l of 16-mm diameter clay beads (Argi-16TBF), 4 l of rhizobia-free substrate (perlite).
- Surface-sterilized seeds (eight in the case of soybean, thinned to four plants after germination) are sown in each container.
- 10^{-1} to 10^{-3} diluted soil suspensions are used as inocula, and, for the positive control, a fresh culture of an efficient rhizobial strain is prepared (Somasegaran and Hoben, 1994) and diluted in sterile water to give an inoculum of 10^6 cells per seed.

- Four containers (i.e. four replicates) are used per treatment (including the two controls).
- Plants are watered (e.g. through a hole at the bottom of the containers) with an N-free mineral nutritive solution, taking care to avoid cross-contamination.
- The experiment follows a randomized block design (four Fisher blocks).
- Plants are grown for 10 weeks in the greenhouse.
- Shoots are harvested, dried at 105°C for 24 h and weighed. Foliage dry matter production is considered a good indicator of total nitrogen uptake (Brockwell *et al.*, 1982), but results may be complemented by analysing total foliage nitrogen (Kjeldahl method).

Calculation

Variance analysis is performed to determine strain efficiency by comparison with the efficient reference strain and the non-inoculated control.

Discussion

A critical point is to include adequate controls, since diluted soil suspensions contain nutrients (including mineral nitrogen) that will not be present in non-inoculated and reference strain treatments. Variance analysis or other statistical treatment of the data should take this bias into account. This method may also be implemented in soil, but plant biomass determinations may be of limited interest under soil conditions where legumes can also acquire significant amounts of combined nitrogen by direct uptake of soil nitrate and/or ammonium. In this case, the amount of plant nitrogen specifically derived from biological fixation can be estimated using the approach presented below.

Assessment of Plant Nitrogen Derived from Symbiotic Fixation

Introduction

The isotopic methods based on ^{15}N variation are the most reliable for evaluating nitrogen fixation by plants under field conditions. Other methods can be used in parallel as useful indicators of nitrogen fixation activity *in situ*, such as acetylene reduction activity by nitrogenase or analysis of the nitrogen content of xylem sap (Dommergues *et al.*, 1999). The latter indirect methods will not be discussed hereafter. Isotopic methods are based on the assumptions that the root systems of the N_2-fixing legume and the non-fixing reference species have similar architectures and explore the same soil horizons, and that they assimilate from the same nitrogen pools (Danso *et al.*, 1993). The occurrence and type (endo- or ectomycorrhizae) of mycorrhizae should also be similar, as they can affect plant $\delta^{15}N$ (Högberg, 1990), and plants should have a similar phenology. These assumptions are rarely fully valid in most field studies, which highlights the importance of the choice of the non-fixing reference plant species. The natural ^{15}N abundance and the ^{15}N enrichment methods are presented here.

Principle of the methods

The natural ^{15}N abundance method is based on the isotopic discrimination process occurring in most chemical reactions involved in gaseous loss, such as ammonia volatilization or denitrification. Indeed, the latter processes favour a progressive disappearance of the lightest, main nitrogen isotope (^{14}N), and thus the heaviest one (^{15}N) tends comparatively to accumulate in soil (Dommergues *et al.*, 1999). Consequently, since the $^{15}N/^{14}N$ ratio of atmospheric N_2 is very low and stable, it is expected that nitrogen immobilized in N_2-fixing legumes will contain a lower proportion of ^{15}N compared with that of plants that do not fix nitrogen and take this element up from soil nitrate and/or ammonium. This means that the comparison of the relative proportion of ^{15}N (i.e. $\delta^{15}N$) in N_2-fixing legumes and non-fixing reference plants can be used to estimate the amount of nitrogen derived from the atmosphere by nitrogen fixation (i.e. % Ndfa) in the former.

The ^{15}N enrichment method, also called isotopic dilution method, is based on the comparison of ^{15}N excess measured in N_2-fixing legumes and non-fixing reference plants grown in soils enriched in ^{15}N through the incorporation of ^{15}N-labelled fertilizers (containing urea, nitrate or ammonium) to the soil. Since the N_2-fixing legumes will both take up combined nitrogen from soil and fix atmospheric nitrogen, their ^{15}N excess will be lower than that of the non-fixing reference plant, which has access to soil nitrogen (enriched in ^{15}N) only.

Materials and apparatus

A highly sensitive mass ratio spectrometer is required for the determination of natural abundance; ^{15}N measurement can be subcontracted to specialized service laboratories equipped with the relevant apparatus.

Procedure

The natural ^{15}N abundance method is only applicable when the soil $\delta^{15}N$ is superior or equal to +2 (Unkovich and Pate, 2001), and when spatial variability of soil $\delta^{15}N$ (as assessed through standard deviation) is not too high. Indeed, soil $\delta^{15}N$ may vary from site to site, according to the successive biological processes that occurred during a given soil history. Therefore, a preliminary sampling of leaves on different local species (including legumes and non-legumes) is required, followed by $\delta^{15}N$ analysis, to determine whether the natural ^{15}N abundance method can be used. If so, a more extensive sampling is then carried out. For herbaceous species, many experimental designs can be chosen, since sampling of plant material is easy to perform on both the N_2-fixing legume and the non-fixing species. The entire plant can be harvested and plant parts (leaves, fruits, stem, roots, nodules) analysed to determine dry weight (after drying for 3 days at 60°C), total nitrogen content and $\delta^{15}N$. $\delta^{15}N_a$ (the $\delta^{15}N$ of the N_2-fixing legume when deriving all its nitrogen from nitrogen fixation) is obtained by growing the legume (after inoculation with the most efficient rhizobial strain available) in an N-free system, e.g. in greenhouse pots containing an artificial substrate such as perlite or vermiculite and watered with an N-free nutrient solution. Plants (usually up to ten plants per treatment) are collected at 3–12 months and oven-dried at 60°C for 3 days. Plant parts are analysed for dry weight, total nitrogen and $\delta^{15}N$ assessments, as above. The best strategy is to compare the $\delta^{15}N$ of different non-fixing reference plant species and to calculate the percentage of N derived from biological fixation (%Ndfa) according to each of them. The different %Ndfa obtained are then presented as different assumptions. The total amount of nitrogen fixed by the legume (Ndfa) is obtained by multiplying %Ndfa with the total N content of the plant (see below for the formula).

In the ^{15}N enrichment method, %Ndfa is calculated from the percentages of ^{15}N excess in the N_2-fixing plant and the non-fixing reference plant (see below for the formula). Several protocols have been described for the ^{15}N enrichment method (Dommergues *et al.*, 1999). The sampling procedure proposed for the natural ^{15}N abundance method can be extrapolated to the ^{15}N enrichment method (except for the determination of $\delta^{15}N_a$, which is not required).

Calculation

In the natural ^{15}N abundance method, $\delta^{15}N$ is computed as follows:

$$\delta^{15}N(‰) = 1000 \times (\%^{15}N_{sample} - 0.3663)/0.3663$$

where 0.3663 is the percentage of ^{15}N in the atmosphere; the value is constant worldwide (Junk and Svec, 1958; Mariotti, 1983).

Then, %Ndfa in the N_2-fixing plant is computed as follows (Amarger *et al.*, 1977; Bardin *et al.*, 1977; Shearer and Kohl, 1986):

$$\%Ndfa = 100 \times (\delta^{15}N_{nf} - \delta^{15}N_f)/(\delta^{15}N_{nf} - \delta^{15}N_a)$$

where: $\delta^{15}N_{nf}$ corresponds to the $\delta^{15}N$ of the non-fixing reference plant; $\delta^{15}N_f$ to the $\delta^{15}N$ of the N_2-fixing legume studied; and $\delta^{15}N_a$ (also known as the β factor of isotopic discrimination) to the $\delta^{15}N$ of the N_2-fixing legume when deriving all its nitrogen from nitrogen fixation.

In the ^{15}N enrichment method, %Ndfa is calculated from the percentages of ^{15}N excess in the N_2-fixing plant and the non-fixing reference plant, as follows:

$$\%\text{Ndfa} = 100 \times (1 - E_i/E_o)$$

where: E_i corresponds to the ^{15}N excess measured in the N_2-fixing plant and E_o to the ^{15}N excess of the non-fixing reference plant. These values are generally determined using an emission spectrometer.

Discussion

The ^{15}N enrichment method displays several drawbacks. First, a correct estimation is only reached when uptake of soil nitrogen by both the N_2-fixing legume and the non-fixing reference plant follows the same kinetics, which is not often the case (Danso *et al.*, 1993). Secondly, gaseous loss (as, for example, N_2O or NH_3) or leaching of the labelled fertilizer can be other sources of error for the final calculation of %Ndfa. Thirdly, it is sometimes difficult to obtain a homogeneous labelling of the soil (especially with woody species), whereas the natural abundance method does not require any handling of soil and is thus less disturbing for the ecosystem. Fourthly, the use of ^{15}N-labelled fertilizer remains costly, making the natural abundance method much cheaper. Fifthly, the measurement is performed over a short period of time, which limits the significance of the results when the method is performed on perennial legumes. Consequently, the enrichment method is generally used in the framework of short-term field experiments, and thus is more adapted to quantification studies on annual crops. Finally, this remains the only applicable method when the soil $\delta^{15}N$ of a given site is too low.

Conclusion

As nitrogen is a key plant nutrient, its availability is a crucial parameter of soil quality. Development and optimization of symbiotic BNF can limit the expensive, energy-consuming and polluting industrial transformation required for production of chemical nitrogen fertilizers. Symbiotic nitrogen fixation performed by rhizobia–legume associations exhibits the highest rate of N fixation, and it is important to detect situations where rhizobial BNF is not optimal and propose solutions to improve it. In particular, inoculation has now been used successfully for years, mainly when the bacterial symbiont is absent from the soil or poorly efficient. Beside BNF, rhizobia are soil and rhizosphere bacteria that could have other roles in improving soil quality. For example, many rhizobia produce plant hormones or modify plant hormonal balance via 1-aminocyclopropane-1-carboxylate deaminase activity, and they might have a potential use as plant-growth-promoting rhizobacteria (PGPR) (Sessitsch *et al.*, 2002; Ma *et al.*, 2003). Certain rhizobia are involved in phosphate solubilization and thus could influence phosphorus nutrition of plants (Chabot *et al.*, 1996). The size and diversity of indigenous rhizobial populations may be affected by soil pollutants, farming practices, etc. (McGrath *et al.*, 1995; Morrissey *et al.*, 2002; Chen *et al.*, 2003; Walsh *et al.*, 2003), which highlights the usefulness of these bacteria as bioindicators of soil quality.

9.3 Contribution of Arbuscular Mycorrhiza to Soil Quality and Terrestrial Ecotoxicology

SILVIO GIANINAZZI,[1] EMMANUELLE PLUMEY-JACQUOT,[1,2] VIVIENNE GIANINAZZI-PEARSON[1] AND CORINNE LEYVAL[2]

[1]UMR INRA1088/CNRS 5184/UB PME, CMSE-INRA, 17 rue Sully, BP 86510, 21065 Dijon cedex, France; [2]LIMOS, Laboratoire des Interactions Microorganismes-Minéraux-Matière Organique dans les Sols, CNRS FRE 2440, 17 rue N.D. des Pauvres, B.P. 5, 54501 Vandoeuvre-les-Nancy cedex, France

Introduction

Although evolution has produced a general state of resistance to 'non-self' in plants, more than 95% of plant taxa in fact form compatible root associations, mycorrhizas, with certain soil fungi. These symbiotic associations are no doubt the most frequent examples of susceptibility of plants to fungi. Consequently, the root systems of a very large number of plants, and in particular those of many cultivated plants, whether they are agricultural, horticultural or fruit crops, do not exist simply as roots, but as complex mycorrhizal associations (Harley and Harley, 1987).

Mycorrhizas are usually divided into three morphologically distinct groups, depending on whether or not there is fungal penetration of the root cells: endomycorrhizas, ectomycorrhizas and ectendomycorrhizas. However, the most widespread plant root symbiosis is represented by (arbuscular) endomycorrhiza (AM) and is formed by more than 80% of plant families. The fungi involved all belong to the Glomeromycota (*Glomus, Acaulospora, Gigaspora, Entrophospora, Sclerocystis, Scutellospora*).

An AM association is generally mutualistic, in that the fungi obtain a carbon source from the host, while the latter benefits from enhanced nutrient uptake through transfer of mineral elements from the soil via the fungal hyphae (Smith and Read, 1997). In fact, there is an external network of ramifying mycelium of the symbiotic fungus into the soil from mycorrhizal root systems. This external mycelium supplies the plant with an extensive supplementary pathway for absorbing mineral nutrients and water from the soil, and it can facilitate mobilization by the plant of poorly mobile elements such as phosphate, ammonium, zinc and copper. The distance over which the fungal hyphae can translocate these nutrients exceeds the radius of any depletion zone likely to develop around an actively absorbing root, enabling plants to better exploit soil resources. It has been estimated that mycorrhizal hyphae can explore volumes of soil that are a hundred to a thousand times greater than the volume exploited by roots alone (Gianinazzi-Pearson and Smith, 1993). Strong evidence supports the existence of a direct link between the level of AM fungal biodiversity in soil and that of the plant species above ground (van der Heijden *et al.*, 1998).

Precolonization of roots by AM fungi can also alleviate stress due to metal and organic acid toxicity in soils of low pH (Leyval *et al.*, 1997), and reduce damage by soil-borne pathogens such as nematodes, *Fusarium*, *Pythium* or *Phytophthora*. The mechanism involved in this protection against a pathogen is complex, but there is strong evidence that AM fungi activate plant defence mechanisms against microbial pathogens (Gianinazzi, 1991; Cordier *et al.*, 1998; Dumas-Gaudot *et al.*, 2000). Furthermore, AM formation stimulates the development in the mycorrhizosphere of microorganisms with antagonistic activity towards soil-borne pathogens (Linderman, 2000). Therefore, the contribution of AM to soil quality and terrestrial ecotoxicology is of primary importance.

Mycorrhizal Inoculation to Improve Plant Health and Growth

The AM association is the most common type of mycorrhiza in agricultural systems. The importance of AM for plant growth and health is now widely demonstrated. It has become clear that they form an integral part of many cultivated plants and positively influence several aspects of plant physiology: mineral nutrition, water uptake, hormone production and resistance to root diseases (Smith and Read, 1997). Plants often grow badly in soils where AM fungi have been eliminated and their presence appears to be a factor determining soil fertility. Because of these characteristics, it is considered that AM fungi can be used as biological tools for increasing plant productivity in the field without creating problems of environment degradation, and by reducing chemical fertilizer or pesticide input (Gianinazzi and Gianinazzi-Pearson, 1988).

Given the effects of AM fungal inoculation on plants, it is generally accepted that appropriate management of this symbiosis should permit reduction of agrochemical inputs, and thus provide for sustainable and low-input plant productivity. Maximum benefits will only be obtained from inoculation with efficient AM fungi and a careful selection of compatible host/fungus/soil combinations (Gianinazzi *et al.*, 2002).

AM fungi cannot be grown in pure culture and must therefore be multiplied on living roots. This requires the use of techniques for inoculum production that are different from those employed for other mycorrhizal fungi. This, usually considered as a major disadvantage, appears in our experience to be an advantage, since the risk of multiplying AM fungi that have lost their symbiotic properties is very low, because culture collections have to be maintained on living host plants. On the other hand, producing symbiotic fungi on living plants raises difficulties because of precautions that have to be taken in order to obtain 'clean' inoculum. It is possible to produce AM inoculum on either excised roots or whole plants under axenic conditions using disinfected spores as inoculum (Gianinazzi *et al.*, 1989). However, for the moment, only one company is proposing commercial inoculum produced *in vitro*.

In order to use any of these types of inoculum rationally and successfully, it is necessary to define a strategy of inoculation, that is, to determine whether for a given situation it is necessary to inoculate and which fungi to use. We have developed biological tests for this. Without going into details about these, the aim is first to estimate the soil mycorrhizal potential, which indicates the number of fungal propagules, including spores, roots or hyphae, present in a given soil or substrate, and secondly, to evaluate the potential effectiveness of indigenous fungi for plant production and determine soil receptivity to fungi that are to be introduced into the system.

Information obtained using these biotests for mycorrhizal potential and fungal effectiveness, together with current knowledge about the mycorrhizal dependency of different plant genera and species, will provide the essential tools for a strategy of inoculation (Gianinazzi *et al.* 2002). The creation of 'La Banque Européenne des Glomales (BEG)' under the impulsion of the EU-COST programme (Dodd *et al.*, 1994), and its recent development into the international bank of Glomeromycota (http://www.kent.ac.uk/bio/beg/), will greatly help in this task. AM biotechnology is feasible for many crop production systems and the recent development of AM fungal-specific molecular probes provides tools for monitoring these microsymbionts in soil and roots (Van Tuinen *et al.*, 1998; Jacquot *et al.*, 2000).

A perspective for the near future should be the development of integrated biotechnologies (plant biotization) in which not only AM fungi, but also other microorganisms capable of promoting plant growth and/or protection – such as symbiotic or associative bacteria, plant-growth-promoting rhizobacteria (PGPR), pathogen antagonists (*Trichoderma, Gliocladium, Bacillus*, etc.), or hypovirulent strains of pathogens – will be used in synergy.

Application of Arbuscular Mycorrhiza as Bioindicators of Soil Quality

Introduction

Arbuscular mycorrhizal fungi are abundant soil fungi and are associated with many plant species. They are an integral part of the plant (Gianinazzi *et al.*, 1982), and many plants are highly dependent on mycorrhizal colonization for their growth. Not only can they be considered as a powerful extension of the plant-root system, providing access to a larger soil volume for nutrient uptake and playing a role in root protection against disease or drought, but they also affect the transfer of pollutants, such as heavy metals, to plants and can be affected by elevated metal concentrations (Gildon and Tinker, 1983; van der Heijden *et al.*, 1998) before toxicity symptoms can be seen on plants. Because of this, and because they are a direct link between the soil and plant roots, AM fungi have been proposed as bioindicators of heavy metal toxicity (Leyval *et al.*, 1995), as a supplement to chemical extraction procedures commonly used to assess availability of metals. Toxicity tests have been performed using techniques based on spore germination (Leyval *et al.*, 1995; Weissenhorn and Leyval, 1996; Jacquot-Plumey *et al.*, 2003), and on AM root colonization using MPN (Leyval *et al.*, 1995; Jacquot-Plumey *et al.*, 2002), or using Ri-T-DNA transformed roots (Wan *et al.*, 1998) or nested PCR, which allows monitoring of the effect of sewage sludges on the diversity of AM fungi *in planta* and in soil (Jacquot *et al.*, 2000; Jacquot-Plumey *et al.*, 2002). Very recently, the potential use of proteomics to study changes in protein expression of AM plants under stress conditions has also been investigated (Bestel-Corre, 2002). Two bioassays have been developed in collaboration with ADEME (Agence de l'Environnement et de la Maîtrise de l'Energie – Contract number 02750021) at LIMOS-CNRS (Nancy, France) and UMR BBCE-IPM, INRA/University of Burgundy (Dijon, France), based on spore germination and root colonization, respectively, and have been submitted to the Technical Committee T95^{E} on Terrestrial Ecotoxicology of the French Association of Normalisation (AFNOR) as a bioassay for ecotoxicological effects of chemicals or wastes and for soil quality.

Principle of the method

These bioassays are based on the germination of spores of an AM fungus (*Glomus mosseae*) sensitive to metals (Weissenhorn *et al.*, 1993), and on colonization of roots of *Medicago truncatula* inoculated with the same fungus. They are complementary since the spore germination assay concerns the AM fungus alone and the initial stage of the symbiotic cycle, while the root colonization assay reflects a further stage of the symbiosis when the fungus interacts with the host plant. Both bioassays are direct contact tests, for acute and chronic toxicity, respectively. Contaminated soils, sewage sludge, wastes or solutions can be tested for their potential toxicity to beneficial microbial activity. AM spore germination and root colonization within a test substrate containing a substance, or a contaminated soil, are compared to a control test substrate alone or a non-contaminated soil comparable to the soil sample to be tested.

Materials and apparatus

- Nitrocellulose filter membranes (0.45 μm, 47 mm in diameter)
- Petri dishes
- Filter paper
- Seeds of *Medicago truncatula* Gaertn (line J5)
- *G. mosseae* inoculum (Biorize, Dijon, France)

- Sand (0.5–1 mm particle size)
- Growth chamber
- Binocular magnifier
- Photonic microscope

Chemical and solutions

- Agar water
- Mycofert nutrient solution (Biorize, Dijon, France)

Procedure

AM spore germination:

1. Thirty *G. mosseae* spores (Nicol. & Gerd.) Gerdemann & Trappe (BEG12) (obtained from Biorize, Dijon, France) are placed between two nitrocellulose filter membranes (0.45 μm, 47 mm in diameter) with gridlines, held together by a slide frame as described by Weissenhorn *et al.* (1993).
2. This so-called sandwich with spores is placed in a Petri dish containing 50 g of test substrate (sand, particle size 0.5–1 mm) mixed with, for example, a chemical compound solution or a contaminated soil to be tested (Fig. 9.3).
3. The sandwich is then covered with another 50 g layer of sand/soil.
4. Soil and sand are allowed to moisten up to water-holding capacity. A filter paper, lining the bottom of the dish and extending outside the dish into a water-filled larger dish, moistens the substrate by capillarity when soil or sludges are used. When sand is used, it is moistened by watering.
5. The Petri dish is sealed and kept at 24°C in the dark for 2 weeks.
6. The spore sandwich is then removed from the soil/sand, stained using tryptan blue for 1 h (Bestel-Corre, 2002) and carefully rinsed with tap water before opening.
7. The number of spores recovered and of germinated spores are recorded at ×32 magnification.
8. Six replicates are made for each sample.
9. The dose effect can be studied using sample dilution in sand and estimation of the mean inhibitory concentration at 50% (MIC50).

Root colonization:

1. Pre-germinated seeds of *M. truncatula* Gaertn. (line J5) on water-agar (0.7%) are transferred individually into pots, and each plant is inoculated with 15 g of a *G. mosseae* inoculum placed in the planting hole of 200 g of a mixture of sand (0.5–1 mm particle size) with the solution, soil or sewage sludge to be tested.
2. Plants are cultivated under controlled conditions (photoperiod 16 h, 19–22°C, relative humidity 60–70%, light intensity 320 $\mu E/m^2/s^1$), and receive weekly 10 ml Mycofert nutrient solution.

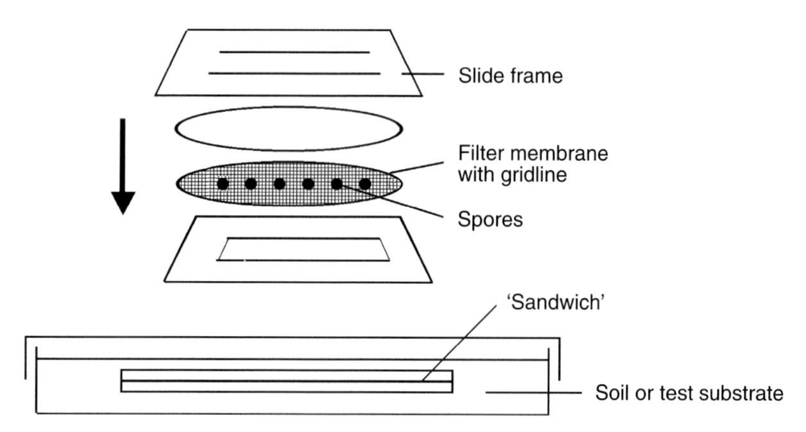

Fig. 9.3. Arbuscular mycorrhiza (AM) spore germination test for pollutants in soil, sewage sludge or wastes.

3. Frequency of root colonization by *G. mosseae* is estimated after 4 weeks according to Trouvelot *et al.* (1985) and a simplified determination is presently made based on a + (fungal detection)/– notation.
4. Six replicate plants are used for each sample.
5. The dose effect can be studied using sample dilution in sand to measure the mean inhibitory concentration at 50% (MIC50).

Discussion

The spore germination bioassay has been used to compare the toxicity of different soils (Leyval *et al.*, 1995) and sewage sludges, and to compare the tolerance to cadmium of different AM fungi (Weissenhorn *et al.*, 1993). The comparison of composted sewage sludges and other wastes from different origins showed a decrease in spore germination with a composted sludge amended with polycyclic aromatic hydrocarbons (PAH) (B31), with combustion ashes (C1) or combusted household refuse (C2), but not when the sludge was unamended or amended with metallic pollutants (B32) (Fig. 9.4A). Dilution of these sludges and ashes showed that C1 and C2 were more toxic to AM spore germination than B31 sludge (Fig. 9.4B).

The root colonization bioassay has been used to compare, *ex situ*, the toxicity of different non-composted sewage sludges and to study, *in situ*, the effects of three successive sewage sludge amendments (Jacquot *et al.*, 2000, Jacquot-Plumey *et al.*, 2002). Comparison of the sewage sludges showed a slight decrease in root colonization with the addition of B3 sewage sludge to sand and even more when the sludge was amended with polycyclic aromatic hydrocarbons (B31) and heavy metals (B32) (Fig. 9.5).

Neither of the bioassays based on AM fungi is specific: a reduction in germination or in root colonization may be due to different pollutants, and possibly also to other soil characteristics, such as very low pH, or high P

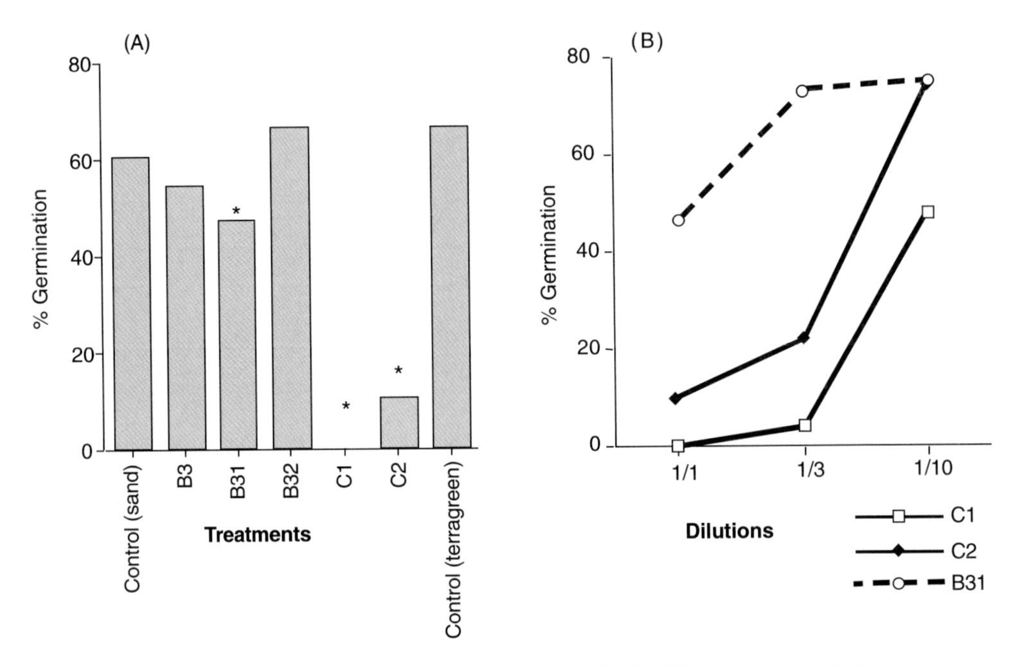

Fig. 9.4. Germination (%) of *Glomus mosseae* spores (A) in different composted sewage sludges and wastes and (B) after dilution of the sludge and wastes in sand using the sandwich bioassay (Fig. 9.3). * indicates significant differences from control (sand or terragreen, chi-squared test, $P \leqslant 0.05$). B3, composted sludge; B31, B3 composted sludge spiked with polycyclic aromatic hydrocarbons (PAH); B32, B3 composted sludge spiked with heavy metals; C1, combustion ash; C2, combusted household refuse.

content. Therefore, when there is such a reduction, further investigations should be performed to identify its origin.

Early in the 1980s, Gildon and Tinker reported the presence of heavy-metal-tolerant AM fungi in polluted soils and their effect on metal uptake by plants (Gildon and Tinker, 1983). Since then, the presence of AM fungi in heavy-metal-polluted sites, and the contribution of AM fungi to the transfer of heavy metals and radionuclides from soil to plants and to the translocation from root to shoots, has been addressed in many studies. They have concerned different metallic trace elements, such as Cd, Zn and Ni, and radionuclides such as Cs (Leyval and Joner, 2001). Results showed that heavy-metal-tolerant AM fungi may reduce metal transfer to plants and protect them against heavy metal toxicity (Leyval *et al.*, 1997). However, results are not always consistent, probably due to different metal concentrations, fungi and availability in soils (Leyval and Joner, 2001; Leyval *et al.*, 2002; Rivera-Becerril *et al.*, 2002). They also cannot be generalized and may differ with plants, with heavy metals/radionuclides and their availability, and possibly, although this has been less investigated, with other microbial components of the soils.

Remediation of heavy metal and radionuclide contaminated soils includes immobilization and extraction techniques. Different plants can be

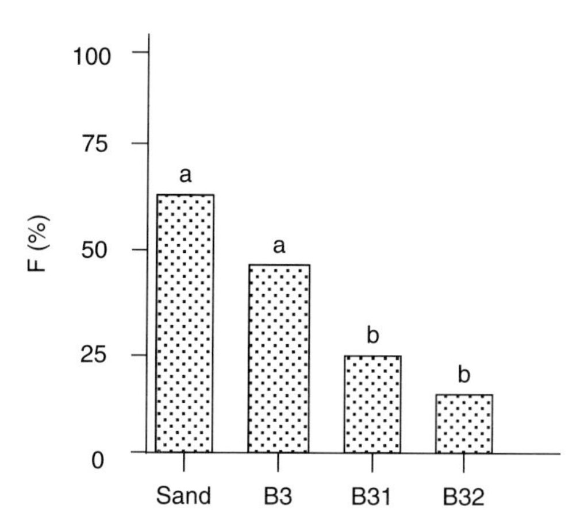

Fig. 9.5. Frequency of root colonization (F%) by *Glomus mosseae* grown in sand and in sand mixed with an unamended sewage sludge (B3), or amended with polycyclic aromatic hydrocarbons (PAH) (B31) and metallic pollutants (B32). Values in columns headed by the same letter are not significantly different ($P \leq 0.05$).

used for phytoextraction (extraction of metals by plants accumulating high metal concentrations) and phytostabilization (use of plants to reduce heavy metal availability, erosion and leaching). For phytoextraction, hyperaccumulative plants from the *Brassicaceae* family are often used, but most plants of this family are reported as non-mycotrophic. However, other plants accumulating lower metal concentrations, but producing higher biomass, such as sunflower and willow, are also receiving attention and these are mycorrhizal. There has been much work done, and much progress made, in the field of AM fungi in soil remediation and restoration. AM inoculum production on a commercial scale is available and is no longer a limit to applied studies (Von Alten *et al.*, 2002). Field studies and field applications of AM fungi have been reported and are currently running, but they concern mainly revegetation studies, horticultural and agricultural applications. There is still a lack of demonstration and of *in situ* projects using AM fungi in polluted sites. However, preliminary field trials have indicated that dual inoculation with a selected AM fungus and *Rhizobium* can improve the yield and nodulation of pea genotypes in a metal contaminated site (Borisov *et al.*, unpublished results; EU project INCO-Copernicus IC15-CT98–0116). The persistence of the inoculated fungi, and the relative diversity of AM fungi in roots *in situ* are poorly known and need to be investigated further.

Arbuscular mycorrhizal fungi provide a direct link between soil and roots and are key components of soil–plant ecosystems. Their presence and beneficial activity can be affected by soil perturbation and contamination, and can be used as an early sign of soil disfunctioning. Besides improving soil fertility and plant productivity, they affect the fate of pollutants such as heavy metals and radionuclides, and therefore bring many advantages to mycorrhizal plants in polluted soils. However, there is still much research needed, especially to predict the conditions in which a beneficial effect of AM fungi can be expected. This requires a better understanding of the mechanisms involved in heavy metal tolerance, taking into account not only AM fungi but also the complexity of the soil–plant–microbe systems.

9.4 Concepts and Methods to Assess the Phytosanitary Quality of Soils

CLAUDE ALABOUVETTE,[1] JOS RAAIJMAKERS,[2] WIETSE DE BOER,[3] RÉGINA NOTZ,[4] GENEVIÈVE DÉFAGO,[4] CHRISTIAN STEINBERG[1] AND PHILIPPE LEMANCEAU[1]

[1]UMR 1229 INRA/Université de Bourgogne 'Microbiologie et Géochimie des Sols', INRA-CMSE, BP 86510, 21065 Dijon cedex, France; [2]Wageningen University, Laboratory of Phytopathology, Binnenhaven 5, 6709 PG Wageningen, The Netherlands; [3]NIOO-CTE, Department of Plant–Microorganisms Interactions, PO Box 40, 6666 ZG Heteren, The Netherlands; [4]Institut für Pflanzenwissenschaften/Phytopathologie, LFW B25, Universitätsstraße 2, ETH Zentrum, CH-8092 Zurich, Switzerland

Concepts

A plant disease results from the intimate interaction between a plant and a pathogen, and today there is a great effort of research devoted to the study of plant–pathogen interactions at the cellular and molecular levels. The importance of these direct interactions should not hide the role of environmental factors that influence disease severity. These indirect interactions are particularly important in the case of diseases induced by soilborne pathogens. The existence of soils that suppress diseases provides an example of biotic and abiotic factors affecting the pathogen, the plant and/or the plant–pathogen interaction. Indeed, in suppressive soils, disease incidence or severity remains low in spite of the presence of a virulent pathogen, a susceptible host plant and climatic conditions favourable for disease development.

Soils suppressive to diseases caused by a range of important soilborne pathogens have been described; they include fungal and bacterial pathogens, but also nematodes (Schneider, 1982; Cook and Baker, 1983; Schippers, 1992). These soils control root rot and wilt diseases induced by: *Aphanomyces euteiches*, *Cylindrocladium* sp., several formae speciales of *Fusarium oxysporum*, *Gaeumannomyces graminis*, *Pythium* spp., *Phytophthora* spp., *Rhizoctonia solani*, *Ralstonia solanacearum*, *Streptomyces scabies*, *Thielaviopsis basicola* (*Chalara elegans*) and *Verticillium dahliae*. This large diversity of pathogens controlled by suppressive soils shows that soil suppressiveness is not a rare phenomenon. On the contrary, every soil has some potential of disease suppression, leading to the concept of *soil receptivity* to diseases.

The receptivity of a soil to soilborne diseases is its capacity to suppress more or less the saprophytic growth and infectious activity of the pathogenic populations present in the soil. Indeed, the soil is not a neutral milieu where pathogenic microorganisms interact freely with the roots of the host plant. On the contrary, the soil interferes in several ways with the relationships between and among microorganisms, pathogens and plants, and it can modify the interactions among microorganisms themselves. Soil receptivity (or soil suppressiveness) is a continuum, going from highly conducive soils to strongly suppressive soils (Alabouvette *et al.*, 1982; Linderman *et al.*, 1983).

Soil suppressiveness to some diseases can be related to another fundamental phenomenon affecting the soil microorganisms: microbiostasis. Well studied by Lockwood (1977) in the case of fungal spores, *fungistasis* is defined as the global effect of the soil that restricts the germination and growth of fungi. In general, the germination and saprophytic growth of fungi are more restricted in soil than would be expected from their behaviour *in vitro* under similar environmental conditions of temperature, moisture and pH. Based on this definition, fungistasis only concerns the saprophytic growth of the fungi, without taking into account their interaction with the plant, but in some cases fungistasis has been associated with soil suppressiveness to diseases.

The concept of soil receptivity to diseases was already evoked in the definition of *inoculum potential* proposed by Garrett (1970) as 'the energy of growth of a parasite available for infection of a host at the surface of the host organ to be infected'. One of the most important words in this definition is 'energy' of growth. It clearly states that the presence of the inoculum, although necessary, is not sufficient to explain the disease. Among the factors that affect the 'energy of growth' from the inoculum, Garrett (1970) pointed to 'the collective effect of environmental conditions', and indicated that 'the endogenous nutrients of the inoculum might be augmented by exogenous nutrients from the environment'.

Applied to soilborne pathogens, this concept of inoculum potential led to that of 'soil inoculum potential', which was at the origin of both theoretical and practical studies. Baker (1968) gave a definition of inoculum potential as the product of inoculum density per capacity. Louvet (1973) proposed to define inoculum capacity as the product of innate inoculum energy and the effects of the environment on this inoculum. Thus, in this definition, the effects of the environment on the inoculum corresponds to what we have defined above as the soil receptivity to diseases.

At the same time, the soil inoculum potential was defined by Bouhot (1979) as the pathogenic energy present in a soil. This inoculum potential depends on three main factors: (i) the inoculum density; (ii) the pathogenic capacity of this inoculum; and (iii) the soil factors which influence both the inoculum density and capacity. This last factor, again, corresponds to the soil receptivity as defined above.

Thus, whatever the definition, all these authors acknowledge that the soil plays a major role in influencing the interactions between a susceptible host plant and its specific pathogens present in soil. It is therefore very important to take into consideration both the inoculum potential of a naturally infested soil and its level of suppressiveness, when elaborating control strategies.

Indeed, the traditional approach to control soilborne disease consisted of trying to eradicate the pathogens from the soil. This led to the use of very dangerous biocides, such as methyl bromide, use of which will be totally banned in the near future. Opposite to that, a new approach consists of either enhancing the natural suppressive potential that exists in every soil, or introducing specific biological control agents. In both cases it is necessary to characterize and, if possible, quantify the soil inoculum potential and the soil receptivity of the soil to the disease, and also the soil receptivity to the biological control agents.

To characterize the phytosanitary quality of a soil, it is important to detect the presence or absence of the main pathogens of the crop to be cultivated. But this knowledge might not be sufficient to predict the risk for the crop to be severely diseased. Therefore, it is preferable to assess the soil inoculum potential, which will give a better view of the capacity of the soil to provoke the disease. A low inoculum potential may result from a very low inoculum density in a conducive environment, or a high inoculum density in a suppressive environment. Therefore, it is very interesting to assess the level of both soil inoculum potential and receptivity, to better estimate the probability of a healthy crop.

Assessment of soil inoculum potential and receptivity requires bioassays, and a few of them will be described in this section. Fungistasis, which is assessed *in vitro*, may be considered as a bioindicator of the phytosanitary quality of the soil, a few examples will be described below. Another strategy to assess the phytosanitary soil quality, which will not be described in detail in this section, is to quantify specific populations and/or specific genes shown to be involved in suppressiveness (Weller *et al.*, 2002), for example, phloroglucinol-producing pseudomonads, shown to be involved in the take-all decline phenomenon (Raaijmakers and Weller, 1998).

Bioassays for Assessing Soil Suppressiveness to *Fusarium* Wilts

Principle of the method

A standardized method has been proposed by Alabouvette *et al.* (1982). It involves infesting soil samples with increasing concentrations of a pathogenic strain of *Fusarium oxysporum*, to grow a susceptible plant under well-controlled conditions, and to establish the disease progress curve in relation to inoculum concentrations. This method is usually applied using flax (*Linum usitatissinum*) and its specific pathogen *F. oxysporum* f. sp. *lini*.

Materials and apparatus

- Seeds of 'Opaline', a highly susceptible cultivar of flax (*Linum usitatissinum*)
- Stock culture of a virulent strain of *F. oxysporum* f. sp. *lini* (Foln3)
- Vessels to produce fungal inoculum in shake culture
- Rotary shaker
- Sintered glass funnel (pore size 40–100 μm)
- Growth chamber to perform the bioassay
- Polystyrene trays (60 × 40 × 5 cm)
- Calcinated clay (Chemsorb®, CONEX Damolin GmbH, Peckhauser Str. 11, D. 40822 Mettmann, Germany)

Chemicals and solution

- Liquid malt extract (10 g/l, pH 5.5)
- Hydrokani® nutrient solution

Procedure

1. The pathogenic strain of *F. oxysporum* f. sp. *lini* (Foln3) is cultivated in liquid malt extract (10 g/l, pH 5.5) for 5 days on a rotary shaker (150 rpm) at 25°C.
2. The whole culture is filtered through a sterile sintered glass funnel (pore size 40–100 μm) and the conidial suspension in the filtrate is adjusted to different concentrations in order to infest the soil samples at concentrations of 1×10^3, 10^4, 10^5 propagules/ml soil.
3. The bioassay is set up in polystyrene trays (60 × 40 × 5 cm). Each tray has 12 rows consisting of eight holes of 7 cm diameter. The holes are closed at the bottom with a plastic grid, topped with a layer of small beads, in order to maintain the soil in the hole and permit excess water to drain.
4. 800 g soil is distributed in two rows of eight holes (i.e. 50 g/hole) and is infested with 4 ml of inoculum suspension at the concentrations needed to reach 1×10^3, 10^4, 10^5 propagules/ml soil.
5. Several seeds of flax cultivar (Opaline) are placed in-between two layers of calcinated clay (Chemsorb®) on the top of the soil, to prevent seeds from damping off. Uninoculated rows (4 ml sterile water) serve as a control.
6. After emergence, the number of seedlings is reduced to one per hole, making a total of 16 plants per treatment.
7. Each treatment is replicated three times.
8. The experiment is carried out in a growth chamber, with 17°C night and 22°C day temperature for 2 weeks, then with 20°C night and 25°C day temperature with a daily 15-h photoperiod of 16,000 lux for the rest of the experiment.

9. The plants are irrigated regularly with nutrient solution (Hydrokani®) and water alternatively, and the number of healthy plants is recorded twice a week from 21 days to 58 days after sowing.

Calculation

Depending on the number of plants used in the bioassay, two procedures of calculation might be used.

When a small number of plants per replicate is used, each plant is identified (as it is the case with the flax, for example) and survival data analysis is used (Hill *et al.*, 1990). This permits estimation of the duration of life of the population and the comparison of the soil receptivity to *Fusarium* wilts (Fig. 9.6). This procedure uses a Kaplan Meier estimate to calculate the survival function $S(t)$, which provides the probability that the failure time (typical wilt symptom appearance) is at least t for each plant, t being the time elapsed since the day of inoculation. A mean survival time (MST) is then evaluated for each subpopulation of plants (i.e. the set of 16 plants making a replicate) inoculated with the respective doses of the pathogen, provided that at least one plant among the 16 of a subpopulation of flax exhibits symptoms (Tu, 1995). In the example given, the MST of the control with no pathogen could not be evaluated because no mortality occurred during the bioassay.

When large numbers of plant are required for the experimental set-up (as might be the case with the assessment of soil suppressiveness to other diseases), or when a disease index is used to evaluate the disease incidence, the area under the disease progress curve (AUDPC) is calculated for each replicate by plotting the cumulative value of the disease index (symptom occurrence, symptom intensity or dead plants, etc.) against time.

Both procedures are applicable to evaluate natural soil suppressiveness to diseases and to evaluate the impact of any treatment on the level of soil suppressiveness or the pathogenic infectious activity (Höper *et al.*, 1995; Migheli *et al.*, 2000). MST values and AUDPC values are then analysed by analysis of variance (ANOVA) and tests of multiple comparison of means (as Fisher's LSD test and Student t-test, for instance, with $P = 0.05$). All statistical analyses are performed using StatView® software issued by SAS Institute Inc. (Cary, North Carolina, USA).

Discussion

Soil suppressiveness to *Fusarium* wilts, whatever the plant and the corresponding forma specialis, can be assessed by this procedure using flax and its specific pathogen, *F. oxysporum* f. sp. *lini* (Abadie *et al.*, 1998). But it is also possible to adapt this procedure to other plant–formae speciales models, such as tomato–*F. oxysporum* f. sp. *lycopersici* or melon–*F. oxysporum* f.

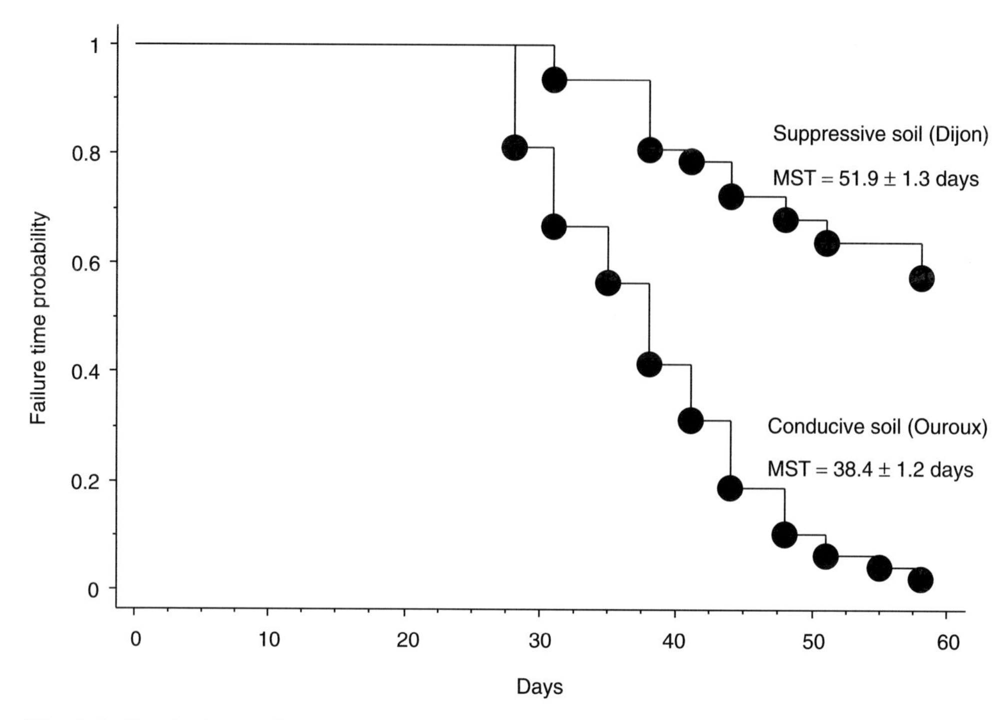

Fig. 9.6. Survival time of populations of flax cultivated in a soil suppressive to *Fusarium* wilt (Dijon) and in a soil conducive to the same disease (Ouroux). Both soils were inoculated with *F. oxysporum* f. sp. *lini* Foln3.

sp. *melonis*. Procedures based on the same principle have been proposed to measure soil suppressiveness to *Pythium* or *Rhizoctonia solani* damping-off (Camporota, 1980), *Aphanomyces* root rot (Persson *et al.*, 1999), take-all (Raaijmakers *et al.*, 1997), *Thielaviopsis basicola* (*Chalara elegans*) root rot (Stutz and Défago, 1985), etc.

Assessment of Soil Inoculum Potential

Principle

Inoculum potential being the pathogenic energy present in a soil, assessment of inoculum potential consists of growing susceptible host plants in the soil under environmental conditions chosen to be very favourable to disease expression.

To quantify this potential, it is necessary to dilute the naturally infested soil in a disinfested soil in different proportions, in order to obtain a dose–response relationship. This principle will be illustrated by the procedure described by Bouhot (1975a,b,c) to estimate the inoculum potential of soils infested by *Pythium* spp.

Materials and apparatus

- Seeds of a variety of cucumber (*Cucumis sativus*), for example 'Le généreux', highly susceptible to *Pythium* damping-off but not to other fungi, such as *Phytophthora* spp., also responsible for damping-off
- A mortar
- A sieve with mesh of 1000 μm
- Pots containing 300 ml of soil
- Sterile soil
- A blender to mix soil and oatmeal
- A growth chamber in which to perform the bioassay

Chemicals and solutions

- Oatmeal

Procedure

1. The soil to be analysed is first air-dried, then ground in a mortar and sieved through a mesh of 1000 μm.
2. This soil is amended with oatmeal at a rate of 20 g soil/litre.
3. The amended soil is diluted into sterile soil to obtain the following concentrations: 30%, 10%, 3%, 1%, 0.3% and 0.1% of soil in sterile soil.
4. Cucumber plants, cultivar 'Le généreux', are produced by sowing ten seeds in steamed, disinfested soil, in 10-cm diameter pots.
5. Plants are cultivated at 25°C under only 4000 lux for 15 h/day.
6. After 5 days, when the hypocotyls are 3–4 cm long, water is added in order to reach the water-holding capacity of the soil.
7. 60 ml of the infested soil amended with oatmeal is spread on the surface of the disinfested soil.
8. This soil layer is adjusted to 80% of its water-holding capacity.
9. The pots with the cucumber plants are then placed for 24 h at 15°C in the dark, then transferred again to 22°C during the day, 19°C at night, with a photoperiod of 15 h under 9000 lux.
10. Damping-off symptoms appear from the second to the seventh day.

Calculations

A correlation can be calculated between the number of dead plants and the concentration of the amended soil. From the regression lines it is possible to determine an inoculum potential unit, defined as the weight of soil needed to provoke the death of 50% of the plants. But, in many cases, the distribution is not normal; it is therefore necessary to utilize different transformations before calculating the inoculum potential units.

Discussion

Based on the same principle, other procedures have been proposed to assess the inoculum potential of soil infested by *Rhizoctonia solani* and *Aphanomyces euteiches* (Camporota, 1982; Williams-Woodward *et al.*, 1998; Reverchon, 2001). It is important to choose the experimental conditions and the test plant carefully, in order to avoid symptoms produced by other fungi than that of interest. Depending on the pathogen studied, the dose–response relationship is not always evident; thus, it might be difficult to quantify the soil inoculum potential.

Soil Fungistasis

Introduction

Following the definition given by Lockwood (1977), the terms 'fungistasis' and 'mycostasis' are used to describe the phenomenon that germination and growth of fungi in most natural soils are more restricted than would be expected from their behaviour under similar environmental conditions of temperature, moisture and pH *in vitro*. Fungistasis affects both plant-pathogenic and saprophytic fungi, but the former are generally more sensitive (Garrett, 1970; De Boer *et al.*, 1998). The high sensitivity of plant-pathogenic fungi to fungistasis can have both positive and negative consequences for disease development (Lockwood, 1977). A clearly negative aspect of fungistasis is that it protects propagules of plant pathogens from germination under unfavourable conditions, e.g. the absence of a host plant. On the other hand, continuing exposure to fungistasis results in loss of vitality of plant-pathogenic propagules. Furthermore, fungistasis limits the distance over which plant pathogens can reach host roots. In general, the intensity of fungistasis for a pathogenic fungal species is found to be positively correlated with the receptivity or disease suppressiveness of soils, but lack of correlation has also been reported (Lockwood, 1977; Hornby, 1983; Larkin *et al.*, 1993; Knudsen *et al.*, 1999; Peng *et al.*, 1999).

The intensity of fungistasis is dependent on several factors, such as physical and chemical soil properties, fungal life-history characteristics, soil microbial activity and soil microbial community composition (Schüepp and Green, 1968; Lockwood, 1977; Hornby, 1983; Toyota *et al.*, 1996; De Boer *et al.*, 1998). The importance of a biological factor is indicated by the relief of fungistasis by (partial) sterilization treatments and addition of antibiotics (Lockwood, 1977; Toyota *et al.*, 1996; De Boer *et al.*, 1998, 2003).

The explanation given most often for the microbial cause of fungistasis is that soil microorganisms limit nutrient availability to the germinating spores or invading hyphae (Ho and Ko, 1986; Lockwood, 1988). Fungistasis has also been attributed to the presence of antifungal compounds of microbiological origin (Romine and Baker, 1973; Liebman and Epstein, 1992; De Boer *et al.*, 1998, 2003). The distinction between nutrient limitation and antibiosis as main mechanisms for fungistasis is not so easily made, as both can be the result of competition for nutrients (Fravel, 1988; Paulitz, 1990). Yet, although the actual mechanism of fungistasis is still open to debate, it is clear that it is a function of activities of soil microorganisms and their interactions with fungi. Therefore, fungistasis can be considered as a soil ecosystem function.

Principle of the methods

The methods used to quantify fungistasis focus on inhibition of either germination of fungal spores or extension of fungal hyphae. To determine differences in fungistasis between soils, it is important to use fixed amounts of soil (dry weight basis) and identical moisture conditions. In general, soil samples are wetted to pF 1 (–1 kPa) or 2 (–10 kPa) to establish optimal conditions for diffusion of nutrients or toxic compounds. For germination tests, spores are mixed directly into the soil or inoculated on a membrane or gel that is in contact with the soil (Schüepp and Green, 1968; Wacker and Lockwood, 1991; Knudsen *et al.*, 1999). After incubation, the amount of germinated spores and/or the length of the germination tubes is determined microscopically. The basic methods have been modified when the aim of the study has been to elucidate the mechanism of fungistasis, e.g. to establish the contribution of toxic solutes or volatiles (Romine and Baker, 1973; Liebman and Epstein, 1992).

Quantification of the effect of soil on hyphal growth has been given much less attention than that on spore germination. The methods used, so far, report on production of fungal biomass, growth rate or colonizing ability (Hsu and Lockwood, 1971; De Boer *et al.*, 1998).

Almost all studies on soil fungistasis have been done using one method, i.e. either spore germination or hyphal extension. For a dune soil isolate of *Fusarium oxysporum*, both methods were compared during incubation of a dune soil that was subjected to a sterilization + soil inoculum treatment (Fig. 9.7; W. de Boer and P. Verheggen, unpublished results). Both methods indicated a rather constant level of fungistasis in the untreated soil and a quick (within a few weeks) return to this level in the re-inoculated sterile soil.

Differences in intensity of fungistasis between soils can remain obscured when undiluted soils are used. To overcome this, dilution series of soil in sterilized quartz sand have been used (Wacker and Lockwood, 1991; Knudsen *et al.*, 1999).

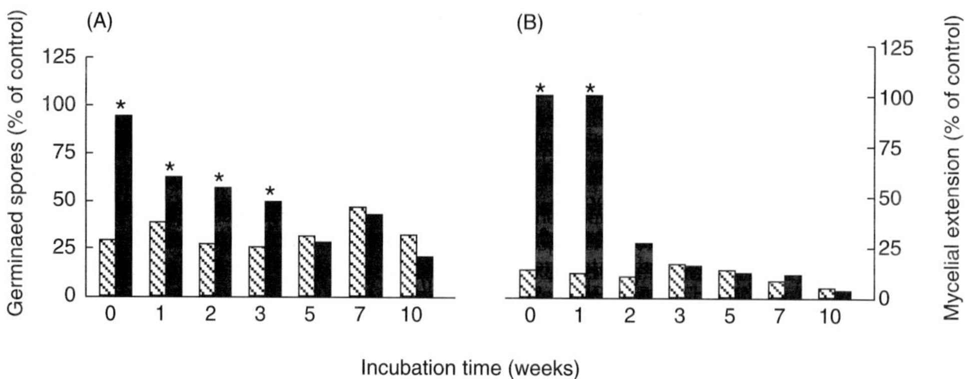

Fig. 9.7. (A) Spore germination and (B) mycelial extension of a dune isolate of the fungus *Fusarium oxysporum* in a dune soil that was incubated for 10 weeks after sterilization + soil inoculation (black bars). Results of similarly incubated untreated dune soil (hatched bars) are also given.* indicates significant differences (two sample *t*-test) between treated and untreated soil. Spore germination and mycelial extension were determined as described in the text.

Materials and apparatus

- Petri dishes (13.5 cm diameter)
- Acid-washed sand (or another nutrient-free, soil-like matrix)
- Glass wool
- Microcentrifuge
- Large membrane filters (polycarbonate; pore 0.2 μm; diameter 9 cm)
- Small membrane filters (polycarbonate or cellulose ester; pore 0.2 μm; diameter 2.5 cm)
- Glass slides
- Binocular microscope
- Epifluorescence microscope

Chemicals and solutions

- Calcofluor White M2R (Sigma Aldricht, St Louis, Missouri, USA) staining solution: 2.3 mg/ml in sterile water
- Potato dextrose agar (Oxoid: UNIPATH S.A., 6 rue de Paisy, 69570 Dardilly cedex, France)

Procedure for the spore germination test

For the sake of comparison, the procedure for isolation and storage of the spores should be standarized, as it has been shown that their germination ability is affected by age, storage conditions, medium composition and strain characteristics (Garrett, 1970; Mondal and Hyakumachi, 1998). Since growth requirements and sporulation conditions differ among fungal species, it is not possible to give one general applicable protocol. Most often, fungi are grown for 2–3 weeks on a solid medium such as potato dextrose agar or oatmeal agar. The spores are isolated by applying a small amount of sterile water or salt solution and gently rubbing the surface with a bent glass rod. The spore suspension is then filtered through sterile glass wool to remove clumps of spores and mycelial fragments. Further cleaning of spores is done by centrifugation. The density of the spores is determined microscopically and adjusted, if necessary. Spore suspensions are used immediately or stored at –80°C in an aqueous solution of glycerol (Liebman and Epstein, 1992).

As indicated above, several methods can be used for the germination test. Here we propose to inoculate the spores on top of a filter, as this is easiest for subsequent microscopic inspection. The following protocol is given as an example:

1. Petri dishes are filled with 90 g moist (–10 kPa) soil, which is spread evenly to obtain a smooth surface (bulk density mineral soil about 1 g/cm^3).
2. A sterile, water-saturated polycarbonate membrane filter (9 cm diameter) is placed on top of the soil. Two sterile glass slides are put on top of the membrane filter (4 cm apart) to keep it in close contact with the soil. The Petri dish is sealed with Parafilm and incubated for 48 h at 20°C to allow diffusion of soil solutes into the membrane filter.
3. Small membrane filters containing the spores (added by vacuum filtration) are placed in the area between the slide glasses. The Petri dish is sealed again and incubated for 16–48 h (depending on fungal species).
4. Spores are stained by floating the membranes for 2 min in a solution of Calcofluor White M2R (Sigma) in demineralized water (2.3 mg/ml). After destaining with demineralized water, at least 100 spores/filter are checked for germination using an epifluorescence microscope.
5. As control for germination ability, soil is replaced by nutrient-free, acid-washed sand.

Procedure for the hyphal extension test

The method described here tests the ability of fungal hyphae to invade soils from nutrient-rich agar (De Boer *et al.*, 1998, 2003).

1. Petri dishes (8.5 cm diameter) are filled with 50 g moist (–10 kPa) soil, which is spread evenly to obtain a smooth surface (bulk density mineral soil about 1 g/cm^3).
2. Potato dextrose agar discs (1 cm diameter; 0.4 cm thick) from the growing margin of the fungal colony are inverted and placed centrally on top of the soil in a Petri dish.

3. After 3 weeks of incubation at 20°C, the extension of the mycelium is determined using a binocular microscope, and the area of hyphal extension is estimated (Fig. 9.8). Image analysis can be used to obtain a more accurate estimate of the mycelial area.
4. The hyphal extension on the soil samples is compared with that on nutrient-free, acid-washed beach sand (–10 kPa).

As a modification, the agar disc containing fungal inoculum may be separated from soil by inert material (e.g. stainless-steel discs) to prevent stimulation of growth of (antagonistic) soil microorganisms by agar.

Calculations

For both tests (germinated spores or invaded area) fungistasis is expressed as the percentage of the control (acid-washed sand).

Discussion

A simple estimate of intensity of soil fungistasis is obtained by measuring the difference between hyphal extension or spore germination on a nutrient-free, soil-like control and on the soil under study. In fact, the nutrient availability in the control should be the same as in the soil, but this is very hard to realize. Therefore, given the fact that most soils are carbon-limited for microbial growth, a nutrient-free control seems a reasonable option. In addition, effect of soil sterilization on growth and germination may be determined to differentiate between biotic and abiotic causes of fungistatis (Dobbs and Gash, 1965; De Boer *et al.*, 1998, 2003). Yet, care should be taken that the sterilization procedure does not introduce undesired (inhibiting) side-effects.

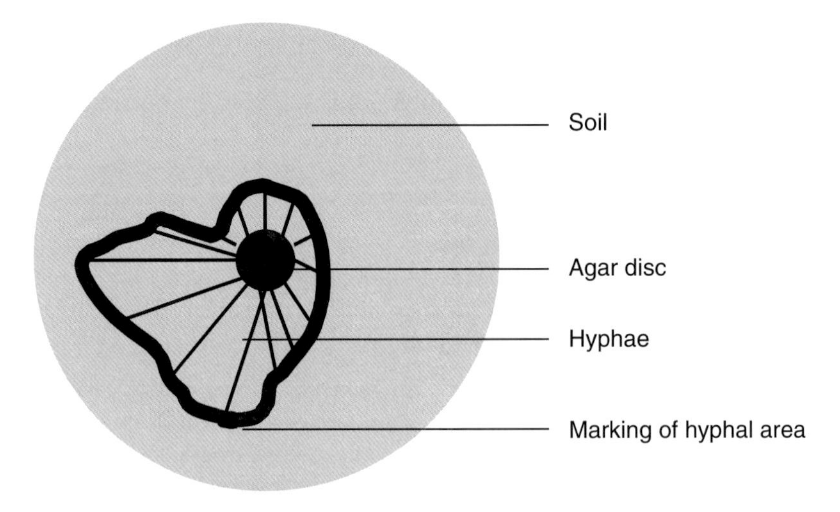

Fig. 9.8. Determination of the area of hyphal extension.

For comparison of fungistastic properties of a wide range of soils, it would be best to use a fixed set of type fungal species/strains. However, from an ecological or phytopathological point of view this has little meaning, as subspecies or strains can be adapted to specific climatic and environmental conditions. Hence, the specific questions to be answered determine what will be the most appropriate fungal strains to use for fungistasis measurement.

It is important to realize that soil fungistasis is dynamic. It can be decreased strongly by enhanced carbon availability, e.g. caused by roots. This explains why even highly fungistatic soils do not necessarily prevent pathogenic fungi from causing disease. Hence, measurements of fungistasis must be completed by bioassays allowing the assessment of receptivity and inoculum potential.

9.5 Free-living Plant-beneficial Microorganisms and Soil Quality

YVAN MOËNNE-LOCCOZ,[1] SHERIDAN L. WOO,[2] YAACOV OKON,[3] RENÉ BALLY,[1] MATTEO LORITO,[2] PHILIPPE LEMANCEAU[4] AND ANTON HARTMANN[5]

[1]UMR CNRS 5557 Ecologie Microbienne, Université Claude Bernard (Lyon 1), 43 bd du 11 Novembre, 69622 Villeurbanne cedex, France; [2]Universita degli Studi di Napoli 'Federico II', Dip. di Arboricoltura, Botanica e Patologia Vegetale, Sez. Patologia Vegetale, Lab. di Lotta Biologica, Portici (NA), Italy; [3]Hebrew University of Jerusalem (HUJI), Department of Plant Pathology and Microbiology, Faculty of Agricultural, Food and Environmental Quality Sciences, Rehovot, Israel; [4]UMR 1229 INRA/Université de Bourgogne 'Microbiologie et Géochimie des Sols', INRA-CMSE, BP 86510, 21065 Dijon cedex, France; [5]GSF-National Research Center for Environment and Health, Department of Rhizosphere Biology, Ingolstädter Landstraße 1, 85764 Neuherberg/München, Germany

Introduction

Many microorganisms living in the rhizosphere and benefiting from root exudates can have positive effects on plant growth and health. The relationship between these plant-beneficial microorganisms and the plant host corresponds to a symbiosis or an associative symbiosis (cooperation). The first case often involves differentiation of one partner or both, which facilitates identification of such symbiotic interactions; in this handbook, symbiotic interactions with the plant are covered in the sections dealing with nodulating, nitrogen-fixing bacteria (Section 9.2) and mycorrhizal fungi (Section 9.3). The second case corresponds to microorganisms designated plant-growth-promoting rhizobacteria (PGPR) and plant-growth-promoting fungi (PGPF), and these are the focus of this section.

PGPR and PGPF can exert positive effects on plants by various mechanisms, some of them implying a directly positive effect on seed germination, root development, mineral nutrition and/or water utilization (i.e. phytostimulation) (Jacoud *et al.*, 1998, 1999; Dobbelaere *et al.*, 2001). Indirect effects can also take place, and typically involve suppression of phytopathogenic bacteria or fungi, and/or phytoparasitic nematodes (i.e. biological control) (Cronin *et al.*, 1997; Walsh *et al.*, 2001; Burdman *et al.*, 2002). In certain cases, the biocontrol effect mediated by indigenous free-living plant-beneficial microorganisms results in the suppression of plant disease (Moënne-Loccoz and Défago, 2004), which is an emerging ecosystemic property (see Section 9.4). In most cases, however, the plant-beneficial effects of indigenous free-living microorganisms remain unnoticed because of the absence of visible differentiation (in contrast to root nodules in the nitrogen-fixing symbiosis, for example), and the fact that these effects add up with a multitude of other effects related to variability in space of plant genotypes, soil composition, farming practices, microclimatic conditions and the composition of the soil biota. This is particularly true for phytostimulatory effects due to indigenous free-living plant-beneficial microorganisms, because they do not necessarily lead to differences in plant health. However, the largely unnoticed contribution of indigenous free-living plant-beneficial microorganisms to plant growth and health is important. Furthermore, since chemical inputs into farming can have deleterious effects on environmental health and food product quality, effective PGPR and PGPF inoculants have the potential to be used as a replacement for, or in combination with reduced rates of, chemical fertilizers (phytostimulation) and pesticides (biocontrol).

Many taxa of free-living plant-beneficial microorganisms are known, for both bacteria and fungi. Some of them, e.g. fluorescent pseudomonads and non-pathogenic *Fusarium oxysporum*, have been discussed in Section 9.4, due to their contribution to disease suppressiveness of certain soils. In this section, the emphasis will be placed on free-living plant-beneficial microorganisms belonging to the bacterial genus *Azospirillum* and the fungal genera *Trichoderma* and *Gliocladium*.

Azospirillum (group 1 of the α-proteobacteria) are PGPR found in close association with the roots of plants, particularly *Gramineae* (Tarrand *et al.*, 1978; Bally *et al.*, 1983), and they mainly colonize the root elongation zone. They exert beneficial effects on plant growth and yield of many crops of agronomic importance (Okon and Labandera-Gonzalez, 1994), and represent one of the best-characterized PGPR (Okon, 1994). During the early days of the investigation on *Azospirillum*–plant associations, plant growth promotion was thought to derive from the contribution of biological N_2 fixation by the bacterial partner. However, further studies demonstrated that the positive effects of *Azospirillum* are due mainly to morphological and physiological changes of the roots of inoculated plants, which lead to an enhancement of water and mineral uptake, especially when plants grow in suboptimal conditions (Dobbelaere *et al.*, 2001). Indeed, inoculation with *Azospirillum* increases the density and length of root hairs, as well as the appearance and elongation rates of lateral roots, thus increasing the root surface area. These effects are linked to the secretion of plant growth hormones such as auxins, gibberellins and cytokinins by the bacterium (Dobbelaere *et al.*, 2001), a property shared with a variety of other plant-beneficial root-colonizing bacteria. In addition to phytostimulation, certain *Azospirillum* strains display biocontrol properties towards phytopathogenic bacteria (Bashan and de-Bashan, 2002) or parasitic plants (Bouillant *et al.*, 1997; Miché *et al.*, 2000).

Trichoderma and *Gliocladium* are fast-growing, spore-producing fungi, commonly found in soil throughout the world (Klein and Eveleigh, 1998). They are resistant to many xenobiotic compounds, and can catabolize a wide range of natural and synthetic organic compounds (including complex polymers), thus having an effect on the nitrogen and carbon cycles (Danielson and Davey, 1973b; Kubicek-Pranz, 1998). Most importantly, biocontrol interactions exist with several phytopathogens, such as *Rhizoctonia*, *Pythium*, *Sclerotinia*, *Fusarium*, *Verticillium*, *Phytophthora*, *Phomopsis*, *Gaeumannomyces* and *Sclerotium* (Weindling, 1932; Jeffries and Young, 1994). The biocontrol mechanisms implicated are diverse. First, *Trichoderma* and *Gliocladium* species are relatively unspecialized, disruptive or necrotrophic mycoparasites, and may also be parasitic towards nematodes. Constitutive secretion of cell-wall-degrading enzymes (CWDEs), such as chitinases and cellulase, plays an important role (Lorito *et al.*, 1996; Lorito, 1998; Zeilinger *et al.*, 1999). The subsequent release of cell-wall degradation products enables chemotactical location of the host (Zeilinger *et al.*, 1999). Physical contact with the phytopathogen triggers coiling, attachment and host penetration by *Trichoderma* (Inbar and Chet, 1992, 1995). Secondly, *Trichoderma* and *Gliocladium* species produce antibiotics, which can affect bacteria and/or pathogenic fungi (Howell, 1998; Kubicek *et al.*, 2001). Thirdly, *Trichoderma* and *Gliocladium* can compete with phytopathogenic species. Although competition is less important than other biocontrol mechanisms, it is a prerequisite for efficient plant colonization (Lo *et al.*, 1996; Harman and Bjorkman, 1998). Fourthly, *Trichoderma* and *Gliocladium* may have direct effects on the plant, e.g. via solubilization of inorganic nutrients or induced resistance in the plant (Windham *et al.*, 1986; Harman, 2000), resulting in better seed germination, enhanced plant growth and development, and increased yield (Lindsey and Baker, 1967; Windham *et al.*, 1986; Harman 2000). The population levels at which one can expect beneficial effects from these indigenous biocontrol fungi are not known, and they probably depend on the phytopathogen, since biocontrol effects are host-dependent. However, there is a link to soil quality, in that environmental conditions that stimulate growth and subsequent colonization and sporulation of biocontrol fungi (e.g. high nutrient availability) will also reduce pathogen populations in the soil. *Trichoderma* and/or *Gliocladium* are used in many commercial inoculants worldwide. Websites that can be consulted for information include http://www.agrobiologicals.com, http://www.oardc.ohio-state.edu/apsbcc/productlist.htm, http://attra.ncat.org/attra-pub/orgfert.html and http://www.epa.gov/pesticides/biopesticides.

In this section, methods available to assess indigenous populations of free-living plant-beneficial microorganisms (*Azospirillum*, *Trichoderma* and *Gliocladium*) are described. Since these microorganisms may be present in different microbial habitats in the soil ecosystem (from bulk soil to root tissues), a strategy to separate the relevant microhabitats/compartments and validated in the case of bacteria is presented (pp. 273–274). The resulting samples can be processed via cultivation-based methods (pp. 274–278 and 281–284), which can only recover a minority of individuals (the culturable ones), but are needed to obtain strains to be used as inoculants. Alternatively, methods for cultivation-independent analysis of free-living plant-beneficial bacteria are described (pp. 291–281).

Extraction of Indigenous Free-living Plant-beneficial Bacteria from Soil and Roots

Introduction

Desorption of bacteria from soil and roots is important for qualitative and quantitative monitoring, regardless of whether indigenous bacteria or inoculants are considered. A standardized protocol for the separation and differentiation of different rhizosphere compartments (e.g. rhizosphere soil/rhizoplane versus root tissues) and the extraction of bacterial cells adsorbed to the root surface has been described in the case of *Medicago sativa* cv. Europae (Mogge *et al.*, 2000; Hartmann *et al.*, 2004). The procedure followed most recommendations made by Macdonald (1986) and Herron and Wellington (1990) and is presented hereafter. Fluorescent *in situ* hybridization (FISH) in combination with confocal scanning laser microscopy (CSLM) is useful to confirm successful desorption of bacteria in root surface studies (see pp. 279–281).

Principle of the method

The method consists of separating roots from soil by physical means, followed by the maceration of root tissues to free endophytic microorganisms. Microbial cells are then extracted from the samples.

Materials and apparatus

- Sterile tweezers
- Stomacher 80 (Seward Medical, Thetford, UK)
- Laboratory glassware (Erlenmeyer flasks, etc.)
- Gauze (40 µm mesh size)
- 5-µm syringe filter (Sartorius No. 17549, Göttingen, Germany)

Chemicals and solutions

- 0.01 M phosphate buffer (Na_2HPO_4/KH_2PO_4, pH 7.4)
- 0.1% sodium cholate buffer
- Polyethylene glycol 6000 (PEG 6000; Sigma, Deisenhofen, Germany)
- Cation exchange polystyrene beads (Chelex 100; Sigma)

Procedure

1. Roots are carefully separated from the soil using sterile tweezers. All steps are conducted with sterile solutions on ice.
2. Non-rhizosphere soil (bulk soil compartment) and root-attached soil particles collected by shaking the roots (rhizosphere compartment) are each suspended in 0.01 M phosphate buffer in a 1:9 (w/v) ratio and dispersed for 1 min at highest speed in a Stomacher 80.
3. To extract rhizoplane and endophytic bacteria (root compartment), 1 g of fresh roots previously cleaned from adhering soil particles (see above) and washed in phosphate buffer are suspended in 20 ml of 0.1% sodium cholate buffer (Macdonald, 1986). The suspension is treated in a Stomacher 80 at highest speed for 4 min to disrupt polymers.
4. After transfer into an Erlenmeyer flask, 0.5 g of PEG 6000 and 0.4 g of cation exchange polystyrene beads (Chelex 100) are added and the suspension is stirred for 1 h at 50 rpm and 4°C.
5. The Stomacher/stirring procedure is repeated three times, the roots being transferred to fresh 0.1% sodium cholate buffer with PEG 6000 and Chelex 100 after each extraction step. The suspensions obtained after each step are pooled.
6. Root and soil particles are removed by filtration through a gauze (40 μm mesh size) and subsequently a 5-μm syringe filter. The resulting sample is used to study bacteria from the root compartment.

Discussion

The suspensions thus obtained from bulk soil, the rhizosphere and the root compartment (i.e. rhizoplane and root tissues) can be used for analysis of indigenous free-living plant-beneficial bacteria (see below) as well as monitoring of bacterial inoculants.

Cultivation Approach to Enumerate Indigenous *Azospirillum* spp.

Introduction

The presence of *Azospirillum* spp. in the rhizosphere can be shown by enrichment and cultivation on semi-solid nitrogen-free media, and this approach will also target other root-associated nitrogen-fixing bacteria (Döbereiner, 1995). Table 9.4 summarizes the different media suitable for the enrichment of *Azospirillum* spp. on the basis of different pH requirements and carbon substrate preferences. *Azospirillum* spp. differ in their physiology, as summarized in Table 9.5, which can be used to identify the isolates using physiological criteria. In addition, routine API® and Biolog™ test systems are used for convenient physiological identification purposes. Here, protocols based on 16S rDNA hybridization to identify *Azospirillum* species are presented.

Table 9.4. Four media used for the isolation and cultivation of *Azospirillum* spp.

Ingredients (per litre)	NFb[a,b]	LGI[a,b]	Modified NFb	Potato agar[c]
DL-Malic acid	5 g	–	5 g	2.5 g
Sucrose	–	5 g	–	2.5 g
K_2HPO_4	0.5 g	0.2 g	0.13 g	–
KH_2PO_4	–	0.6 g	–	–
$MgSO_4.7H_2O$	0.2 g	0.2 g	0.25 g	–
NaCl	0.1 g	–	1.2 g	–
$CaCl_2.2H_2O$	0.02 g	0.02 g	0.25 g	–
$Na_2MoO_4.2H_2O$	–	0.002 g	–	–
Na_2SO_4	–	–	2.4 g	–
$NaHCO_3$	–	–	0.22 g	–
Na_2CO_3	–	–	0.09 g	–
K_2SO_4	–	–	0.17 g	–
Minor element solution[d]	2 ml	–	2 ml	2 ml
Bromothymol blue solution[e]	2 ml	2 ml	–	–
Fe-EDTA, 1.64%	4 ml	4 ml	4 ml	–
pH-value (adjusted with KOH)	6.8	6.0	8.5	6.8
Vitamin solution[f]	1 ml	1 ml	1 ml	1 ml
Agar	1.75 g	1.75 g	1.75 g	15 g

[a] Ingredients should be added to the medium in the stated order; [b] Semi-solid medium; [c] 200 g fresh potatoes are peeled, cooked for 30 min and filtered through cotton before other ingredients are added; [d] $CuSO_4.5H_2O$, 0.4 g; $ZnSO_4.7H_2O$, 0.12 g; H_2BO_3, 1.4 g; $Na_2MoO_4.2H_2O$, 1 g; $MnSO_4.H_2O$, 1.5 g; H_2O, 1000 ml; [e] 0.5% bromothymol blue in 0.2 N KOH; [f] Biotin, 10 mg; Pyridoxol/HCl, 20 mg; H_2O, 100 ml.

Principle of the methods

Colony hybridization with 16S rDNA-targeted oligonucleotides (Kabir *et al.*, 1994, 1995; Chotte *et al.*, 2002) can be used to identify *Azospirillum* species, and the procedure is usually carried out on nitrogen-fixing isolates able to fix the dye Congo Red, because this is a typical *Azospirillum* attribute (Rodriguez Caceres, 1982). Species-specific probes include Al (for *A. lipoferum*, 5′-CGTCGGATTAGGTAGT-3′), used at a hybridization/washing temperature of 43°C, and Aba (for both *A. brasilense* and *A. amazonense*, 5′-CGTCCGATTAGGTAGT-3′), used at a hybridization/washing temperature of 51°C (Chotte *et al.*, 2002).

Materials and apparatus

- Sterile membranes (GeneScreen Plus; Nen Life Science Products, Boston, Massachusetts, USA)
- 3MM paper
- UV table
- Hybridization tubes and oven (65°C)
- Hyper film TM-MP (Amersham Labs, Amersham, UK)
- Autoradiography set-up

Table 9.5. Main physiological characteristics of *Azospirillum* spp.

	A. doebereinerae	*A. lipoferum*	*A. largimobile*	*A. brasilense*	*A. amazonense*	*A. irakense*	*A. halopraeferens*
Carbon utilization test (API)							
N-Acetylglucosamine	-[a]	+	+	−	d	+	ND
D-Glucose	d	+	+	d	+	+	−
Glycerol	+[b]	+	+	+	−	−	+
D-Mannitol	+	+	+	−	−	−	+
D-Ribose	−	+	+	−	+	d	+
D-Sorbitol	+	+	+	−	−	−	−
Sucrose	−	−	ND[d]	−	+	+	−
Acid formation (API 50 anaerobe)							
From glucose	d[c]	+	+	−	−	−	−
From fructose	+	+	+	−	−	−	+
Miscellaneous							
Biotin requirement	−	+	−	−	−	−	+
Optimum growth temperature	30°C	37°C	28°C	37°C	35°C	33°C	41°C
Optimum pH for growth	6.0–7.0	5.7–6.8	ND	6.0–7.8	5.7–6.5	5.5–8.5	6.8–8.0
Occurrence of pleomorphic cells	+	+	+	−	+	+	+

[a] sign (−) means less than 10% of the investigated strains showed a positive response; [b] sign (+) means more than 90% of the investigated strains showed a positive response; [c] d (depends) means between 11% and 89% of the investigated strains showed a positive response; [d] Not determined.

For the alternative hybridization protocol:

- Glass slides
- Epifluorescence microscope

Chemicals and solutions

- Tryptone-yeast extract (TY; per litre: tryptone, 5 g; yeast extract, 3 g; $CaCl_2.H_2O$, 0.5 g)
- 10% (w/v) sodium dodecyl sulphate (SDS)
- Denaturing solution (NaOH 0.5 M and NaCl 1.5 M)
- Neutralizing solution (NaCl 1.5 M, Tris 1 M, pH 7.4)
- 2× standard saline citrate (SSC; NaCl 0.1 M, sodium citrate 15 mM, pH 7.0)
- Pre-hybridization solution (i.e. 16.5 ml sterile distilled water, 3 ml dextran 50%, 1.5 ml SDS 10%, 0.58 g NaCl)

- ^{32}P
- Denatured herring sperm DNA

For the alternative hybridization protocol:

- Phosphate-buffered saline (PBS; 0.13 M NaCl, 7 mM Na_2HPO_4 and 3 mM NaH_2PO_4 (pH 7.2))
- Paraformaldehyde
- Agarose
- Ethanol
- FISH hybridization buffer (20 mM Tris-hydrochloride, pH 7.2, 0.01% SDS and 5 mM EDTA)
- NaCl
- Formamide
- DAPI
- Citifluor AF1 (Citifluor Ltd, London, UK)

Procedure

1. The bacteria are grown for 48 h at 28°C on sterile membranes (GeneScreen Plus) previously placed on TY plates. Two membranes are used for each probe.
2. The membranes are then treated successively for 6 min in 10% (w/v) SDS, 10 min in denaturing solution, 9 min in neutralizing solution and 9 min in 2× standard saline citrate to promote cell lysis.
3. The membranes are left drying for 1 h on 3MM paper, followed by a 4-min UV treatment for DNA binding.
4. The hybridization procedure consists of moistening the membranes in 2× SSC, rolling and transferring them into hybridization tubes containing pre-hybridization solution.
5. Each probe is labelled at the 5′ end with ^{32}P (as described by Sambrook *et al.*, 1989).
6. The hybridization tubes are then put for 2 h in an oven (65°C) before adding the labelled probe and 600 μl of denatured herring sperm DNA.
7. After hybridization for 12 h, the membranes are rinsed at room temperature for 5 min (twice) using 2× SSC, 30 min (twice) using 2× SSC and 1% SDS, and 30 min (three times) using 0.1× SSC.
8. The membranes are left to dry for 1 h at room temperature on 3MM paper sheets and are then packed with Hyper film TM-MP for at least 12 h at –80°C for autoradiography.

An alternative hybridization protocol involves:

1. Fixing bacteria overnight at 4°C in PBS containing 3% paraformaldehyde.
2. Then they are washed in PBS, mixed with 0.3% agarose, dropped on to glass slides and dried at room temperature.

3. These glass slides are immersed successively in 50%, 80% and 96% ethanol for 3 min each and stored at room temperature.
4. Oligonucleotide probes (Table 9.6) labelled with Cy3, Cy5 or 5(6)-carboxyfluorescein-*N*-hydroxysuccinimide ester (FLUOS) at the 5′ end are used.
5. The oligonucleotides are stored in distilled water at a concentration of 50 ng/μl (Amann *et al.*, 1990).
6. FISH is performed (Wagner *et al.*, 1993) at 46°C for 90 min in FISH hybridization buffer containing 0.9 M NaCl and formamide at the percentages shown in Table 9.6.
7. Hybridization is followed by a stringent washing step at 48°C for 15 min.
8. The washing buffer is removed by rinsing the slides with distilled water.
9. Counterstaining with DAPI and mounting in Citifluor AF1 is performed as described previously (Aßmus *et al.*, 1995).
10. Observations are made by epifluorescence microscopy.

Table 9.6. 16S rRNA-targeted oligonucleotide probes for FISH analysis of the *Azospirillum* cluster (Stoffels *et al.*, 2001).

Probes and competitors	Sequence (5′–3′)	Stringency[a]	Specificity
AZO440a +	GTCATCATCGTCGCGTGC	50	*Azospirillum* spp.
AZO440b	GTCATCATCGTCGTGTGC	50	*Conglomeromonas* spp., *Rhodocista* spp.
AZOI665	CACCATCCTCTCCGGAAC	50	*Azospirillum* species cluster[b]
Abras1420	CCACCTTCGGGTAAAGCCA	40	*A. brasilense*
Alila1113	ATGGCAACTGACGGTAGG	35	*A. lipoferum*, *A. largimobile*
Adoeb587	ACTTCCGACTAAACAGGC	30	*A. doebereinerae*
Ahalo1115	ATGGTGGCAACTGGCAGCA	45	*A. halopraeferens*
Aama1250	CACGAGGTCGCTGCCCAC	50	*A. amazonense*
Airak985	TCAAGGCATGCAAGGGTT	35	*A. irakense*
Rhodo654	ACCCACCTCTCCGGACCT	65	*Rhodocista centenaria*
Sparo84	CGTGCGCCACTAGGGGCG	20	*Skermanella parooensis*
Abras1420C	CACCTTCGGGTAAAACCA	40	Competitor[c]
Alila113C	ATGGCAACTGGCGGTAGG	35	Competitor
Ahalo1115C	ATGATGGCAACTGGCAGTA	45	Competitor

[a] Amount of formamide (%, v/v) in hybridization buffer; [b] *lipoferum*, *brasilense*, *halopraeferens*, *doebereinerae*, and *largimobile*; [c] the competitor oligonucleotide (without fluorescent label) is used in the FISH analysis to prevent false-positive hybridizations, which could be possible in rare cases of indigenous bacteria harbouring very close oligonucleotide similarity according to the sequence analysis.

Cultivation-independent Approach to Monitor Indigenous *Azospirillum* spp. in Extracted Root Compartments or in the Rhizosphere

Introduction

Culture-dependent techniques may not always enable recovery of all targeted bacteria, even when considering a particular strain belonging to an easily culturable taxon (Défago *et al.*, 1997; Mascher *et al.*, 2003). Even the efficacy of the PCR-based method may depend on the physiological status of the cells (Rezzonico *et al.*, 2003). Here, a FISH method enabling detailed localization of cells and an *in situ* approach is presented.

Principle of the methods

Concomitant staining with the general DNA stain DAPI and FISH enables counting total and hybridizing bacteria in the three compartments outlined in Fig. 9.9, after collection on polycarbonate filters. The main advantage of FISH analysis is that the bacteria can also be identified and localized directly in the rhizosphere, provided that they are present in a physiologically active state (i.e. harbouring a high ribosome content).

Materials and apparatus

- 0.2 μm polycarbonate filters
- Zeiss Axiophot 2 epifluorescence microscope (Zeiss, Jena, Germany)
- Filter sets F31–000, F41–001 and F41–007 (Chroma Tech. Corp., Battleboro, Vermont, USA)

Chemicals and solutions

For analysis of cell suspensions:

- Formaldehyde
- Ethanol
- Citifluor AF1 (Citifluor Ltd, London, UK)
- DAPI
- Species-specific DNA probes (see Table 9.6)

For analysis *in situ*:

- PBS (0.13 M NaCl, 7 mM Na_2HPO_4 and 3 mM NaH_2PO_4 (pH 7.2))
- Formaldehyde
- A confocal laser scanning microscope, e.g. microscopes LSM 410 or LSM 510 (Zeiss, Jena, Germany).

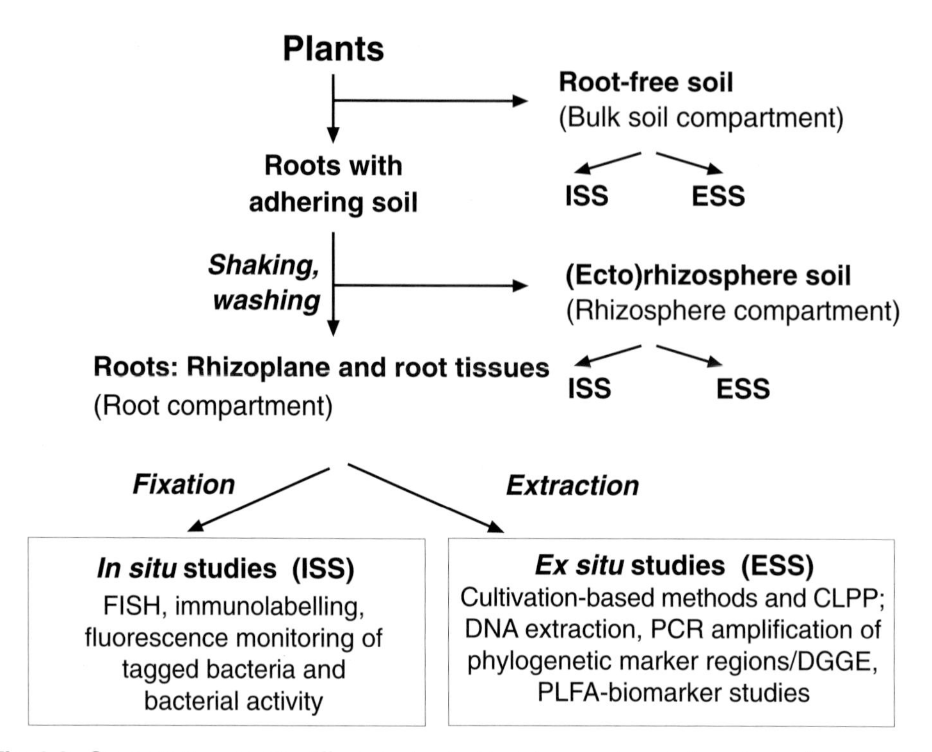

Fig. 9.9. Strategy to separate different microhabitats/compartments in soil–plant systems for detailed analysis of indigenous free-living plant-beneficial microorganisms and monitoring of microbial inoculants by *in situ* studies (ISS) and *ex situ* studies (ESS).

Procedure

Concomitant staining with DAPI and FISH is done as follows.

1. The cell suspensions (for extraction details, see pp. 273–274) are fixed overnight at 4°C with 3% formaldehyde and concentrated on to 0.2 μm polycarbonate filters (100 μl aliquots).
2. Dehydration of cells is performed successively with 50%, 80% and 96% ethanol for 3 min each.
3. The slides are mounted with Citifluor AF1 to reduce photobleaching.
4. A Zeiss Axiophot 2 epifluorescence microscope equipped with filter sets F31–000, F41–001 and F41–007 can be used for the enumeration of bacteria on filters.
5. Total cell counts (DAPI) and hybridizing bacteria using a set of domain-specific to species-specific probes (see Table 9.6) are determined by evaluating at least ten microscopic fields with 20–100 cells per field.

When FISH is used for *in situ* assessment, root samples are fixed overnight at 4°C in PBS containing 3% paraformaldehyde, washed in PBS and treated as described above. For *in situ* identification of *Azospirillum* on the root surface, the autofluorescence problem (Hartmann *et al.*, 1998b) makes it necessary to use a confocal laser scanning microscope, such as the inverted Zeiss microscopes LSM 410 or LSM 510, which are equipped with lasers (Ar-ion UV; Ar-ion visible; HeNe) supplying excitation wavelengths at 365 nm, 488 nm, 543 nm and 633 nm. Usually, a general cell DNA staining with DAPI is combined with FISH using probes specific for the domain bacteria, group-specific probes (Amann *et al.*, 1995), and genus- or species-specific probes.

Calculation

For *in situ* assessment, sequentially recorded images are assigned to the respective fluorescence colour and then merged to obtain a true colour display. All image combining and processing is performed with the standard software provided by Zeiss.

Discussion

These analyses may be completed as follows. When cell suspensions are used, PCR amplification of the 16S rDNA from the samples and subsequent electrophoretic fingerprinting of the amplification products or clone bank analysis can be performed (Weidner *et al.*, 1996; Muyzer and Smalla, 1998). Structural and functional microbial diversity aspects can also be assessed using community-level fatty acid analysis (Zelles, 1997; White and Ringelberg, 1998) and physiological profiling (Garland *et al.*, 1997).

When the assessment is performed *in situ*, other specific oligonucleotide probes available for a number of root-associated and symbiotic bacteria (Kirchhof *et al.*, 1997; Ludwig *et al.*, 1998; Hartmann *et al.*, 2000) can also be used, which enables extension of the characterization of rhizosphere bacteria to other root-associated bacteria of interest.

Cultivation Approach to Monitoring Indigenous *Trichoderma* and *Gliocladium* Species

Introduction

A range of different methods have been assessed for monitoring of *Trichoderma* and *Gliocladium* species, including direct PCR (Lieckfeldt *et al.*, 1998) and methods based on the use of fluorogenic substrates (Miller *et al.*, 1998) and specific monoclonal antibodies (Thornton *et al.*, 2002). Hereafter, only cultivation methods will be presented. *Trichoderma* and *Gliocladium* species are ubiquitous in soil, and they can be readily isolated due to their rapid growth and profuse sporulation. Samples may be obtained from agricultural soils growing the crop of interest for a biocontrol application, or from soils naturally suppressive to the target pathogen. Particular soil microhabitats can be targeted (see pp. 273–274).

Principle of the methods

The fungi are isolated by homogenizing soil in water, diluting the soil suspension and plating on to solid media containing various compounds to selectively suppress growth of bacteria, oomycetes, mucorales and other fungi. Many different media have been formulated for selective isolation of *Trichoderma* and *Gliocladium* (Davet, 1979; Elad *et al.*, 1981; Johnson *et al.*, 1987; Park *et al.*, 1992; Askew and Laing, 1993), and examples of commonly used media are provided below. Identification of fungal colonies is performed by microscopic analysis of key morphological properties.

Materials and apparatus

- Analytical balance
- Magnetic stirrer and stir bars
- pH meter
- Blender
- Autoclave
- 50°C water bath
- Vortexer
- Laminar flow hood
- Bunsen burner
- Incubator or growth chamber with a constant temperature of 25°C and light
- 180°C oven
- Light microscope
- Glassware: Erlenmeyer flasks (1 l) with autoclavable lids or aluminium foil, beakers (250 ml), 10–15 ml tubes in rack, graduated cylinder (1 l)
- Parafilm
- Pipettes (P10, P200, P1000) and sterile disposable tips
- Petri plates (90 mm) and plate spreader
- Spatulas
- Microscope slides and coverslips

Chemicals and solutions

Petri plates are prepared in advance:

- *Trichoderma* Selective Medium (TSM; Elad and Chet, 1983) is prepared from a 1 l solution containing 200 mg $MgSO_4.7H_2O$, 900 mg KH_2PO_4, 150 mg KCl, 1 g NH_4NO_3, 3 g glucose, 20 g agar. Autoclave and cool to 50°C, then add 250 mg chloramphenicol, 300 mg fenaminosulf, 200 mg pentachloronitrobenzene, 200 mg Rose Bengal, 20 mg Captan (50% wettable powder).

- Modified *Trichoderma* Selective Medium (Smith *et al.*, 1990) is prepared from a 1 l solution containing 260 mg KNO_3, 260 mg $MgSO_4.7H_2O$, 120 mg KH_2PO_4, 50 mg citric acid, 1 g $Ca(NO_3)_2$, 1 g $CaCl_2.2H_2O$, 2 g sucrose, 20 g agar, Igepal CA-630 (pH adjusted to 4.5). Autoclave and cool to 50°C, then add 50 mg chlortetracycline, 40 mg Captan (50% wettable powder), 2.5 mg Vinclozolin.
- Selective Basal Medium (Papavizas and Lumsden, 1982) is prepared by mixing 200 ml V-8, 800 ml water, 1 g glucose, 20 g agar. Autoclave and cool to 50°C, then add 500 μg sodium propionate, 100 μg neomycin sulphate, 100 μg Bacitran, 100 μg penicillin G, 100 μg chloroneb, 25 μg chlortetracycline, 20 μg nystatin, 2 ml alkylaryl polyether alcohol.

Procedure

1. Soil (10 g) is added to 100 ml distilled water and homogenized for 1 min.
2. A dilution series (0, 10^{-1}, 10^{-2}, 10^{-3}) is prepared by transferring 1 ml of the soil suspension into 9 ml distilled water and vortexing well before each pipetting. If the medium contains Igepal CA-630, the dilution series can be reduced by one.
3. Add 100 μl of each dilution per plate, spread the sample evenly over the surface with a sterile spreader and cover the plates (three plates per dilution).
4. The plates are incubated 5–7 days with light, at 25°C. In addition, particular temperature or pH conditions or the presence of chemical pesticides may be used as additional selective criteria for obtaining potential inoculant for, for example, post-harvest applications (Johnson *et al.*, 1987; Widden and Hsu, 1987; Faull, 1988).
5. Sporulating fungal colonies are isolated from the plates and identified. Identification of fungal isolates is confirmed microscopically by morphology examination using taxonomic keys such as those of Gams and Bissett (1998).
6. The *Trichoderma* or *Gliocladium* isolates can be purified. They are maintained on potato dextrose, oatmeal or cornmeal agar at 25°C or as a spore suspension prepared in 20% glycerol for storage at –20°C.

Calculation

Colony-forming units (CFU) are computed and expressed per g of soil.

Discussion

The CFUs obtained from direct soil plating are more likely indicative of the number of dormant spores than the amount of active mycelial biomass in the soil. In general, *Trichoderma* and *Gliocladium* species are relatively adaptable to diverse ecological habitats. However, the distribution of diverse species and isolates depends upon varying abiotic environmental factors, such as temperature, moisture availability and nutrients, as well as different biotic factors, such as crop type and the presence of other microorganisms (Hjeljord and Tronsmo, 1998). *Trichoderma* spp. were found to comprise up to 3% and 1.5% of the total quantity of fungal propagules obtained from a range of forest and pasture soils, respectively (Brewer *et al.*, 1971; Danielson and Davey, 1973a).

Conclusion

Populations of free-living plant-beneficial microorganisms are naturally present in soil ecosystems, colonizing plant roots and benefiting the plant. Therefore, they make a very significant contribution to soil quality, even if this contribution is often difficult to quantify. The only case where the effects of indigenous free-living plant-beneficial microorganisms can be 'visualized' corresponds to disease-suppressive soils, but even then these soils are not so easy to identify and therefore most of them remain unnoticed. Free-living plant-beneficial microorganisms can be found in a wide range of bacterial and fungal taxa, and their effect on the plant can involve many different modes of action. Indigenous free-living plant-beneficial microorganisms can be monitored in the soil ecosystem, and this has been illustrated in the case of the PGPR *Azospirillum*, and the PGPF *Trichoderma* and *Gliocladium*. Although culture-independent methods such as FISH are available to detect certain free-living plant-beneficial microorganisms, much remains to be done to develop these approaches (e.g. PCR methods) for detection and quantification of a wider range of free-living plant-beneficial bacteria and fungi.

Normative references for Chapter 9

Results from a nationally organized evaluation of the bioassays in several laboratories has led to an experimental national standardization of the germination bioassay in France (XP X 31–205–1).

References for Chapter 9

Abadie, C., Edel, V. and Alabouvette, C. (1998) Soil suppressiveness to fusarium wilt: influence of a cover-plant on density and diversity of *Fusarium* populations. *Soil Biology and Biochemistry* 30, 643–649.

Alabouvette, C., Couteaudier, Y. and Louvet, J. (1982) Comparaison de la réceptivité de différents sols et substrats de culture aux fusarioses vasculaires. *Agronomie* 2, 1–6.

Amann, R.I., Binder, B.J., Olson, R.J., Chisholm, S.W., Devereux, R. and Stahl, D.A. (1990) Combination of 16S rRNA-targeted oligonucleotide probes with flow cytometry for analyzing mixed microbial populations. *Applied and Environmental Microbiology* 56, 1919–1925.

Amann, R.I., Ludwig, W. and Schleifer, K.-H. (1995) Phylogenetic identification and *in situ* detection of individual microbial cells without cultivation. *FEMS Microbiology Reviews* 59, 43–169.

Amarger, N. (1980) Aspect microbiologique de la culture des légumineuses. *Le Sélectionneur Français* 28, 61–66.

Amarger, N. (2001) Rhizobia in the field. *Advances in Agronomy* 73, 109–168.

Amarger, N., Mariotti, A. and Mariotti, F. (1977) Essai d'estimation du taux d'azote fixé symbiotiquement chez le lupin par le traçage isotopique naturel (^{15}N). *Comptes-Rendus de l'Académie des Sciences série D* 284, 2179–2182.

Askew, D.J. and Laing, M.D. (1993) An adapted selective medium for the quantitative isolation of *Trichoderma* species. *Plant Pathology* 42, 686–690.

Aßmus, B., Hutzler, P., Kirchhof, G., Amann, R.I., Lawrence, J.R. and Hartmann, A. (1995) *In situ* localization of *Azospirillum brasilense* in the rhizosphere of wheat with fluorescently labeled, rRNA-targeted oligonucleotide probes and scanning confocal laser microscopy. *Applied and Environmental Microbiology* 61, 1013–1019.

Bais, H.P., Park, S.-W., Weir, T.L., Callaway, R.M. and Vivanco, J.M. (2004) How plants communicate using the underground information superhighway. *Trends in Plant Science* 9, 26–32.

Baker, R. (1968) Mechanisms of biological control of soilborne pathogens. *Annual Review of Phytopathology* 6, 263–294.

Bally, R., Thomas-Bauzon, D., Heulin, T., Balandreau, J., Richard, C. and de Ley, J. (1983) Determination of the most frequent N_2-fixing bacteria in the rice rhizosphere. *Canadian Journal of Microbiology* 29, 881–887.

Bardin, R., Domenach, A.M. and Chalamet, A. (1977) Rapports isotopiques naturels de l'azote. II. Application à la mesure de la fixation symbiotique de l'azote *in situ*. *Revue d'Ecologie et de Biologie des Sols* 14, 395–402.

Bashan, Y. and de-Bashan, L.E. (2002) Protection of tomato seedlings against infection by *Pseudomonas syringae* pv. tomato by using the plant growth-promoting bacterium *Azospirillum brasilense*. *Applied and Environmental Microbiology* 68, 2637–2643.

Bestel-Corre, G. (2002) La protéomique: un outil d'étude des interactions dans la rhizosphère-Identification de protéines reliées à la mycorhization et à la nodulation et évaluation de l'impact de boues d'épuration sur les symbioses. Doctoral thesis, University of Burgundy, France.

Bouhot, D. (1975a) Recherches sur l'écologie des champignons parasites dans le sol. V. Une technique sélective d'estimation du potentiel infectieux des sols, terreaux et substrats infestés par *Pythium* sp. Etudes qualitatives. *Annales de Phytopathologie* 7, 9–18.

Bouhot, D. (1975b) Recherches sur l'écologie des champignons parasites dans le sol. VII. Quantification de la technique d'estimation du potentiel infectieux des sols, terreaux et substrats infestés par *Pythium* sp. *Annales de Phytopathologie* 7, 147–154.

Bouhot, D. (1975c) Technique sélective et quantitative d'estimation du potentiel infectieux des sols, terreaux et substrats infestés par *Pythium* sp. Mode d'emploi. *Annales de Phytopathologie* 7, 155–158.

Bouhot, D. (1979) Estimation of inoculum density and inoculum potential: techniques and their values for disease prediction. In: Schippers, B. and Gams, W. (eds) *Soil-borne Plant Pathogens*. Academic Press, London, pp. 21–34.

Bouillant, M.L., Miché, L., Ouedraogo, O., Alexandre, G., Jacoud, C., Sallé, G. and Bally, R. (1997) Inhibition of *Striga* seed germination associated with sorghum growth promotion by soil bacteria. *Comptes-Rendus de l'Académie des Sciences* 320, 159–162.

Brewer, D., Calder, F.W., MacIntyre, T.M. and Taylor, A. (1971) Ovine ill-thrift in Nova Scotia: I. The possible regulation of the rumen flora in sheep by the fungal flora of permanent pasture. *Journal of Agriculture Science* 76, 465–477.

Brockwell, J., Diatloff, A., Roughley, R.J. and Date, R.A. (1982) Selection of rhizobia for inoculants. In: Vincent, J.M. (ed.) *Nitrogen Fixation in Legumes*. Academic Press, Sydney, pp. 173–191.

Bromfield, E.S.P., Barran, L.R. and Wheatcroft, R. (1995) Relative genetic structure of a population of *Rhizobium meliloti* isolated directly from soil and from nodules of lucerne (*Medicago sativa*) and sweet clover (*Melilotus alba*). *Molecular Ecology* 4, 183–188.

Burdman, S., Kadouri, D., Jurkevitch, E. and Okon, Y. (2002) Bacterial phytostimulators in the rhizosphere: from research to application. In: Bitton, G. (ed.) *Encyclopedia of Environmental Microbiology: Soil Microbiology* (Volume 1). John Wiley & Sons, New York, pp. 343–354.

Camporota, P. (1980) Recherches sur l'écologie des champignons parasites dans le sols. XV.-Choix d'une plante piège et caractérisation de souches de *Rhizoctonia solani* pour la mesure du potentiel infectieux des sols et substrats. *Annales de Phytopathologie* 12, 31–44.

Camporota, P. (1982) Recherches sur l'écologie des champignons parasites dans le sol. XVII: Mesure du potentiel infectieux de sols et substrats infestés par *Rhizoctonia solani* Kühn, agent de fontes de semis. *Agronomie* 2, 437–442.

Chabot, R., Antoun, H. and Cescas, M.P. (1996) Growth promotion of maize and lettuce by phosphate-solubilizing *Rhizobium leguminosarum* biovar *phaseoli*. *Plant and Soil* 184, 311–321.

Chen, Y.X., He, Y.F., Yang, Y., Yu, Y.L., Zheng, S.J., Tian, G.M., Luo, Y.M. and Wong, M.H. (2003) Effect of cadmium on nodulation and N_2-fixation of soybean in contaminated soils. *Chemosphere* 50, 781–787.

Chotte, J.-L., Schwartzmann, A., Bally, R. and Jocteur Monrozier, L. (2002) Changes in bacterial communities and *Azospirillum* diversity in soil fractions of a tropical soil under 3 or 19 years of natural fallow. *Soil Biology and Biochemistry* 34, 1083–1092.

Clays-Josserand, A., Lemanceau, P., Philippot, L. and Lensi, R. (1995) Influence of two plant species (flax and tomato) on the distribution of nitrogen dissimilative abilities within fluorescent *Pseudomonas* spp. *Applied and Environmental Microbiology* 61, 1745–1749.

Cook, R. and Baker, K.F. (1983) *The Nature and Practice of Biological Control of Plant Pathogens*. APS, St Paul, Minnesota.

Cordier, C., Pozo, M.J., Barea, J.M., Gianinazzi, S. and Gianinazzi-Pearson, V. (1998) Cell defense responses associated with localized and systemic resistance to *Phytophthora parasitica* induced in tomato by an arbuscular mycorrhizal fungus. *Molecular Plant Microbe Interactions* 11, 1017–1028.

Cronin, D., Moënne-Loccoz, Y., Dunne, C. and O'Gara, F. (1997) Inhibition of egg hatch of the potato cyst nematode *Globodera rostochiensis* by chitinase-producing bacteria. *European Journal of Plant Pathology* 103, 433–440.

Curl, E.A. and Truelove, B. (1986) *The Rhizosphere*. Advanced series in agricultural sciences, vol. 15. Springer-Verlag, Berlin.

Danielson, R.M. and Davey, C.B. (1973a) The abundance of *Trichoderma* propagules and distribution of species in forest soils. *Soil Biology and Biochemistry* 5, 485–494.

Danielson, R.M. and Davey, C.B. (1973b) Carbon and nitrogen nutrition of *Trichoderma*. *Soil Biology and Biochemistry* 5, 505–515.

Danso, S.K.A., Hardarson, G. and Zapata, F. (1993) Misconceptions and practical problems in the use of ^{15}N soil enrichment techniques for estimating N fixation. *Plant and Soil* 152, 25–52.

Davet, P. (1979) Technique pour l'analyse de populations de *Trichoderma* et de *Gliocladium virens* dans le sol. *Annales de Phytopathologie* 11, 529–534.

De Boer, W., Klein Gunnewiek, P.J.A. and Woldendorp, J.W. (1998) Suppression of hyphal growth of soil-borne fungi by dune soils from vigorous and declining stands of *Ammophila arenaria*. *New Phytologist* 138, 107–116.

De Boer, W., Verheggen, P., Klein Gunnewiek, P.J.A., Kowalchuk, G.A. and van Veen, J.A. (2003) Microbial community composition affects soil fungistasis. *Applied and Environmental Microbiology* 69, 835–844.

Défago, G., Keel, C. and Moënne-Loccoz, Y. (1997) Fate of released *Pseudomonas* bacteria in the soil profile: implications for the use of genetically-modified microbial inoculants. In: Zelikoff, J.T., Lynch, J.M. and Shepers, J. (eds) *EcoToxicology: Responses, Biomarkers and Risk Assessment*. SOS Publications, Fair Haven, New Jersey, pp. 403–418.

Dobbelaere, S., Croonenborghs, A., Thys, A., Ptacek, D., Vanderleyden, J., Dutto, P., Labandera-Gonzalez, C., Caballero-Mellado, J., Aguirre, J.F., Kapulnik, Y., Brener, S., Burdman, S., Kadouri, D., Sarig, S. and Okon, Y. (2001) Responses of agronomically important crops to inoculation with *Azospirillum*. *Australian Journal of Plant Physiology* 28, 871–879.

Dobbs, C.G. and Gash, M.J. (1965) Microbial and residual mycostasis in soils. *Nature* (London) 207, 1354–1356.

Döbereiner, J. (1995) Isolation and identification of aerobic nitrogen-fixing bacteria from soil and plants. In: Alef, K. and Nannipieri, P. (eds) *Methods in Applied Soil Microbiology and Biochemistry*. Academic Press, London, pp. 134–141.

Dodd, J.C., Gianinazzi-Pearson, V., Rosendahl, S. and Walker, C. (1994) European Bank of *Glomales* – an essential tool for efficient international and interdisciplinary collaboration. In: Gianinazzi, S. and Schüepp, H. (eds) *Impact of Arbuscular Mycorrhizas on Sustainable Agriculture and Natural Exosystems*. Birkhäuser Verlag, Basel, Switzerland, pp. 41–45.

Dommergues, Y.R., Duhoux, E. and Diem, H.G. (1999) In: Ganry, F. (ed.) *Les Arbres Fixateurs d'Azote. Caractéristiques Fondamentales et Rôle dans l'Aménagement des Ecosystèmes Méditerranéens et Tropicaux*. CIRAD/Editions Espaces 34/FAO/IRD, Montpellier, France, pp. 163–165.

Duffy, B., Schouten, A. and Raaijmakers, J.M. (2003) Pathogen self-defense: mechanisms to counteract microbial antagonism. *Annual Review of Phytopathology* 41, 501–538.

Dumas-Gaudot, E., Gollotte, A., Cordier, C., Gianinazzi, S. and Gianinazzi-Pearson V. (2000) Modulation of host defense systems. In: Kapulnik, Y. and Douds, D.D. Jr (eds) *Arbuscular Mycorrhizas: Physiology and Function*. Kluwer Academic, Dordrecht, The Netherlands, pp. 173–200.

Edel, V., Steinberg, C., Gautheron, N. and Alabouvette, C. (1997) Populations of nonpathogenic *Fusarium oxysporum* associated with roots of four plant species compared to soilborne populations. *Phytopathology* 87, 693–697.

Elad, Y. and Chet, I. (1983) Improved selective media for isolation of *Trichoderma* spp. or *Fusarium* spp. *Phytoparasitica* 11, 55–58.

Elad, Y., Chet, I. and Henis, Y. (1981) A selective medium for improving quantitative isolation of *Trichoderma* spp. from soil. *Phytoparasitica* 9, 59–68.

Faull, J.L. (1988) Competitive antagonism of soil-borne plant pathogens. In: Burge, M.N. (ed.) *Fungi in Biological Control Systems*. Manchester University Press, Manchester, UK, pp. 125–140.

Fettell, N.A., O'Connor, G.E., Carpenter, D.J., Evans, J., Bamforth, I., Oti-Boateng, C., Hebb, D.M. and Brockwell, J. (1997) Nodulation studies on legumes exotic to Australia: the influence of soil populations and inocula of *Rhizobium leguminosarum* bv. viciae on nodulation and nitrogen fixation by field peas. *Applied Soil Ecology* 5, 197–210.

Fisher, R.A. and Yates, F. (1963) *Statistical Tables for Biological, Agricultural and Medical Research*, 6th edn. Oliver and Boyd, Edinburgh, UK.

Fravel, D.R. (1988) Role of antibiosis in the biocontrol of plant diseases. *Annual Review of Phytopathology* 26, 75–91.

Frey, P., Frey-Klett, P., Garbaye, J., Berge, O. and Heulin, T. (1997) Metabolic and genotypic fingerprinting of fluorescent pseudomonads associated with the Douglas Fir–*Laccaria bicolor* mycorrhizosphere. *Applied and Environmental Microbiology* 63, 1852–1860.

Gams, W. and Bissett, J. (1998) Morphology and identification of *Trichoderma*. In: Kubicek, C.P. and Harman, G.E. (eds) Trichoderma *and* Gliocladium*: Basic Biology, Taxonomy and Genetics* (Volume 1). Taylor & Francis, London, pp. 3–34.

Garland, J.L., Cook, K.L., Loader, C.A. and Hungate, B.A. (1997) The influence of microbial community structure and function on community-level physiological profiles. In: Insam, H. and Rangger, A. (eds) *Microbial Communities: Functional versus Structural Approaches*. Springer, New York, pp. 171–183.

Garrett, S.D. (1970) *Pathogenic Root-Infecting Fungi*. Cambridge University Press, Cambridge, UK.

Gianinazzi, S. (1991) Vesicular-arbuscular endomycorrhizas: cellular, biochemical and genetic aspects. *Agriculture, Ecosystems and Environment* 35, 105–119.

Gianinazzi, S. and Gianinazzi-Pearson, V. (1988) Mycorrhizae: a plant's health insurance. *Chimica Oggi*, pp. 56–58.

Gianinazzi, S., Gianinazzi-Pearson, V. and Trouvelot, A. (1982) Mycorrhizae, an integral part of plants: biology and perspectives for their use. *Les colloques de l'INRA*, no. 13.

Gianinazzi, S., Gianinazzi-Pearson, V. and Trouvelot, A. (1989) Potentialities and procedures for the use of endomycorrhizas with special emphasis on high value crops. In: Whipps, J.M. and Lumsden, R.D. (eds) *Biotechnology of Fungi for Improving Plant Growth*, vol. 3. Cambridge University Press, Cambridge, UK, pp. 41–54.

Gianinazzi, S., Schüepp, H., Barea, J.M. and Haselwandter, K. (2002) *Mycorrhizal Technology in Agriculture: From Genes to Bioproducts*. Birkhäuser Verlag, Basel, Switzerland.

Gianinazzi-Pearson, V. and Smith, S.E. (1993) Physiology of mycorrhizal mycelia. In: Ingram, D.S., Williams, P.H. and Tommerup, I.C. (eds) *Advances in Plant Pathology*. Academic Press, London, pp. 55–82.

Gibson, A.H. (1963) Physical environment and symbiotic nitrogen fixation. I. The effect of root temperature on recently nodulated *Trifolium subterraneum* L. plants. *Australian Journal of Biological Sciences* 16, 28–42.

Gildon, A. and Tinker, P.B. (1983) Interactions of vesicular-arbuscular mycorrhizal infection and heavy metal in plants: I: The effect of heavy metals on the development of vesicular arbuscular mycorrhizas. *New Phytolologist* 95, 247–261.

Harley, J.L. and Harley, E.L. (1987) A check list of mycorrhiza in the British Flora. *New Phytologist* 105, 1–102.

Harman, G.E. (2000) Myths and dogmas of biocontrol – changes in perceptions derived from research on *Trichoderma harzianum* T-22. *Plant Disease* 84, 377–393.

Harman, G.E. and Bjorkman, T. (1998) Potential and existing uses of *Trichoderma* and *Gliocladium* for plant disease control and plant growth enhancement. In: Harman, G.E. and Kubicek, C.P. (eds) Trichoderma *and* Gliocladium*: Enzymes, Biological Control and Commercial Applications* (Volume 2). Taylor & Francis, London, pp. 229–265.

Hartmann, A., Giraud, J.J. and Catroux, G. (1998a) Genotypic diversity of *Sinorhizobium* (formerly *Rhizobium*) *meliloti* strains isolated directly from a soil and from nodules of lucerne (*Medicago sativa*) grown in the same soil. *FEMS Microbiology Ecology* 25, 107–116.

Hartmann, A., Lawrence, J.R., Aßmus, B. and Schloter, M. (1998b) Detection of microbes by laser confocal microscopy. In: Akkermans, A.D.L., van Elsas, J.D. and de Bruijn, F.J. (eds) *Molecular Microbial Ecology Manual*. Kluwer Academic Publishers, Dordrecht, The Netherlands, pp. 1–34.

Hartmann, A., Stoffels, M., Eckert, B., Kirchhof, G. and Schloter, M. (2000) Analysis of the presence and diversity of diazotrophic endophytes. In: Triplett, E.W. (ed.) *Prokaryotic Nitrogen Fixation: A Model System for Analysis of a Biological Process*. Horizon Scientific Press, Wymondham, UK, pp. 727–736.

Hartmann, A., Pukall, R., Rothballer, M., Gantner, S., Metz, S., Schloter, M. and Mogge, B. (2004) Microbial community analysis in the rhizosphere by *in situ* and *ex situ* application of molecular probing, biomarker and cultivation techniques. In: Varma, A., Abbott, L., Werner, D. and Hampp, R. (eds) *Plant Surface*

Microbiology. Springer-Verlag, Berlin, pp. 449–469.

Hellriegel, H. and Wilfarth, H. (1888) *Untersuchungen über die Stickstoff-nahrung der Gramineen und Leguminosen. Beilageheft zu der Zeitschrift des Vereins für die Rübenzucker-Industrie des Deutschen Reiches, Buchdruckerei der "Post"*. Kayssler & Co., Berlin.

Herron, P.R. and Wellington, E.M.H. (1990) New method for the extraction of streptomycete spores from soil and application to the study of lysogeny in sterile amended and nonsterile soil. *Applied and Environmental Microbiology* 56, 1406–1412.

Hill, C., Com-Nougué, C., Kramar, A., Moreau, T., O'Quigley, J., Senoussi, R. and Chastang, C. (1990) *Analyse Statistique des Données de Survie*. Edited by M.-S. INSERM. Flammarion, Paris.

Hiltner, L. (1904) Über neuere erfahrungen und problem auf dem gebeit der bodenbakteriologie und unter besonderer berucksichtigung der grundungung und brache. *Arbeit und Deutsche Landwirschaft Gesellschaft* 98, 59–78.

Hirsch, A.M., Bauer, W.D., Bird, D.M., Cullimore, J., Tyler, B. and Yodder, J.I. (2003) Molecular signals and receptors: controlling rhizosphere interactions between plants and other organisms. *Ecology* 84, 858–868.

Hjeljord, L. and Tronsmo, A. (1998) *Trichoderma* and *Gliocladium* in biological control: an overview. In: Harman, G.E. and Kubicek, C.P. (eds) Trichoderma *and* Gliocladium*: Enzymes, Biological Control and Commercial Applications* (Volume 2). Taylor & Francis, London, pp. 131–151.

Ho, W.C. and Ko, W.H. (1986) Microbiostasis by nutrient deficiency shown in natural and synthetic soils. *Journal of General Microbiology* 132, 2807–2815.

Högberg, P. (1990) ^{15}N natural abundance as a possible marker of the ectomycorrhizal habit of trees in mixed African woodlands. *New Phytologist* 115, 483–486.

Höper, H., Steinberg, C. and Alabouvette, C. (1995) Involvement of clay type and pH in the mechanisms of soil suppressiveness to fusarium wilt of flax. *Soil Biology and Biochemistry* 27, 955–967.

Hornby, D. (1983) Suppressive soils. *Annual Review of Phytopathology* 21, 65–85.

Howell, C.R. (1998) The role of antibiosis in biocontrol. In: Harman, G.E. and Kubicek, C.P. (eds) Trichoderma *and* Gliocladium*: Enzymes, Biological Control and Commercial Applications* (Volume 2). Taylor & Francis, London, pp. 173–183.

Hsu, S.C. and Lockwood, J.L. (1971) Responses of fungal hyphae to soil fungistasis. *Phytopathology* 61, 1355–1362.

Inbar, J. and Chet, I. (1992) Biomimics of fungal cell–cell recognition by use of lectin-coated nylon fibers. *Journal of Bacteriology* 174, 1055–1059.

Inbar, J. and Chet, I. (1995) The role of recognition in the induction of specific chitinases during mycoparasitism by *Trichoderma harzianum*. *Microbiology* 141, 2823–2829.

Jacoud, C., Faure, D., Wadoux, P. and Bally, R. (1998) Development of a strain-specific probe to follow inoculated *Azospirillum lipoferum* CRT1 under field conditions and enhancement of maize root development by inoculation. *FEMS Microbiology Ecology* 27, 43–51.

Jacoud, C., Job, D., Wadoux, P. and Bally, R. (1999) Initiation of root growth stimulation by *Azospirillum lipoferum* CRT1 during maize seed germination. *Canadian Journal of Microbiology* 45, 339–342.

Jacquot, E., van Tuinen, D., Gianinazzi, S. and Gianinazzi-Pearson, V. (2000) Monitoring species of arbuscular mycorrhizal fungi *in planta* and in soil by nested PCR: application to the study of the impact of sewage sludge. *Plant and Soil* 226, 179–188.

Jacquot-Plumey, E., van Tuinen, D., Chatagnier, O., Gianinazzi, S. and Gianinazzi-Pearson, V. (2002) Molecular monitoring of arbuscular mycorrhizal fungi in sewage sludge treated field plots. *Environmental Microbiology* 3, 525–531.

Jacquot-Plumey, E., Caussanel, J.P., Gianinazzi, S., van Tuinen, D. and Gianinazzi-Pearson, V. (2003) Heavy metals in sewage sludges contribute to their adverse effects on the arbuscular mycorrhizal fungus *Glomus mosseae*. *Folia Geobotanica* 38, 167–176.

Jeffries, P. and Young, T.W.K. (1994) *Interfungal Parasitic Relationships.* CAB International, Wallingford, UK.

Jensen, H.L. (1942) Nitrogen fixation in leguminous plants. I. General characters of root nodule bacteria isolated from species of *Medicago* and *Trifolium* in Australia. *Proceedings of the Linnean Society of New South Wales* 66, 98–108.

Johnson, L.F., Bernard, E.C. and Peiyuan, Q. (1987) Isolation of *Trichoderma* spp. at low temperatures from Tennessee and Alaska (USA) soils. *Plant Disease* 71, 137–140.

Junk, G. and Svec, H.J. (1958) The absolute abundance of the nitrogen isotopes in the atmosphere and compressed gas from various sources. *Geochimica et Cosmochimica Acta* 14, 234–243.

Kabir, M.M., Chotte, J.L., Rahman, M., Bally, R. and Jocteur Monrozier, L. (1994) Distribution of soil fractions and location of soil bacteria in a vertisol under cultivation and perennial grass. *Plant and Soil* 163, 243–255.

Kabir, M.M., Faure, D., Haurat, J., Jacoud, C., Normand, P., Wadoux, P. and Bally, R. (1995) Oligonucleotide probes based on 16S rRNA sequences for the identification of four *Azospirillum* species. *Canadian Journal of Microbiology* 41, 1081–1087.

Kirchhof, G., Schloter, M., Aßmus, B. and Hartmann, A. (1997) Molecular microbial ecology approaches applied to diazotrophs associated with non-legumes. *Soil Biology and Biochemistry* 29, 853–862.

Klein, D. and Eveleigh, D.E. (1998) Ecology of *Trichoderma.* In: Kubicek, C.P. and Harman, G.E. (eds) Trichoderma *and* Gliocladium*: Basic Biology, Taxonomy and Genetics* (Volume 1). Taylor & Francis, London, pp. 57–74.

Knudsen, I.M.B., Debisz, K., Hockenhull, J., Jensen, D.F. and Elmholt, S. (1999) Suppressiveness of organically and conventionally managed soils towards brown foot rot of barley. *Applied Soil Ecology* 12, 61–72.

Kubicek, C.P., Mach, R.L., Peterbauer, C.K. and Lorito, M. (2001) *Trichoderma*: from genes to biocontrol. *Journal of Plant Pathology* 83, 11–23.

Kubicek-Pranz, E.M. (1998) Nutrition, cellular structure and basic metabolic pathways in *Trichoderma* and *Gliocladium.* In: Kubicek, C.P. and Harman, G.E. (eds) Trichoderma *and* Gliocladium*: Basic Biology, Taxonomy and Genetics* (Volume 1). Taylor & Francis, London, pp. 95–119.

Laguerre, G., Bardin, M. and Amarger, N. (1993) Isolation from soil of symbiotic and nonsymbiotic *Rhizobium leguminosarum* by DNA hybridization. *Canadian Journal of Microbiology* 39, 1142–1149.

Larkin, R.P., Hopkins, D.L. and Martin, F.N. (1993) Ecology of *Fusarium oxysporum* f. sp. *niveum* in soils suppressive and conducive to *Fusarium* wilt of watermelon. *Phytopathology* 83, 1105–1116.

Lemanceau, P., Samson, R. and Alabouvette, C. (1988) Recherches sur la résistance des sols aux maladies. XV. Comparaison des populations de *Pseudomonas* fluorescents dans un sol résistant et un sol sensible aux fusarioses vasculaires. *Agronomie* 8, 243–249.

Lemanceau, P., Corberand, T., Gardan, L., Latour, X., Laguerre, G., Boeufgras, J.-M. and Alabouvette, C. (1995) Effect of two plant species, flax (*Linum usitatissinum* L.) and tomato (*Lycopersicon esculentum* Mill.), on the diversity of soilborne populations of fluorescent pseudomonads. *Applied and Environmental Microbiology* 61, 1004–1012.

Leyval, C. and Joner, E. (2001) Bioavailability of heavy metals in the mycorrhizosphere. In: Gobran, G.R., Wenzel, W.W. and Lombi, E. (eds) *Trace Elements in the Rhizosphere.* CRC Press, Boca Raton, Florida, pp. 165–185.

Leyval, C., Singh, B.R. and Joner, E. (1995) Occurrence and infectivity of arbuscular-mycorrhizal fungi in some Norwegian soils influenced by heavy metals and soil properties. *Water Air Soil Pollution* 84, 203–216.

Leyval, C., Turnau, H. and Haselwandter, K. (1997) Effect of heavy metal pollution on mycorrhizal colonization and function: physiological, ecological and applied aspects. *Mycorrhiza* 7, 139–153.

Leyval, C., Joner, E.J., del Val, C. and Haselwandter, K. (2002) Potential of arbuscular mycorrhizal fungi for bioremediation. In: Gianinazzi, S., Barea, J.M., Schuepp, H. and Haselwandter, K. (eds) *Mycorrhizal Technology in Agriculture: From Genes to Bioproducts*. Birkhäuser Verlag, Basel, Switzerland, pp. 175–186.

Liebman, J.A. and Epstein, L. (1992) Activity of fungistatic compounds from soil. *Phytopathology* 82, 147–153.

Lieckfeldt, E., Kuhls, K. and Muthumeenakshi, S. (1998) Molecular taxonomy of *Trichoderma* and *Gliocladium* and their teleomorphs. In: Kubicek C.P. and Harman, G.E. (eds) Trichoderma *and* Gliocladium*: Basic Biology, Taxonomy and Genetics* (Volume 1). Taylor & Francis, London, pp. 35–56.

Linderman, R.G. (2000) Effects of mycorrhizas on plant tolerance to diseases – mycorrhiza–disease interactions. In: Kapulnik, Y. and Douds, D.D. Jr (eds) *Arbuscular Mycorrhizas: Physiology and Function*. Kluwer Academic, Dordrecht, The Netherlands, pp. 345–365.

Linderman, R.G., Moore, L.W., Baker, K.F. and Cooksey, D.A. (1983) Strategies for detecting and characterizing systems. *Plant Disease* 67, 1058–1064.

Lindsey, D.L. and Baker, R. (1967) Effect of certain fungi on dwarf tomatoes grown under gnotobiotic conditions. *Phytopathology* 57, 1262–1263.

Lo, C.T., Nelson, E.B. and Harman, G.E. (1996) Biological control of turfgrass diseases with a rhizosphere competent strain of *Trichoderma harzianum*. *Plant Disease* 80, 736–741.

Lockwood, J.L. (1977) Fungistasis in soils. *Biological Reviews* 52, 1–43.

Lockwood, J.L. (1988) Evolution of concepts associated with soilborne plant pathogens. *Annual Review of Phytopathology* 26, 93–121.

Lorito, M. (1998) Chitinolytic enzymes and their genes. In: Harman, G.E. and Kubicek, C.P. (eds) Trichoderma *and* Gliocladium*: Enzymes, Biological Control and Commercial Applications* (Volume 2). Taylor & Francis, London, pp. 87–115.

Lorito, M., Woo, S.L., D'Ambrosio, M., Harman, G.E., Hayes, C.K., Kubicek, C.P. and Scala, F. (1996) Synergistic interaction between cell wall degrading enzymes and membrane affecting compounds. *Molecular Plant–Microbe Interactions* 9, 206–213.

Louvet, J. (1973) Les perspectives de lutte biologique contre les champignons parasites des organes souterrains des plantes. *Perspectives de lutte biologique contre les champignons parasites des plantes cultivées et des tissus ligneux. Symposium International, Lausanne*, pp. 48–56.

Ludwig, W., Amann, R., Martinez-Romero, E., Schönhuber, W., Bauer, S., Neef, A. and Schleifer, K.-H. (1998) rRNA based identification and detection systems for rhizobia and other bacteria. *Plant and Soil* 204, 1–19.

Lynch, J.M. and Whipps, J.M. (1990) Substrate flow in the rhizosphere. *Plant and Soil* 129, 1–10.

Ma, W., Sebestianova, S.B., Sebestian, J., Burd, G.I., Guinel, F.C. and Glick, B.R. (2003) Prevalence of 1-aminocyclopropane-1-carboxylate deaminase in *Rhizobium* spp. *Antonie van Leeuwenhoek* 83, 285–291.

Macdonald, R.M. (1986) Sampling soil microfloras: dispersion of soil by ion exchange and extraction of specific microorganisms from suspension by elutriation. *Soil Biology and Biochemistry* 18, 399–406.

Mariotti, A. (1983) Atmospheric nitrogen is a reliable standard for natural ^{15}N abundance measurements. *Nature* 303, 685–687.

Mascher, F., Schnider-Keel, U., Haas, D., Défago, G. and Moënne-Loccoz, Y. (2003) Persistence and cell culturability of biocontrol *Pseudomonas fluorescens* CHA0 under plough pan conditions in soil and influence of the anaerobic regulator gene *anr*. *Environmental Microbiology* 5, 103–115.

Mavingui, P., Laguerre, G., Berge, O. and Heulin, T. (1992) Genetic and phenotypic diversity of *Bacillus polymyxa* in soil and in the wheat rhizosphere. *Applied and Environmental Microbiology* 58, 1894–1903.

Mazzola, M., Cook, R.J., Thomashow, L.S., Weller, D.M. and Pierson, L.S.D. (1992)

Contribution of phenazine antibiotic biosynthesis to the ecological competence of fluorescent pseudomonads in soil habitats. *Applied and Environmental Microbiology* 58, 2616–2624.

McGrath, S.P., Chaudri, A.M. and Giller, K.E. (1995) Long-term effects of metals in sewage sludge on soils, microorganisms and plants. *Journal of Industrial Microbiology* 14, 94–104.

Miché, L., Bouillant, M.-L., Rohr, R. and Bally, R. (2000) Physiological and cytological studies of the inhibitory effect of soil bacteria of the genus *Azospirillum* on striga seeds germination. *European Journal of Plant Pathology* 106, 347–351.

Migheli, Q., Steinberg, C., Davière, J.M., Olivain, C., Gerlinger, C., Gautheron, N., Alabouvette, C. and Daboussi, M.J. (2000) Recovery of mutants impaired in pathogenicity after transposition of *impala* in *Fusarium oxysporum* f. sp. *melonis*. *Phytopathology* 90, 1279–1284.

Miller, M., Palojärvi, A., Rangger, A., Reeslev, M. and Kjoller, A. (1998) The use of fluorogenic substrates to measure fungal presence and activity in soil. *Applied and Environmental Microbiology* 64, 613–617.

Moënne-Loccoz, Y. and Défago, G. (2004) Life as a biocontrol pseudomonad. In: Ramos, J.L. (ed.) Pseudomonas*: Genomics, Life Style and Molecular Architecture* (Volume 1). Kluwer Academic/Plenum Publishers, New York, pp. 457–476.

Moënne-Loccoz, Y., Powell, J., Higgins, P., McCarthy, J. and O'Gara, F. (1998) An investigation of the impact of biocontrol *Pseudomonas fluorescens* F113 on the growth of sugarbeet and the performance of subsequent clover–*Rhizobium* symbiosis. *Applied Soil Ecology* 7, 225–237.

Mogge, B., Lebhuhn, M., Schloter, M., Stoffels, M., Pukall, R., Stackebrandt, E., Wieland, G., Backhaus, H. and Hartmann, A. (2000) Erfassung des mikrobiellen Populationsgradienten vom Boden zur Rhizoplane von Luzerne (*Medicago sativa*). In: Hartmann, A. (ed.) *Biologische Sicherheit: Biomonitor und Molekulare Mikrobenökologie*. Projektträger BEO, Jülich, Germany, pp. 217–224.

Mondal, S.N. and Hyakumachi, M. (1998) Carbon loss and germinability, viability, and virulence of chlamydospores of *Fusarium solani* f. sp. *phaseoli* after exposure to soil at different pH levels, temperatures, and matric potentials. *Phytopathology* 88, 148–155.

Morrissey, J.P., Walsh, U.F., O'Donnell, A., Moënne-Loccoz, Y. and O'Gara, F. (2002) Exploitation of genetically modified inoculants for industrial ecology applications. *Antonie van Leeuwenhoek* 81, 599–606.

Moulin, L., Munive, A., Dreyfus, B. and Boivin-Masson, C. (2001) Nodulation of legumes by members of the β-subclass of Proteobacteria. *Nature* 411, 948–950.

Muyzer, G. and Smalla, K. (1998) Application of denaturing gradient gel electrophoresis (DGGE) and temperature gradient gel electrophoresis (TGGE) in microbial ecology. *Antonie van Leuwenhook* 73, 127–141.

Nguyen, C. (2003) Rhizodeposition of organic C by plants: mechanisms and controls. *Agronomie* 23, 375–396.

Okon, Y. (ed.) (1994) Azospirillum/*Plant Associations*. CRC Press, Boca Raton, Florida.

Okon, Y. and Labandera-Gonzalez, C.A. (1994) Agronomic applications of *Azospirillum*: an evaluation of 20 years world-wide field inoculation. *Soil Biology and Biochemistry* 26, 1591–1601.

Papavizas, G.C. and Lumsden, R.D. (1982) Improved medium for isolation of *Trichoderma* spp. from soil. *Plant Disease* 66, 1019–1020.

Park, Y.-H., Stack, J.P. and Kenerly, C.M. (1992) Selective isolation and enumeration of *Gliocladium virens* and *G. roseum* from soil. *Plant Disease* 76, 230–235.

Paulitz, T.C. (1990) Biochemical and ecological aspects of competition in biological control. In: Baker, R. and Dunn, P.E. (eds) *New Directions in Biological Control: Alternatives for Suppressing Agricultural Pests and Diseases*. Alan R. Liss, New York, pp. 713–724.

Peng, H.X., Sivasithamparam, K. and Turner, D.W. (1999) Chlamydospore germination and *Fusarium* wilt of banana plantlets in suppressive and conducive soils are affected by physical and chemical factors. *Soil Biology and Biochemistry* 31, 1363–1374.

Persson, L., Larsson Wikström, M. and Gerhardson, B. (1999) Assessment of soil suppressiveness to *Aphanomyces* root rot of pea. *Plant Disease* 83, 1108–1112.

Peters, N.K. (1986) A plant flavone, luteolin, induces expression of *Rhizobium meliloti* genes. *Science* 233, 977–980.

Raaijmakers, J.M. and Weller, D.M. (1998) Natural protection by 2,4-diacetylphloroglucinol-producing *Pseudomonas* spp. in take-all decline soils. *Molecular Plant Microbe Interactions* 11, 144–152.

Raaijmakers, J.M., Weller, D.M. and Thomashow, L.S. (1997) Frequency of antibiotic producing *Pseudomonas* spp. in natural environments. *Applied and Environmental Microbiology* 63, 881–887.

Revellin, C., Pinochet, X., Beauclair, P. and Catroux, G. (1996) Influence of soil properties and soya bean cropping history on the *Bradyrhizobium japonicum* population in some French soils. *European Journal of Soil Science* 47, 505–510.

Reverchon, S. (2001) Un diagnostic adapté aux champignons du sol. *Phytoma – La défense des Végétaux* 542, 20–23.

Rezzonico, F., Moënne-Loccoz, Y. and Défago, G. (2003) Effect of stress on the ability of a *phlA*-based quantitative competitive PCR assay to monitor biocontrol strain *Pseudomonas fluorescens* CHA0. *Applied and Environmental Microbiology* 69, 686–690.

Rivera-Becerril, F., Calantzis, C., Turnau, K., Caussanel, J.P., Belimov, A.A., Gianinazzi, S., Strasser, R.J. and Gianinazzi-Pearson, V. (2002) Cadmium accumulation and buffering of cadmium-induced stress by arbuscular mycorrhiza in three *Pisum sativum* L. genotypes. *Journal of Experimental Botany* 53, 1–9.

Rodriguez Caceres, E.A. (1982) Improved medium for isolation of *Azospirillum* spp. *Applied and Environmental Microbiology* 44, 990–991.

Romine, M. and Baker, R. (1973) Soil fungistasis: evidence for an inhibitory factor. *Phytopathology* 63, 756–759.

Rovira, A.D. (1965) Interactions between plant roots and soil microorganisms. *Annual Review of Microbiology* 19, 241–266.

Sambrook, J., Fritsch, E.F. and Maniatis, T. (1989) *Molecular Cloning: A Laboratory Manual.* Cold Spring Harbor Laboratory Press, Cold Spring Harbor, New York.

Schippers, B. (1992) Prospects for management of natural suppressiveness to control soilborne pathogens. In: Tjamos, E.C., Papavizas, G.C. and Cook, R.J. (eds) *Biological Control of Plant Diseases.* Plenum Press, New York, pp. 21–34.

Schneider, R.W. (1982) *Suppressive Soils and Plant Disease.* American Phytopathological Society, St Paul, Minnesota, p. 96.

Schüepp, H. and Green, R.J. Jr (1968) Indirect assay methods to investigate soil fungistasis with special consideration of soil pH. *Phytopathologische Zeitschrift* 61, 1–28.

Sessitsch, A., Howieson, J.G., Perret, X., Antoun, H. and Martínez-Romero, E. (2002) Advances in *Rhizobium* research. *Critical Reviews in Plant Sciences* 21, 323–378.

Shearer, G. and Kohl, D.H. (1986) N_2-fixation in field settings: estimations based on natural ^{15}N abundance. *Australian Journal of Plant Physiology* 13, 669–756.

Smith, S.E. and Read, D.J. (1997) *Mycorrhizal Symbiosis.* Academic Press, London, 605 pp.

Smith, V.L., Wilcox, W.F. and Harman, G.E. (1990) Potential for biological control of *Phytophthora* root and crown rots of apple by *Trichoderma* and *Gliocladium* spp. *Phytopathology* 80, 880–885.

Somasegaran, P.H. and Hoben, J. (1994) *Handbook for Rhizobia: Methods in Legume–Rhizobium Technology.* Springer, New York.

Stock, J.B., Ninfa, A.J. and Stock, A.M. (1989) Protein phosphorylation and regulation of adaptive responses in bacteria. *Microbiological Reviews* 53, 450–490.

Stoffels, M., Castellanos, T. and Hartmann, A. (2001) Design and application of new 16S rRNA-targeted oligonucleotide probes for the *Azospirillum–Skermanella–Rhodocista* cluster. *Systematic and Applied Microbiology* 24, 83–97.

Stutz, E.W. and Défago, G. (1985) Effect of parent materials derived from different geological strata on suppressiveness of soils to black root rot of tobacco. In: Parker, C.A., Rovira, A.D., More, K.J. and Wong, P.T.W. (eds) *Ecology and Management of Soilborne Plant Pathogens*. APS, St Paul, Minnesota, pp. 215–217.

Sy, A., Giraud, E., Jourand, P., Garcia, N., Willems, A., de Lajudie, P., Prin, Y., Neyra, M., Gillis, M., Boivin-Masson, C. and Dreyfus, B. (2001) Methylotrophic *Methylobacterium* bacteria nodulate and fix nitrogen in symbiosis with legumes. *Journal of Bacteriology* 183, 214–220.

Tarrand, J.J., Krieg, N.R. and Dobereiner, J. (1978) A taxonomic study of the *Spirillum lipoferum* group, with descriptions of a new genus, *Azospirillum* gen. nov. and two species, *Azospirillum lipoferum* (Beijerinck) comb. nov. and *Azospirillum brasilense* sp. nov. *Canadian Journal of Microbiology* 24, 967–980.

Thornton, C.R., Pitt, D., Wakley, G.E. and Talbot, N.J. (2002) Production of a monoclonal antibody specific to the genus *Trichoderma* and closely related fungi, and its use to detect *Trichoderma* spp. in naturally infested composts. *Microbiology* 148, 1263–1279.

Tilman, D. (1998) The greening of the green revolution. *Nature* 396, 211–212.

Toyota, K., Ritz, K. and Young, I.M. (1996) Microbiological factors affecting the colonisation of soil aggregates by *Fusarium oxysporum* f. sp. *raphani*. *Soil Biology and Biochemistry* 28, 1513–1521.

Trouvelot, A., Kough, J.L. and Gianinazzi-Pearson, V. (1985) Mesure du taux de mycorhization VA d'un système radiculaire. Recherche de méthodes d'estimation ayant une signification fonctionnelle. In: *Physiological and Genetical Aspects of Mycorrhizae*. Proceedings of the 1st European Symposium on Mycorrhizae, INRA, Paris, pp. 217–221.

Tu, X.M. (1995) Nonparametric estimation of survival distribution with censored initiating time, and censored and truncated terminating time: application to transfusion data for acquired immune deficiency syndrome. *Applied Statistics* 44, 3–16.

Unkovich, M. and Pate, J.S. (2001) Assessing N_2 fixation in annual legumes using ^{15}N natural abundance. In: Unkovich, M., Pate, J., NcNeill, A. and Gibbs, D.J. (eds) *Stable Isotope Techniques in the Study of Biological Processes and Functioning of Ecosystems*. Kluwer Academic, Dordrecht, The Netherlands, pp. 103–118.

van der Heijden, M.G.A., Klironomos, J.N., Ursic, M., Moutoglis, P., Streitwolfengel, R., Boller, T., Wiemken, A. and Sanders, I.R. (1998) Mycorrhizal fungal diversity determines plant biodiversity, ecosystem variability and productivity. *Nature* 396, 69–72.

van Tuinen, D., Jacquot, E., Zhao, B., Gollotte, A. and Gianinazzi-Pearson, V. (1998) Characterization of root colonization profiles by a microcosm community of arbuscular mycorrhizal fungi using 25S rDNA-targeted nested PCR. *Molecular Ecology* 7, 879–887.

Velázquez, E., Mateos, P.F., Velasco, N., Santos, F., Burgos, P.A., Villadas, P.J., Toro, N. and Martínez-Molina, E. (1999) Symbiotic characteristics and selection of autochthonous strains of *Sinorhizobium meliloti* populations in different soils. *Soil Biology and Biochemistry* 31, 1039–1047.

Villadas, P.J., Velázquez, E., Martínez-Molina, E. and Toro, N. (1995) Identification of nodule-dominant *Rhizobium meliloti* strains carrying pRmeGR4b-type plasmid within indigenous soil populations by PCR using primers derived from specific DNA sequences. *FEMS Microbiology Ecology* 17, 161–168.

Vincent, J.M. (1970) *A Manual for the Practical Study of Root Nodule Bacteria.* I.B.P. Handbook No. 15. Blackwell Scientific Publications, Oxford, UK.

Von Alten, H., Blal, B., Dodd, J.C., Feldmann, F. and Vosatka, M. (2002) Quality control of arbuscular mycorrhizal fungi inoculum in Europe. In: Gianinazzi, S., Barea, J.M., Schuepp, H. and Haselwandter, K. (eds) *Mycorrhizal Technology in Agriculture: From Genes to Bioproducts*. Birkhäuser Verlag, Basel, Switzerland, pp. 281–296.

Wacker, T.L. and Lockwood, J.L. (1991) A comparison of two assay methods for assessing fungistasis in soils. *Soil Biology and Biochemistry* 23, 411–414.

Wagner, M., Amann, R., Lemmer, H. and Schleifer, K.-H. (1993) Probing activated sludge with oligonucleotides specific for proteobacteria: inadequacy of culture-dependent methods for describing microbial community structure. *Applied and Environmental Microbiology* 59, 1520–1525.

Walsh, U.F., Morrissey, J.P. and O'Gara, F. (2001) *Pseudomonas* for biocontrol of phytopathogens: from functional genomics to commercial exploitation. *Current Opinion in Biotechnology* 12, 289–295.

Walsh, U.F., Moënne-Loccoz, Y., Tichy, H.-V., Gardner, A., Corkery, D.M., Lorkhe, S. and O'Gara, F. (2003) Residual impact of the biocontrol inoculant *Pseudomonas fluorescens* F113 on the resident population of rhizobia nodulating a red clover rotation crop. *Microbial Ecology* 45, 145–155.

Wan, M.T., Rahe, J.E. and Watts, R.G. (1998) A new technique for determining the sublethal toxicity of pesticides to the vesicular-arbuscular mycorrhizal fungus *Glomus intraradices*. *Environmental Toxicological Chemistry* 17, 1421–1428.

Wardle, D.A. (1992) A comparative assessment of factors which influence microbial biomass carbon and nitrogen levels in soil. *Biological Reviews* 67, 321–342.

Weidner, S., Arnold, W. and Pühler, A. (1996) Diversity of uncultured microorganisms associated with the seagrass *Halophila stipulacea* estimated from restriction fragment length polymorphism analysis of PCR-amplified 16S rRNA genes. *Applied and Environmental Microbiology* 62, 766–771.

Weindling, R. (1932) *Trichoderma lignorum* as a parasite of other fungi. *Phytopathology* 22, 837–845.

Weissenhorn, I. and Leyval, C. (1996) Spore germination of arbuscular-mycorrhizal (AM) fungi in soils differing in heavy metal content and other physicochemical properties. *European Journal of Soil Biology* 32, 165–172.

Weissenhorn, I., Leyval, C. and Berthelin, J. (1993) Cd-tolerant arbuscular mycorrhizal (AM) fungi from heavy-metal polluted soils. *Plant and Soil* 157, 247–256.

Weller, D.M., Raaijmakers, J.M., McSpadden-Gardener, B.B. and Thomashow, L.S. (2002) Microbial populations responsible for specific soil suppressiveness to plant pathogens. *Annual Review of Phytopathology* 40, 309–348.

White, D.C. and Ringelberg, D.B. (1998) Signature lipid biomarker analysis. In: Burlage, R.S., Atlas, R., Stahl, D., Geesey, G. and Sayler, G. (eds) *Techniques in Microbial Ecology*. Oxford University Press, New York, pp. 255–272.

Whitehead, N.A., Barnard, A.M.L., Slater, H., Simpson, N.J.L. and Salmond, G.P.C. (2001) Quorum-sensing in Gram-negative bacteria. *FEMS Microbiology Review* 25, 365–404.

Widden, P. and Hsu, D. (1987) Competition between *Trichoderma* species: effects of temperature and litter type. *Soil Biology and Biochemistry* 19, 89–94.

Williams-Woodward, J.L., Pfleger, F.L., Allmaras, R.R. and Fritz, V.A. (1998) *Aphanomyces euteiches* inoculum potential: a rolled-towel bioassay suitable for fine-textured soils. *Plant Disease* 82, 386–390.

Windham, M.T., Elad, Y. and Baker, R. (1986) A mechanism for increased plant growth induced by *Trichoderma* spp. *Phytopathology* 76, 518–521.

Young, J.M., Kuykendall, L.D., Martinez-Romero, E., Kerr, A. and Sawada, H. (2001) A revision of *Rhizobium* Frank 1889, with an amended description of the genus, and the inclusion of all species of *Agrobacterium* Conn 1942 and *Allorhizobium undicola* de Lajudie *et al.* 1998 as new combinations: *Rhizobium radiobacter*, *R. rhizogenes*, *R. rubi*, *R. undicola* and *R. vitis*. *International Journal of Systematic and Evolutionary Microbiology* 51, 89–103.

Zeilinger, S., Galhaup, C., Payer, K., Woo, S.L., Mach, R.L., Fekete, C., Lorito, M. and Kubicek, C.P. (1999) Chitinase gene expression during mycoparasitic interaction of *Trichoderma harzianum* with its host. *Fungal Genetics and Biology* 26, 131–140.

Zelles, L. (1997) Phospholipid fatty acid profiles in selected members of soil microbial communities. *Chemosphere* 35, 275–294.

10 Census of Microbiological Methods for Soil Quality

OLIVER DILLY

Lehrstuhl für Bodenökologie, Technische Universität München, D-85764 Neuherberg and Ökologie-Zentrum, Universität Kiel, Schauenburgerstraße 112, D-24118 Kiel, Germany; Present address: *Lehrstuhl für Bodenschutz und Rekultivierung, Brandenburgische Technische Universität, Postfach 101344, D-03013 Cottbus, Germany*

Introduction

An extensive spectrum of soil microbiological methods has been developed to investigate and evaluate soil quality. Classical soil microbiological methods referring to activity rates and biomass content can be separated from modern methods related to molecular techniques and isotope determinations. Several textbooks containing methods of the two groups have been published. These textbooks frequently contain an enormous set of methods selected by authors or editors without clearly indicating the field of application, or they apply more specifically to modern techniques that are still under development for routine analyses. This handbook aims to give a selection of methods for soil quality determination. Some information should also be given to rank the use of the methods. Therefore, a questionnaire was developed in order to evaluate the use of these methods. In addition, the database should provide information to enable location of research partners and laboratory expertise; thus, the questionnaire was not anonymous. Scientists involved in COST Action 831, and also other laboratories working in this area, were encouraged to complete the questionnaire.

Method

The questionnaire was developed to be completed via the Internet and was connected to a database. The database expertise was provided by http://www.kaufraum.de. The Internet address of the questionnaire was located at http://www.soiltechnology.de (accessed 26 October 2004) and was available for more than 1 year; evaluation was carried out at the end of 2003.

The questionnaire asked for information on country, city, laboratory, scientist, postal address, phone, fax and e-mail address. Using a radio button, it was possible to select 1 of 18 methods dealing with biomass, community structure and activity, carried out either in the laboratory or in the field. For each method, one form was to be filled in. Whether the methods were used as routine analysis, in monitoring programmes or for research only was to be noted. In addition, information was collected about application of the method with reference to: (i) soil use; (ii) soil type according ISSS/ISRIC/FAO (1998); (iii) soil texture; (iv) data use; (v) specific comments; and (vi) references. In some categories, 'unspecified' could be selected. The field 'unspecified' referred either to broad variability or to vague knowledge.

Results

This homepage was initiated by, and announced at, meetings or via e-mail to those on the distribution lists of COST Action 831, so it was visited mainly by scientists actively involved in this Action. Only a few people not actively involved in COST Action 831 found the homepage and filled in the questionnaire.

Overall, 159 questionnaires were completed by 32 laboratories in 14 countries. The highest number of questionnaires were returned from Germany, followed by Austria and Italy (Table 10.1). Most methods refer to bacterial community structure, unspecified activity measures and microbial biomass estimated by fumigation–extraction (Table 10.2). The methods were used mainly for research; approximately 50% were also used for monitoring and one-third for routine analysis (Table 10.2).

The soils studied were mainly under agricultural land use. One-third were industrial or urban soils. Eighteen different soil types were studied with the techniques. However, most of the scientists did not specify soil type when completing the questionnaire. In contrast, soil texture was specified in more detail, showing that the contributors were dealing mainly with loamy or sandy soil samples and, to a lesser extent, clay soils.

Conclusions

Although the database represents only an initial picture of European laboratories dealing with microbiological methods for evaluating soil quality, it is clear that several methods for bacterial community structure, a variety of activity methods, and microbial biomass estimated by fumigation–extraction are mainly used. Furthermore, agricultural soils and soils with loamy texture were investigated most frequently and information on soil units was generally not given.

The questionnaire was not completed by all scientists involved in COST Action 831 over the 12 months that it was available, most likely due to the fact that these scientists did not visit the website or found the procedure

Table 10.1. Number of questionnaires returned from the countries involved in COST Action 831.

Country	Amount
Germany	36
Austria	27
Italy	18
United Kingdom	17
Switzerland	12
The Netherlands	9
Spain	9
Sweden	9
Slovenia	8
Denmark	6
Belgium	2
Norway	2
France	2
Hungary	2

difficult and time consuming, e.g. each method needed to be reported in a separate questionnaire. In contrast, some laboratories offered a broad spectrum of methods, because they found the information useful.

Perspectives

To obtain more completed questionnaires for a more representative census, we hope that readers of this book will communicate the Internet address further to their scientific communities, or to societies dealing in this field, such as soil science societies, societies in the field of microbiology and private laboratories.

To enhance the acceptance of, and interest in, completing the questionnaire, some incentives may be provided, e.g. raffle of books or laboratory tools. Interest may also be encouraged when the webpage is officially supported by national authorities and the European Commission. The ability to refer to personal homepages, or to obtain information about specific contributing scientists, may further stimulate interest.

On the technical side, the actual display of graphics with statistics of methods used for the respective purposes of monitoring, research or routine analysis may be of interest. The simultaneous selection of several methods with the respective information would make the repeated filing of the data related to one laboratory unnecessary. Finally, a button for the submission of comments for improving the questionnaire may be included.

Regarding the content of the questionnaire, some questions related to soil–plant interactions, such as microorganisms and plants (in *Latin*), should be considered in the questionnaire, since these aspects are included in this book. Furthermore, the following categories may be included.

Table 10.2. Results of the questionnaire on methods used to study soil quality involved in COST Action 831.

Method		Soil Use		Soil type[a]		Soil texture		Use for	
Microbial community structure: Bacteria	21	Agricultural	133	Not specified	77	Loam	119	Research	140
Microbial activity: Others	18	Forest	79	Cambisol	35	Sand	96	Monitoring	74
Microbial biomass: Fumigation–extraction	18	Grassland	73	Luvisol	25	Clay	77	Routine analysis	43
Microbial activity: Basal respiration	16	Urban/industrial	53	Histosol	24	Silt	59		
Microbial biomass: Substrate-induced respiration	15	Not specified	20	Podzol	22	Others	53		
Microbial activity: N mineralization	13			Gleysol	21				
Molecular tools: Bacteria	12			Fluvisol	19				
Microbial biomass: ATP	10			Anthrosol	16				
Microbial biomass: Fungi	6			Regosol	11				
Microbial activity: C mineralization	5			Acrisol	9				
Molecular tools: Others	5			Arenosol	9				
Field experiments: Others	4			Leptosol	9				
Microbial biomass: Bacteria	4			Andosol	8				
Microbial community structure: Fungi	3			Umbrisol	8				
Microbial biomass: Others	3			Chernozem	7				
Field experiments: Litter bag	2			Ferralsol	7				
Field experiments: *in situ* C mineralization	2			Phaeozem	2				
Field experiments: *in situ* N mineralization	2			Vertisol	1				
				Durisol	1				

[a] According to ISSS/ISRIC/FAO (1998).

Soils under extreme environmental factors:

- soils under cold or high temperature;
- soils under limited or excess of water;
- soil polluted with heavy metals;
- soil polluted with organic compounds;
- salty soils;
- soil with extreme soil pH value;

and information may be included on soil pH value, for example:

- highly acidic soil;
- acid soil;
- neutral soil;
- alkaline soil.

To get more insight into the use of the respective methods, it would be interesting to know how many samples per year are analysed with each technique.

References

ISSS/ISRIC/FAO (1998) *World Reference Base for Soil Resources*. World Soil Resources, Report 84. FAO, Rome.

Index